Optimal Estimation in Approximation Theory

THE IBM RESEARCH SYMPOSIA SERIES

Computational Methods in Band Theory
Editors: P.M. Marcus, J.F. Janak, and A.R. Williams

Computational Solid State Physics
Editors: F. Herman, N.W. Dalton, and T.R. Koehler

Sparse Matrices and Their Applications
Editors: D.J. Rose and R.A. Willoughby

Complexity of Computer Computations
Editors: R.E. Miller and J.W. Thatcher
Associate Editor: J.D. Bohlinger

Computational Methods for Large Molecules and Localized States in Solids
Editors: F. Herman, A.D. McLean, and R.K. Nesbet

Ion Implantation in Semiconductors and Other Materials
Editor: Billy L. Crowder

Stiff Differential Systems
Editor: Ralph A. Willoughby

Optimal Estimation in Approximation Theory
Editors: Charles A. Micchelli and Theodore J. Rivlin

A Continuation Order Plan is available for this series. A continuation order will bring delivery of each new volume immediately upon publication. Volumes are billed only upon actual shipment. For further information please contact the publisher.

Optimal Estimation in Approximation Theory

Edited by

Charles A. Micchelli and Theodore J. Rivlin

IBM
Yorktown Heights, New York

PLENUM PRESS · NEW YORK AND LONDON

Library of Congress Cataloging in Publication Data

International Symposium on Optimal Estimation in Approximation Theory, Freudenstadt, Ger., 1976.
Optimal estimation in approximation theory.

(The IBM research symposia series)
Includes index.
1. Approximation theory—Congresses. 2. Mathematical optimization—Congresses. 3. Computational complexity—Congresses. I. Micchelli, Charles A. II. Rivlin, Theodore J., 1926- III. Title. IV. Series: International Business Machines Corporation. IBM research symposia series.
QA297.5.I57 1976 511'.4 77-4329
ISBN 0-306-31049-X

Proceedings of an International Symposium on Optimal Estimation in Approximation Theory held in Freudenstadt, Federal Republic of Germany, September 27–29, 1976

A Division of Plenum Publishing Corporation
227 West 17th Street, New York, N.Y. 10011

Printed in the United States of America

PREFACE

The papers in this volume were presented at an International Symposium on Optimal Estimation in Approximation Theory which was held in Freudenstadt, Federal Republic of Germany, September 27-29, 1976.

The symposium was sponsored by the IBM World Trade Europe/Middle East/Africa Corporation, Paris, and IBM Germany. On behalf of all the participants we wish to express our appreciation to the sponsors for their generous support.

In the past few years the quantification of the notion of complexity for various important computational procedures (e.g. multiplication of numbers or matrices) has been widely studied. Some such concepts are necessary ingredients in the quest for optimal, or nearly optimal, algorithms. The purpose of this symposium was to present recent results of similar character in the field or approximation theory, as well as to describe the algorithms currently being used in important areas of application of approximation theory such as: crystallography, data transmission systems, cartography, reconstruction from x-rays, planning of radiation treatment, optical perception, analysis of decay processes and inertial navigation system control. It was the hope of the organizers that this confrontation of theory and practice would be of benefit to both groups.

Whatever success the symposium had is due, in no small part, to the generous and wise scientific counsel of Professor Helmut Werner, to whom the organizers are most grateful.

Dr. T.J. Rivlin
IBM T.J. Watson Research Center
Yorktown Heights, N. Y.
Symposium Chairman

Dr. P. Schweitzer
IBM Germany
Scientific and Education Programs
Symposium Manager

Dr. C.A. Micchelli
IBM T.J. Watson Research Center
Yorktown Heights, N. Y.

CONTENTS

A SURVEY OF OPTIMAL RECOVERY

C. A. Micchelli and T.J. Rivlin

Research Staff Members of IBM

IBM - Yorktown Heights, New York 10598

ABSTRACT: The problem of optimal recovery is that of approximating as effectively as possible a given map of any function known to belong to a certain class from limited, and possibly error-contaminated, information about it. In this selective survey we describe some general results and give many examples of optimal recovery.

TABLE OF CONTENTS

1. THE SETTING, BASIC BOUNDS AND RELATIONSHIP TO PREVIOUS WORK

Let X be a linear space and Y, Z normed linear spaces. Suppose K to be a subset of X and U to be a given linear operator from X into Z. Our object is to approximate Ux for $x \in K$ using limited information about x. A linear operator, I, the information operator, maps X into Y, and we assume that Ix for $x \in K$ is known possibly only with some error. Thus in our attempt to recover Ux we actually know $y \in Y$ satisfying $||Ix-y|| \le \varepsilon$ for some given $\varepsilon \ge 0$. An algorithm, A, is any transformation - not necessarily a linear one - from $IK+\varepsilon S$, where $S = \{y \in Y: ||y|| \le 1\}$, into Z. The process is schematized in Figure 1.

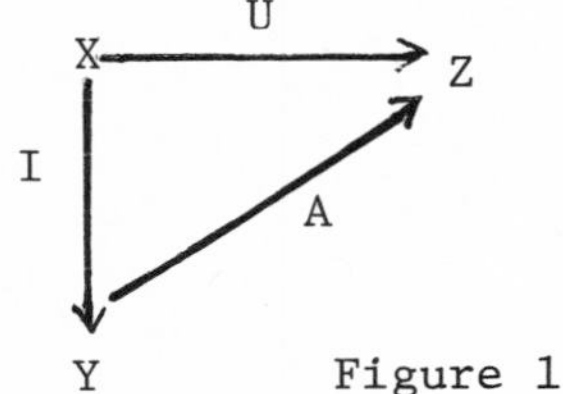

Figure 1

The algorithm A produces an error

$$E_A(K,\varepsilon) = \sup_{\substack{x \in K \\ ||Ix-y|| \le \varepsilon}} ||Ux - Ay||. \tag{1}$$

We call

$$E(K,\varepsilon) = \inf_A E_A(K,\varepsilon) \tag{2}$$

the intrinsic error in the recovery problem, and any algorithm A satisfying

$$E_A(K,\varepsilon) = E(K,\varepsilon) \tag{3}$$

is called an optimal algorithm.

Our interest is in determining $E(K,\varepsilon)$ for various choices of X,Y,Z,U,I,K and ε and finding optimal or nearly optimal algorithms.

A lower bound for $E(K,\varepsilon)$ which is most useful is given by the following simple result. Recall that K is balanced if $x \in K$ implies $-x \in K$.

Theorem 1. If K is a balanced convex subset of X then

$$e(K,\varepsilon) = \sup_{\substack{x \in K \\ ||Ix|| \le \varepsilon}} ||Ux|| \le E(K,\varepsilon). \tag{4}$$

Proof. Let 0 denote the zero vector in Y. For every $x \in K$ satisfying $||Ix|| \le \varepsilon$ we have, for any algorithm A,

$$||Ux-A0|| \leq E_A(K,\varepsilon)$$

and

$$||U(-x)-A0|| = ||Ux+A0|| \leq E_A(K,\varepsilon)$$

in view of (1). Therefore,

$$||2Ux|| = ||Ux-A0+Ux+A0|| \leq 2E_A(K,\varepsilon)$$

and (4) follows, since A was arbitrary.

Remark. This result is still valid if U and I are possibly nonlinear transformations with $U(-x)=-Ux$ and $||Ix|| = ||I(-x)||$ for all $x \in X$.

An upper bound for $E(K,\varepsilon)$ in terms of $e(K,\varepsilon)$ can also be obtained.

Theorem 2. If K is a balanced convex subset of X then

$$E(K,\varepsilon) \leq 2e(K,\varepsilon).$$

Proof. First we prove the theorem when $\varepsilon=0$ and then reduce the general result to this special case.

Suppose $\varepsilon=0$, so that an algorithm is a transformation from IK into Z. Consider the algorithm, A, defined on IK as follows: For each $y \in Y$ put

$$B(y) = \inf\{c : c>0, \{z \in X : Iz=y\} \cap cK \neq \emptyset\}.$$

Note that when K is a unit ball in X, $B(y)=\inf\{||z|| : Iz=y\}$, the distance of the hyperplane $Iz=y$ from the origin in X. Also observe that $B(Ix) \leq 1$, if $x \in K$. Hence given any fixed $\delta>0$ and any $x \in K$ there is a $z(Ix) \in X$ such that $I(z(Ix)) = Ix$ and $z(Ix) \in cK$ for some c satisfying $0<c<1+\delta$. Define the algorithm A by

$$A: Ix \to Uz(Ix).$$

This algorithm produces an error

$$E_A(K,0) = \sup_{x \in K} ||Ux-Uz(Ix)|| = \sup_{x \in K} ||U(x-z(Ix))||.$$

But since if $x \in K$ also $(-1/c)z(Ix) \in K$, and hence $x-z(Ix) \in (1+c)K$. Thus, because $I(x-z(Ix)) = 0$ we have

$$E_A(K,0) \leq (1+c) \sup_{\substack{x \in K \\ Ix=0}} ||Ux|| \leq (2+\delta)e(K,0),$$

and since $\delta>0$ was arbitrary the theorem holds when $\varepsilon=0$.

Next we show that optimal recovery with error is, at least formally, equivalent to recovery with exact information. This fact is just the observation that knowing both y and the actual error in computing Ix is the same as having exact information. Thus we introduce the space $\hat{X} = X \times Y$, and in $\hat{X}$ the convex balanced subset

$$\hat{K} = \{(x,y) : x \in \hat{K}, ||y|| \leq 1\}.$$

We extend U to $\hat{X}$ by defining $\hat{U}(x,y) = Ux$, and finally define the information operator as

$$\hat{I}(x,y) = Ix + \varepsilon y.$$

In the context we are now considering Figure 1 is replaced by

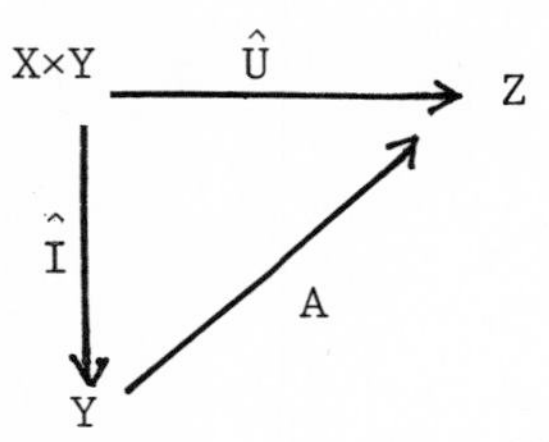

Figure 2

Now we observe that

$$\begin{aligned} E(\hat{K},0) &= \inf_{A} \sup_{(x,y)\in\hat{K}} ||\hat{U}(x,y)-A(\hat{I}(x,y))|| \\ &= \inf_{A} \sup_{\substack{x\in K \\ ||y||\le 1}} ||Ux-A(Ix+\varepsilon y)|| \\ &= \inf_{A} \sup_{\substack{x\in K \\ ||Ix-y||\le\varepsilon}} ||Ux-Ay|| = E(K,\varepsilon). \end{aligned}$$

Hence, applying the part of the theorem we have already proved, we have

$$E(K,\varepsilon) \le 2e(\hat{K},0).$$

But

$$\begin{aligned} e(\hat{K},0) &= \sup_{\substack{\hat{I}(x,y)=0 \\ (x,y)\in\hat{K}}} ||\hat{U}(x,y)|| = \sup_{\substack{Ix+\varepsilon y=0 \\ ||y||\le 1,\ x\in K}} ||Ux|| \\ &= \sup_{\substack{||Ix||\le\varepsilon \\ x\in K}} ||Ux|| = e(K,\varepsilon), \end{aligned}$$

and the theorem is proved.

It is possible, in some fairly general circumstances, to close the gap between Theorems 1 and 2 and solve the optimal recovery problem. For example, suppose K is convex and balanced, and

$$K_o = \{x\in K : Ix = 0\}.$$

We have

Theorem 3. Suppose there exists a transformation

$$G : IX \to X$$

such that $x-GIx \in K_o$ for all $x \in K$, then $e(K,0) = E(K,0)$ and $D = UG$ is an optimal algorithm.

Proof.

$$\begin{aligned} E_D(K,0) &= \sup_{x\in K} \|Ux-UGIx\| \\ &= \sup_{x\in K} \|U(x-GIx)\| \\ &\le \sup_{x\in K_o} \|Ux\| = e(K,0). \end{aligned}$$

Remark. See Morozov and Grebennikov [36] for related material.

Let us look at some examples of Theorem 3.

Example 1.1. Suppose X to be a Hilbert space, K the unit ball in X and $K_1 = \{x \in X : Ix=0\}$ to be closed. For $x \in X$ let Q be the projection of X on the subspace K_1, that is,

$$\|x-Qx\| = \min_{z\in K_1} \|x-z\|.$$

Note that since $\|Qx\| \le \|x\|$, $Qx \in K_o$ for $x \in K$. The transformation

$$G : Ix \to x - Qx \tag{5}$$

is now seen to be a well-defined linear operator of IX into X which satisfies the hypothesis of Theorem 3. Thus UG, where G is defined in (5) is seen to be an optimal algorithm and $e(K,0) = E(K,0)$.

For example, if we assume that I is a bounded linear operator from X onto a Hilbert space, Y, so that, in particular, K_1 is closed, then if I^* denotes the adjoint of I we define

$$G = I^*(II^*)^{-1} \tag{6}$$

and it is easy to verify that (5) is satisfied.

In particular, suppose Y to be $\mathcal{R}^n$, let $z_1,\ldots,z_n$ be linearly independent elements of X and define I by $Ix=((x,z_1),\ldots,(x,z_n))$. Then II^* is the Gram matrix $g(z_1,\ldots,z_n)$ and (6) reduces to a familiar formula,

$$G(Ix) = \sum_{j=1}^{n} a_j z_j$$

where $\underline{a} = g^{-1}(Ix)$.

Remark. In this example we do not require that Z be a Hilbert space.

Example 1.2. Let X be $\ell^p(\mathbb{C})$, $p\geq 1$, the space of sequences of complex numbers, $\{c_j\}$, $-\infty<j<\infty$, satisfying

$$\Sigma|c_j|^p < \infty.$$

Let $a:\{a_j\}$ be any given doubly infinite sequence of complex numbers, and choose

$$K = \{c\in\ell^p : \Sigma|a_j|^p|c_j|^p \leq 1\}.$$

Fix integers m,n, m<n and put $Ic = (c_m, c_{m+1},\ldots,c_n)$ so that $Y = \mathbb{C}^{n-m+1}$. Consider

$$G: (c_m,\ldots,c_n) \to (\ldots 0,\ldots,0,c_m,\ldots,c_n,0,\ldots,0,\ldots).$$

Since $x-GIx = (\ldots,c_{m-1},0,\ldots,0,c_{n+1},\ldots)$ is in K_o if $x\in K$, we have another illustration of Theorem 3.

Example 1.3. Suppose $n\geq 1$ and X is the space of functions having n-1 continuous derivatives on [0,1] having an nth derivative in $L^2[0,1]$ and equipped with the L^2 norm on [0,1]. Z=X and U is the identity operator. Suppose $0\leq x_1\leq x_2\leq\ldots\leq x_{n+r}\leq 1$, with $x_i<x_{i+n}$, $i=1,\ldots,r$. For $f\in X$ define I by

$$If = (f(x_1),\ldots,f(x_{n+r}))$$

with the convention that if $x_{i-1}<x_i=x_{i+1}=\ldots=x_{i+s}<x_{i+s+1}$ then $f(x_{i+j}) = f^{(j)}(x_i)$, $j=1,\ldots,s$. In other words, our information is obtained by sampling the functions at a given set of points. Put

$$K = \{f\in X : ||f^{(n)}||_2 \leq 1\}.$$

We now construct the required operator G. Let S_x be the set of natural splines of degree 2n-1 with respect to the knots $x:(x_1,\ldots,x_{n+r})$. Given $f\in X$ let Sf be the unique element of S_x satisfying

$$(Sf)(x_i) = f(x_i),\ i = 1,\ldots,n+r.$$

Note that $Sf\in X$ and, as is well known,

$$||f^{(n)} - (Sf)^{(n)}||_2 \leq ||f^{(n)}||_2. \tag{7}$$

Thus if we define G by

$$G: If \to Sf, \tag{8}$$

$I(f-G(If)) = If-I(Sf) = If-If = 0$ and in view of (7), if $f\in K$ also $f-GIf\in K$. G defined by (8) satisfies our requirements, and Theorem 3 tells us that natural spline interpolation is an optimal algorithm.

For another optimal method of recovery for this example see the end of Section 5.

Whenever our information is obtained by sampling at a given set of points, as in the present case, it is natural to seek the best set of points. That is, the intrinsic error, $E(K,0)$, depends on the points x. For what choice of x is it minimal? Melkman and Micchelli [23] show that in the present case one considers the differential equation

$$y^{(2n)}(t) = \mu y(t),$$
$$y^{(j)}(0) = y^{(j)}(1) = 0,\ j = n,\ldots,2n-1.$$

Its eigenvalues satisfy

$$0 = \mu_1 = \mu_2 = \ldots = \mu_n < \mu_{n+1} < \ldots .$$

and if $y_i(t)$ is the ith orthonormal eigenfunction of this system, then $y_{n+r+1}(t)$ has exactly $n+r$ simple zeros satisfying

$$0 < \eta_1 < \eta_2 < \ldots < \eta_{n+r} < 1,$$

and they are the optimal sampling points.

Generalizations of this problem are considered by Melkman and Micchelli [23] and its relationship to the n-widths of the appropriate spaces is revealed by them. In this connection see also Melkman [22].

One may be tempted to conjecture on the basis of these examples (as well as other examples to be discussed) that the lower bound of Theorem 1 is always sharp (as was asserted by S. Winograd in [49], see also the article of M. Schultz [43], in the same proceedings for material related to Theorems 1 and 2 and Example 1.1.) We will provide a counterexample to this conjecture and also point out that the upper bound in Theorem 2 may be reduced to $\sqrt{2}e(K,\epsilon)$ when Z is a Hilbert space.

First, we discuss the relationship of the problem of optimal recovery to the notion of Chebyshev center of a set and to the hypercircle inequality (Cf. [12], [14], [21]).

We have already seen in the proof of Theorem 2 that the problem of optimal recovery with error may be reduced to the case when $\varepsilon=0$. Thus we confine our attention to this case.

Recall that if M is a bounded subset of Z then an element $z_o\in Z$ is called a (Chebyshev) center for M if

(9) $$\sup_{a\in M} \|z_0 - a\| = \inf_{z\in Z} \sup_{a\in M} \|z-a\|.$$

The number on the right-hand side in (9) is called the Chebyshev radius of M and is denoted by r(M), see [14, p. 177]. We define the "hypercircle" determined by the intersection of the "ball" UK and the hyperplane Ix = y as

$$H(y) = \{Ux : x\in K,\ Ix=y\}.$$

Clearly, $H(y) \neq \emptyset$ if, and only if, $y\in IK$.

Let us assume that for every $y\in IK$, H(y) has a center, $c(y)\in Z$. Then we have

<u>Theorem 4</u>. $A^*: y \to c(y)$ is an optimal algorithm and

(10) $$E(K,0) = \sup_{y\in IK} r(H(y)).$$

<u>Proof</u>. Suppose that A is any algorithm. Then for all $x\in K$

$$\|Ux - A(Ix)\| \le E_A(K,0).$$

In particular

$$\sup_{z\in H(y)} \|w-z\| \le E_A(K,0)$$

where $w = A(y)$. Hence $r(H(y)) \le E_A(K,0)$ and therefore

(11) $$\sup_{y\in IK} r(H(y)) \le \inf_{A} E_A(K,0) = E(K,0).$$

To establish the reverse inequality and the optimality of A^* we observe that for any $y\in IK$ and $x\in K$ such that Ix=y

$$\|Ux - c(y)\| \le r(H(y)) \le \sup_{y\in IK} r(H(y)) .$$

Hence

$$E(K,0) \le \sup_{x\in K} \|Ux - A^*(Ix)\| \le \sup_{y\in IK} r(H(y)) ,$$

and the theorem is proved.

<u>Remark</u>. The validity of (10) can be proved without assuming that H(y) has a center for every $y\in IK$. Indeed our proof of (11) did not use c(y). But if

$$\sup_{y\in IK} r(H(y)) < \infty,$$

the only case of interest, choose $\delta>0$ and then for every $y\in IK$ there is a $c_\delta(y)\in Z$ such that

$$\|Ux-c_\delta(y)\| \le \delta + r(H(y)) \le \delta + \sup_{y\in IK} r(H(y)),$$

for any $x\in K$ with Ix = y. Letting $\delta \to 0^+$ gives the desired equality.

Remark. Theorem 1, in the case $\varepsilon = 0$, is a consequence of Theorem 4 since, as is easily verified

$$r(H(0)) = \sup_{\substack{x \in K \\ Ix=0}} ||Ux||.$$

Remark. In the above discussion, when $\varepsilon > 0$, $H(y)$ is replaced by $H_\epsilon(y) = \{Ux : x \in K, ||Ix-y|| \le \epsilon\}$ and therefore $r(H_\epsilon(0)) = e(K,\epsilon)$.

Our next example shows that the lower bound in Theorem 1 is not sharp. We owe our appreciation to Dr's. A. Melkman and I. Meilijson for their help in providing this example.

Example 1.4. Let Δ_n be the n-simplex defined as

$$\Delta_n = \{(x_1,\ldots,x_n) : \sum_{i=1}^{n} x_i = 1,\ x_i \ge 0\},\ n \ge 2,$$

and suppose K_n is the smallest convex balanced subset of R^n which contains Δ_n. Hence K_n is the ℓ^1 unit ball $\{x : \sum_{i=1}^{n} |x_i| \le 1\}$ and is the convex hull of the points $\pm e_1,\ldots,\pm e_n$,

$$(e_i)_j = \begin{cases} 0, & i \ne j \\ 1, & i = j \end{cases}.$$

For our example we choose $X = R^n$, $K = K_n$, $Y = R^1$ $Z = E^n$ (Euclidean n-space) and $Ux = x$, $Ix = \sum_{i=1}^{n} x_i$.

Thus we wish to recover a vector x which is known to be in K_n from the sum of its components. The hypercircle

$$H(d) = \{x : \sum_{i=1}^{n} x_i = d,\ x \in K_n\}$$

is the convex hull of the points

$$C = \{\frac{d+1}{2} e_i + \frac{d-1}{2} e_j : i \ne j\}$$

and its center of gravity $\frac{1}{n}(d,\ldots,d)$, being equally spaced from all the points of C is also the Chebyshev center of H(d). Hence

$$r(H(d)) = 2^{-1/2}(d^2(1-2/n)+1)^{1/2}$$

and

$$E(K_n,0) = \max_{|d| \le 1} r(H(d)) = r(H(1))$$

$$= (1-1/n)^{1/2}.$$

However $e(K_n,0) = r(H(0)) = 2^{-1/2}$ which is, for $n \ge 3$, strictly less than $E(K_n,0)$.

The motivation for this example is a classical result of Jung (Cf. [14]) that states that for any bounded subset M of E^n

$$r(M) \le \left(\frac{n}{2n+2}\right)^{1/2} \operatorname{diam}(M)$$

where

$$\operatorname{diam}(M) = \sup_{x,y \in M} ||x-y||$$

and equality occurs if and only if M is a regular simplex. For any (infinite) Hilbert space we have

$$r(M) \le 2^{-1/2} \operatorname{diam}(M)$$

(Routledge, Cf. [14]). Note that the face of K_n which determines I is a regular simplex.

Clearly, for any $y \in IK$

$$\operatorname{diam}(H(y)) \le 2e(K,0).$$

Hence, if Z is a Hilbert space we obtain the estimate

$$E(K,0) \le \sqrt{2}\, e(K,0)$$

(an improvement over Theorem 2). If, in addition, we have $\operatorname{diam}(H(y)) \le n$, all $y \in IK$, e.g. $Y = R^s$, $Z = E^{n+s}$, then Jung's theorem tells us that

$$E(K,0) \le \left(\frac{2n}{n+1}\right)^{1/2} e(K,0).$$

Now, the previous example show that this inequality is sharp for all n. Since, in this case, $\dim(H(y)) = n-1$, $Y = R^1$, $Z = E^n$, and therefore in general we have

$$E(K,0) \le \sqrt{2}\,(1-1/n)^{1/2} e(K,0).$$

However, $r(0) = e(K,0)$ and thus equality is achieved for $K = K_n$.

Let us next interpret the solution to the optimal recovery problem that is provided in Example 1.1 by (6) in the context of Theorem 4. Thus in the setting of the first sub-example of Example 1.1 we are required to find the center of the hypercircle

$$H(y) = \{Ux\colon Ix = y,\ ||x|| \le 1\},\ y \in IK.$$

The result is a minor variation on [14, p. 197].

Theorem 5. Let X be a Hilbert space, K the unit ball in X and I a bounded linear operator of X onto the Hilbert space Y. Then $UI^*(II^*)^{-1}y$ is the center of H(y) and its radius is

$$r(H(y)) = (1-||I^*(II^*)^{-1}y||^2)^{1/2} \sup_{\substack{Ix=0\\ x\in K}} ||Ux|| .$$

Proof. $x_o = I^*(II^*)^{-1}y$ satisfies $Ix_o = y$ and $(x,x_o) = 0$ whenever $Ix = 0$. Thus $||x_o-x||^2=||x||^2-||x_o||^2\le 1-||x_o||^2$ if $Ix = y$ and $||x||\le 1$. Consequently

$$\sup_{\substack{Ix=y\\ ||x||\le 1}} ||Ux-Ux_o|| = \sup_{\substack{Ix=y\\ ||x||\le 1}} ||U(x-x_o)||$$

$$\le (1-||x_o||^2)^{1/2} \sup_{\substack{Ix=0\\ ||x||\le 1}} ||Ux||$$

and we see that $r(H(y)) \le (1-||x_o||^2)^{1/2}\ e(K,0)$.

To establish the reverse inequality put $s = (1-||x_o||^2)^{1/2}$. Then if $I\bar{x}=0$ and $||\bar{x}||\le 1$

$$||x_o\pm s\bar{x}||^2 \le ||x_o||^2+s^2 = 1$$

and also $I(x_o\pm s\bar{x}) = y$. Thus for any $z\in Z$

$$||U(x_o+s\bar{x}) - z|| \le \sup_{\substack{Ix=y\\ ||x||\le 1}} ||Ux-z||$$

and

$$||U(x_o-s\bar{x})-z|| \le \sup_{\substack{Ix=y\\ ||x||\le 1}} ||Ux-z||.$$

Hence by the triangle inequality,

$$s||U\bar{x}|| \le \sup_{\substack{Ix=y\\ ||x||\le 1}} ||Ux-z||$$

and thus we conclude that

$$(1-||x_o||^2)^{1/2} \sup_{\substack{Ix=0\\ ||x||\le 1}} ||Ux|| = r(H(y)),$$

and the proof is complete.

Remark. When $Z = \mathbb{R}$, the real numbers, and $U\in X^*$ the set $H(y)$ is an interval and the specification of this interval given by the theorem is sometimes called the hypercircle inequality.

If $Z = \mathcal{R}$, while X is not necessarily a Hilbert space, the center of our hypercircle is

$$c(y) = \frac{1}{2}\left(\sup_{\substack{Ix=y\\ x\in K}} Ux + \inf_{\substack{Ix=y\\ x\in K}} Ux\right)$$

and $c(Ix)$ provides an optimal algorithm by Theorem 4 (Cf. Bojanov [3]). In general $c(y)$ is a nonlinear function of y. In the next section we show that when $Z = \mathcal{R}$ we can, in fairly general circumstances, find optimal algorithms which are linear. Before turning to this let us observe that when $Z = \mathcal{R}$ it is easy to see that we have equality in (4) of Theorem 1. Since in this case the hypercircle $H(y)$ is an interval and thus its diameter is twice its radius. However, we already pointed out that the diameter of $H(y)$ is $\le 2r(H(0))$ and therefore $r(H(y)) \le r(H(0))$.

Let us now show that sometimes an optimal algorithm may be obtained by variational methods.

We assume that $E(K,0) < \infty$. Then there is an algorithm, A, such that

$$\sup_{x\in K} |Ux - A(Ix)| < M < \infty.$$

Hence if we put $K(y) = \{x\in K: Ix = y\}$ then for all $y\in IK$

$$\sup_{x\in K(y)} Ux < M + A(y)$$

and we define

$$\phi(y;K) = \phi(y) = \sup_{x\in K(y)} Ux.$$

Then $r(H(y)) = (\phi(y) + \phi(-y))/2$ and $c(y) = (\phi(y) - \phi(-y))/2$.

Let us observe that $\phi(y)$ is a concave function of y. For, if $y_i \in IK$, $i = 1,2$ and $x_i \in K(y_i)$, $i = 1,2$ then $\lambda y_1 + (1-\lambda)y_2 \in IK$ and $\lambda x_1 + (1-\lambda)x_2 \in K(\lambda y_1 + (1-\lambda)y_2)$ for $0 \le \lambda \le 1$. Thus $U(\lambda x_1 + (1-\lambda)x_2) \le \phi(\lambda y_1 + (1-\lambda)y_2)$ for <u>any</u> $x_i \in K(y_i)$, $i = 1,2$. Hence $\lambda\phi(y_1) + (1-\lambda)\phi(y_2) \le \phi(\lambda y_1 + (1-\lambda)y_2)$.

Choosing $\lambda = 1/2$ and $y_1 = -y_2 = y$ we obtain

$$r(H(y)) = \frac{1}{2}(\phi(y) + \phi(-y)) \le \phi(0) = \sup_{\substack{Ix=0\\ x\in K}} Ux,$$

and in view of Theorem 4 we have again shown that we have equality in (4).

Next we note that the inequalities

$$-\phi(-y) \le Ux \le \phi(y)$$

are sharp for $x \in K(y)$. Furthermore, the concavity of ϕ implies that, for $\mu > 1$, $\phi(\mu y_1 + (1-\mu) y_2) \le \mu\phi(y_1) + (1-\mu)\phi(y_2)$. Hence we have for $x \in K(y)$ and $0 < \lambda \le 1$

$$|Ux - \frac{\phi(\lambda y) - \phi(-\lambda y)}{2\lambda}| \le \frac{\phi(\lambda y) + \phi(-\lambda y)}{2} \le \phi(0) = e(K,0).$$

Therefore, for any λ, $0 < \lambda \le 1$

$$c_\lambda(Ix) = \frac{\phi(\lambda Ix) - \phi(-\lambda Ix)}{2\lambda}$$

is also an optimal algorithm. But, since ϕ is concave $c_\lambda(Ix)$ converges to a limit as $\lambda \to 0^+$. If, in addition, ϕ is differentiable at zero then by definition

$$\lim_{\lambda \to 0} c_\lambda(Ix) = F(Ix)$$

where F is a linear functional which is necessarily bounded on IK. Thus the hypothesis of differentiability of ϕ at zero insures the existence of an optimal algorithm which is linear.

As it turns out optimal algorithms which are linear exist even when this condition fails. The existence of linear optimal algorithms is treated in the next section. We will also demonstrate in this section that the differentiability of $\phi(y)$ at zero is equivalent to the uniqueness of the optimal linear algorithm.

Remark.

For $\epsilon > 0$, $\phi(y)$ is replaced by

$$\phi_\epsilon(y) = \sup_{x \in K_\epsilon(y)} Ux$$

$$K_\epsilon(y) = \{x \in K \; : \; ||Ix - y|| \le \epsilon\}.$$

2. OPTIMAL ESTIMATION OF LINEAR FUNCTIONALS

In this section we shall suppose that X is a linear space over $\mathcal{R}$ (the real numbers), that $Z = \mathcal{R}$ and that K is a balanced convex subset of X. At the end of our discussion we shall indicate how our methods can be adjusted in the case that $\mathcal{R}$ is replaced by $\mathbb{C}$ (the complex numbers). Our major aim is to show under what circumstances there is an optimal algorithm which is continuous and linear, that is, an element of Y^*.

For any $L \in Y^*$, $x \in K$ and y satisfying $||Ix - y|| \le \varepsilon$ we have

$$|Ux - Ly| = |Ux - LIx + L(Ix - y)|$$

$$\leq \sup_{\substack{x\in K \\ ||y||\leq 1}} |Ux-LIx+\varepsilon Ly| \leq \sup_{x\in K} |Ux-LIx| + \varepsilon||L||.$$

Thus, recalling (4) we have

$$e(K,\varepsilon) \leq E(K,\varepsilon) \leq \inf_{L\in Y^*} \sup_{\substack{x\in K \\ ||y||\leq 1}} |Ux-LIx+\varepsilon Ly| \leq d(K,\varepsilon) \tag{12}$$

where we put

$$d(K,\varepsilon) = \inf_{L\in Y^*} \{\sup_{x\in K} |Ux-LIx| + \varepsilon||L||\}. \tag{13}$$

The inequality between $E(K,\varepsilon)$ and $d(K,\varepsilon)$ in (12) will be useful when we are dealing with linear spaces over $\mathbb{C}$.

A basic question for us is: when is $e(K,\varepsilon) = d(K,\varepsilon)$? Clearly this is true if $d(K,\varepsilon) = 0$. Suppose, then, that $d(K,\varepsilon) > 0$. Put

$$C = \{(Ux,Ix) : x\in K\}$$

and note that C is a convex subset of $\mathbb{R}\times Y$. Then $e(K,\varepsilon) = d(K,\varepsilon)$ holds if, and only if, given any positive $d < d(K,\varepsilon)$ there is an $x\in K$ satisfying $||Ix|| \leq \varepsilon$ such that

$$d \leq Ux \leq d(K,\varepsilon),$$

that is, for every positive $d < d(K,\varepsilon)$ the convex subset of $\mathbb{R}\times Y$

$$C_d = \{(t,y) : d \leq t \leq d(K,\varepsilon), ||y|| \leq \varepsilon\}$$

has a point in common with C.

We are thus led to consider linear separation of convex sets.

<u>Definition</u>. Two convex sets A and B of $\mathbb{R} \times Y$ are <u>separated</u> if there exists a non-trivial linear functional on $\mathbb{R} \times Y$ and a real constant b such that

$$Hz \leq b, \quad z\in A,$$
$$Hz \geq b, \quad z\in B.$$

A and B are strictly separated if $Hz \leq b-\delta$ for $z\in A$ and $Hz \geq b+\delta$ for $z\in B$, for some $\delta>0$.

<u>Lemma 1</u>. H is a linear functional on $\mathbb{R}\times Y$ if, and only if, $H(t,y) = at+Ly$, $a\in\mathbb{R}$, L a linear functional on Y.

<u>Lemma 2</u>. i) If $\varepsilon>0$ and $0<d<d(K,\varepsilon)$ then C_d and C cannot be separated. ii) If $\epsilon\geq 0$ and $d=d(K,\varepsilon)$, C_d and C cannot be strictly separated.

<u>Proof</u>. i) Suppose C_d and C can be separated. Then there exist $b,c\in\mathbb{R}$ and a linear functional L on Y such that

$$cd + Ly \leq b \quad , \quad ||y|| \leq \varepsilon \tag{14}$$

and

$$cUx + LIx \geq b \;, \quad x\in K. \tag{15}$$

Setting $y = 0$ in (14) gives $cd \leq b$ and putting $x = 0$ in (15) yields $b \leq 0$ and therefore $c \leq 0$. Moreover, if $c = 0$ then $b = 0$ and (14) implies that $L = 0$. This is impossible because $H(d,y) = cd+Ly$ is a non-trivial linear functional. Thus $c < 0$ and we have

$$\sup_{x\in K} (Ux+c^{-1}LIx) \le \frac{b}{c}$$

and

$$\varepsilon||L|| \le b-cd,$$

so that, noting that $|b/c| = b/c$,

$$\begin{aligned} d(K,\varepsilon) &\le \sup_{x\in K} |Ux+c^{-1}LIx| + \varepsilon||c^{-1}L|| \\ &\le \frac{b}{c} - \frac{1}{c}(b-cd) = d < d(K,\varepsilon), \end{aligned}$$

a contradiction which establishes i).

ii) Suppose C_d and C are strictly separated. (14) and (15) are now replaced by

(16) $$cd(K,\varepsilon) + Ly \le b-\delta, \quad ||y|| \le \varepsilon$$

and

(17) $$cUx + LIx \ge b+\delta \quad , \quad x\in K$$

for some $\delta>0$. Setting $y=0$ in (16) and $x=0$ in (17) yields $cd(K,\varepsilon) \le b-\delta$ and $b+\delta \le 0$. Hence $cd(K,\varepsilon) < b+\delta \le 0$ and so $c < 0$. Thus, as in i)

$$\begin{aligned} d(K,\varepsilon) &\le \sup_{x\in K} |Ux+c^{-1}LIx| + \varepsilon||c^{-1}L|| \\ &\le \frac{b+\delta}{c} - \frac{1}{c}(b-\delta-cd(K,\varepsilon)) = d(K,\varepsilon) + \frac{2\delta}{c} \\ &< d(K,\varepsilon) \quad , \end{aligned}$$

and the contradiction proves the Lemma.

The value of Lemma 2 is that if we can show that C_d meets C by an argument based on a standard separation theorem, then we will have established that $e(K,\varepsilon) = d(K,\varepsilon)$. Before we carry out this program let us give a condition under which i) of Lemma 2 remains valid for $\varepsilon=0$.

A convex set, K, is <u>absorbing</u> if for every $x\in X$ there is a $\lambda>0$ such that $\lambda x\in K$.

<u>Lemma 3</u>. If $IK = \{Ix:x\in K\}$ is absorbing then Lemma 2 i) is true when $\varepsilon = 0$.

<u>Proof</u>. We need only prove that $c\neq 0$ in (14) and (15). But if $c=0$ then $b=0$ and hence $LIx \ge 0$ for $x\in K$, and since IK is balanced and absorbing $L = 0$, which is impossible.

We recall the following:

<u>Definition</u>. If K is a convex subset of the linear space X, $x\in K$ is an <u>internal point</u> of K if given any $y\in X$ there is a $\delta>0$ such that $x+\lambda y\in K$ for all $|\lambda|<\delta$.

Note that if K is a balanced absorbing convex set then 0 is an internal point of K. We now quote (Cf. Royden [40]).

Theorem A. Let K_1 and K_2 be two disjoint convex subsets of the linear space X, and suppose that one of them has an internal point. Then K_1 and K_2 can be separated.

We now have

Theorem 6. If $\varepsilon>0$ then $e(K,\varepsilon) = d(K,\varepsilon)$. If IK is absorbing then $e(K,0) = d(K,0)$.

Proof. Suppose $d(K,\varepsilon)>0$ and there exists a positive $d<d(K,\varepsilon)$ such that C_d and C are disjoint. If $\varepsilon>0$, C_d has an internal point, while if IK is absorbing then C is a balanced absorbing set. The theorem now follows from Theorem A, Lemma 2 i) and Lemma 3.

Our next result will be used to compute the directional derivative of the function $\phi_\epsilon(y)$ which was introduced at the end of the previous section.

Theorem 7. Let $\epsilon\geq 0$. If $\epsilon>0$ or IK is absorbing and y is an internal point of IK then

$$\phi_\epsilon(y) = \sup_{\substack{||Ix-y||\leq\epsilon \\ x\in K}} Ux = \inf_{L\in Y^*}\{\sup_{x\in K} |Ux-LIx| + \epsilon||L|| + Ly\}$$

Proof. Let $I_y x = Ix-y$ then the above equation reads

$$\sup_{\substack{||I_y x||\leq\epsilon \\ x\in K}} Ux = \inf_{L\in Y^*}\{\sup_{x\in K} (Ux-LI_y x) + \epsilon||L||\}$$

This equation is similar to the duality relation we proved in Theorem 6. However, in the present case I_y is not linear but rather affine, i.e. preserves convex combinations. Nevertheless the previous proof applies in this case as well provided we make the following additional observations. The set C previously defined is still convex when I is affine. Hence for $\epsilon>0$ there is no change in the proof. When $\epsilon=0$ we must show that if $L(I_y x) \geq 0$, $x\in K$, and y an internal point of IK then $L \equiv 0$. This is easily done. Given $z\in Y$ there is a $\delta>0$ such that $y+\lambda z\in IK$ for $|\lambda|<\delta$. Therefore it follows that $Lz = 0$.

Note that Theorem 6 concludes that

$$\sup_{\substack{x\in K \\ ||Ix||\leq\varepsilon}} |Ux| = \inf_{L\in Y^*}\{\sup_{x\in K}|Ux-LIx| + \varepsilon||L||\}$$

$$= \inf_{L \in Y^*} \sup_{\substack{x \in K \\ ||Ix-y|| \le \varepsilon}} |Ux-Ly| = E(K,\varepsilon),$$

but that nothing is said about whether the various limits involved are attained. We turn next to an examination of this question in which topological considerations are important.

We require a <u>feasibility condition</u>: There exists $L \in Y^*$ such that

$$\sup_{x \in K} |Ux-LIx| < \infty,$$

and also need

<u>Theorem B</u> (Banach-Alaoglu (Cf. [41])). Let X be a linear topological space. If N is a neighborhood of 0 and if

$$V = \{g \in X^* : |gx| \le 1, \text{ all } x \in N\}$$

then V is weak* compact.

<u>Theorem 8</u>. If $\varepsilon>0$ an optimal linear algorithm exists. If $\varepsilon=0$ and IK is a neighborhood of the origin then an optimal linear algorithm exists.

<u>Proof</u>. The feasibility condition implies that there exists $L_o \in Y^*$ such that

$$\sup_{x \in K} |Ux-L_o Ix| + \varepsilon||L_o|| \le B < \infty.$$

Thus a linear optimal algorithm must lie in the set

$$\{L \in Y^* : |(L-L_o)Ix| \le 2B; x \in K\}.$$

This set is weak* compact when IK is a neighborhood of the origin. The function

$$G(L) = \sup_{x \in K} |Ux-LIx|$$

is a weak* lower semi-continuous function of L, and so the minimum exists when $\varepsilon=0$. When $\varepsilon>0$, a linear optimal algorithm must lie in the weak* compact set

$$\{L \in Y^* : ||L|| \le \frac{B}{\varepsilon}\}$$

and the proof is complete.

The first part of Theorems 6 and 8 were proved using a version of the min-max theorem by Micchelli [27]. In the case that Y is a finite dimensional space see also Marchuk and Osipenko [20], and when $\epsilon=0$, Bakhvalov [1] and Smolyak [44].

The "off center" version of Theorem 8 is

Theorem 9. Let $\epsilon \geq 0$. If $\epsilon > 0$ or IK is an absorbing neighborhood of y then there exists a weak* compact set $M \subseteq Y^*$ such that

$$\phi_\epsilon(y) = \min_{L \in M} \{\sup_{x \in K} |Ux - LIx| + \epsilon ||L|| + Ly\}.$$

Proof. Let L_o and B be as defined in Theorem 8. For $\epsilon > 0$ we choose $M = \{L \in Y^* : ||L|| \leq \epsilon^{-1}(B + L_o y)\}$. When $\epsilon = 0$, the set $\{L \in Y^* : |(L - L_o)I_y x| \leq 2B\}$ is a weak* compact set in which the minimum lies.

Next we examine the question of the existence of a worst function, that is, an $x \in K$ such that $||Ix|| \leq \varepsilon$ and $Ux = e(K,\varepsilon) = d(K,\varepsilon)$. Existence of a worst function is equivalent to having

$$C_{d(K,\varepsilon)} \cap C \neq \emptyset .$$

According to Lemma 2 $C_{d(K,\varepsilon)}$ and C cannot be strictly separated, but we recall

Theorem C. (Cf. [41]). If A and B are disjoint non-empty convex sets in a locally convex topological linear space X, and A is compact and B is closed, then A and B can be strictly separated.

Thus, for example, we have

Theorem 10. i) If C is norm compact a worst function exists.
ii) If C is weakly closed (equivalently, since C is convex, C is norm closed) and Y is reflexive then a worst function exists.

Proof. i) $C_{d(K,\varepsilon)}$ is closed in the norm topology on Y.
ii) $C_{d(K,\varepsilon)}$ is weakly compact.

We turn next to the question of characterizing an optimal linear algorithm.

Theorem 11. Suppose x_o is a worst function, then $L_o \in Y^*$ is an optimal algorithm if, and only if

i) $\varepsilon ||L_o|| = L_o I x_o$

ii) $\max_{x \in K} |Ux - L_o Ix| = Ux_o - L_o I x_o$.

Proof. If i) and ii) hold then

$$\max_{x \in K} |Ux - L_o Ix| + \varepsilon ||L_o|| = Ux_o = d(K,\varepsilon).$$

Conversely, if L_o is an optimal algorithm

$$\max_{x \in K} |Ux - L_o Ix| + \varepsilon ||L_o|| = Ux_o, \tag{18}$$

hence $Ux_o - L_o Ix_o + \varepsilon||L_o|| \le Ux_o$, or $\varepsilon||L_o|| \le L_o Ix_o$.
But since $L_o Ix_o \le \varepsilon||L_o||$, i) holds and ii) follows from (18).

Remark. i) and ii) may also be used to characterize worst functions. If L_o is an optimal estimator then x_o is a worst function if and only if i) and ii) hold.

Before turning to the case that X is a linear space over the complex numbers we settle the question of the differentiability of $\phi_\epsilon(y)$ at zero. Recall, that the right directional derivative of $\phi_\epsilon(y)$ at y_o in the direction w is defined as

$$\lim_{\lambda\to 0^+} \frac{\phi_\epsilon(y_o+\lambda w)-\phi_\epsilon(y_o)}{\lambda} = (\phi'_\epsilon)_+(y_o,w).$$

The left derivative is defined similarily and will be denoted by $(\phi'_\epsilon)_-(y_o,w)$. We will say that ϕ_ϵ is differentiable at y_o, if its right and left derivative exist for any direction and are equal.

Since $\phi_\epsilon(y)$ is a concave function its right and left derivative exist. We will compute these quantities by using Theorem 9 which expresses $\phi_\epsilon(y)$ as a min function. The study of the differentiability of min (max) functions apparently originated with Danskin [8]. The result we will use is proved in Micchelli [24].

Theorem D. Let T be a compact space and E a linear topological space. Suppose G(t,x) is a real-valued function defined on T×E such that for each $x \in E$, G(t,x) is continuous on T and for each $t \in T$, G(t,x) is a convex function on E. Suppose $g(x) = \max_{t\in T} G(t,x)$.

Then

$$g_+(x,y) = \max_{t\in S(x)} G_+(t,x,y)$$

where $S(x) = \{t : t\in T, g(x) = G(t,x)\}$ and $G_+(t,x,y)$ is the right derivative of G with respect to its second argument in the direction y.

Proof. If $t \in S(x)$ then for $\lambda>0$

$$\frac{g(x+\lambda y)-g(x)}{\lambda} \le \frac{G(t,x+\lambda y)-G(t,x)}{\lambda} .$$

Letting $\lambda\to 0^+$ we obtain $g_+(x;y) \ge G_+(t,x,y)$ for all $t\in S(x)$ which implies $g_+(x,y) \ge \max_{t\in S(x)} G_+(t,x,y)$.

To prove the reverse inequality we use the fact that $f(\lambda)/\lambda$ is a nondecreasing function for $\lambda>0$ whenever $f(0) = 0$ and f is a convex function on $[0,\infty)$.

Set $g_+(x,y) = \delta$ then from our previous remark it follows that

$$g(x+\lambda y) \le \lambda\delta + g(x), \qquad \lambda>0.$$

Define

$$h(\lambda,t) = \frac{G(t,x+\lambda y)-g(x)}{\lambda}, \quad \lambda>0,$$

then for each $t\in T$, $h(\lambda,t)$ is a nondecreasing function of λ because

$$h(\lambda,t) = \frac{G(t,x+\lambda y)-G(t,x)}{\lambda} - \frac{g(x)-G(t,x)}{\lambda}.$$

Therefore the sets $A_\lambda=\{t:t\in T,\ h(\lambda,t)\le\delta\}$ are nonempty, closed and nested for $\lambda>0$. Thus the compactness of T implies that there exists a $t_o\in\bigcap_{\lambda>0}A_\lambda$, i.e.,

$$G(t_o,x+\lambda y) \ge \lambda\delta+g(x), \quad \lambda>0.$$

Thus $t_o\in S(x)$, $G_+(t_o,x,y) \ge \delta$, and therefore

$$g_+(x,y) \le G_+(t_o,x,y).$$

Remark. If max is replaced by min and convex by concave Theorem D is obviously still valid.

Theorem 12. Let $\epsilon\ge0$. If $\epsilon>0$ or IK is a neighborhood of the origin then the function $\phi_\epsilon:IK\to R$ is differentiable at the origin if and only if there is a unique optimal linear algorithm L_o. In this case, $\phi_\epsilon'(0) = L_o$.

Proof. Given any $y\in Y$ for $|\lambda|$ sufficiently small, $w = \lambda y$ is an interior point of IK. Hence by Theorem 9

$$\phi_\epsilon(w) = \min_{L\in M}\ (\sup_{x\in K}|Ux-LIx|+Lw+\epsilon||L||).$$

The function $H(L,w) = \sup_{x\in K}|Ux-LIx|+Lw+\epsilon||L||$ is weak* continuous in L and linear in w. Thus by Theorem D if we set B = the set of all optimal linear algorithms then

$$(\phi_\epsilon')_+(0,y) = \lim_{\lambda\to0^+}\frac{\phi_\epsilon(\lambda y)-\phi_\epsilon(0)}{\lambda} = \min_{L\in B} Ly$$

and

$$(\phi_\epsilon)_-'(0,y) = \max_{L\in B} Ly \quad .$$

The theorem easily follows from these formulas.

For a related result see Osipenko [39].

Finally, as promised we examine the complex case. First note that for any $L \in Y^*$

$$\sup_{\substack{x \in K \\ ||y|| \le 1}} (Ux - LIx + \varepsilon Ly) \ge \sup_{\substack{x \in K \\ ||Ix|| \le \varepsilon}} (Ux - LIx + \varepsilon L(\tfrac{Ix}{\varepsilon})) \ge \sup_{\substack{x \in K \\ ||Ix|| \le \varepsilon}} Ux \quad .$$

Then if the hypothesis of Theorem 3 is assumed, (12) and Theorem 6 imply

$$d(K,\varepsilon) = \sup_{\substack{x \in K \\ ||Ix|| \le \varepsilon}} Ux \le \inf_{L \in Y^*} \sup_{\substack{x \in K \\ ||y|| \le 1}} (Ux - LIx + \varepsilon Ly) \le d(K,\varepsilon).$$

Thus,

$$\sup_{\substack{x \in K \\ ||Ix|| \le \varepsilon}} |Ux| = \inf_{L \in Y^*} \sup_{\substack{x \in K \\ ||y|| \le 1}} (Ux - LIx + \varepsilon Ly).$$

Suppose now that X is a linear space over $\mathbb{C}$, the complex numbers, that K is balanced relative to $\mathbb{C}$, i.e. $\lambda x \in K$ for $x \in K$ and $|\lambda| = 1$, $\lambda \in \mathbb{C}$. Also let $Z = \mathbb{C}$ and let $Y^*(\mathbb{C})$ be the complex valued continuous linear functionals on Y, that is, $L \in Y^*(\mathbb{C})$ if, and only if, $L = L_1 + iL_2$, $L_1, L_2 \in Y^*$.

Clearly,

$$\sup_{\substack{x \in K \\ ||Ix|| \le \varepsilon}} |Ux| = \sup_{\substack{x \in K \\ ||Ix|| \le \varepsilon}} \operatorname{Re} Ux \quad .$$

But by applying our results in the real case with U replaced by ReU we obtain

$$\sup_{\substack{x \in K \\ ||Ix|| \le \varepsilon}} \operatorname{Re}Ux = \inf_{L_1 \in Y^*} \sup_{\substack{x \in K \\ ||y|| \le 1}} (\operatorname{Re} Ux - L_1 Ix + \varepsilon L_1 y)$$

$$= \inf_{L \in Y^*(\mathbb{C})} \sup_{\substack{x \in K \\ ||y|| \le 1}} \operatorname{Re}(Ux - LIx + \varepsilon Ly),$$

$$= \inf_{L \in Y^*(\mathbb{C})} \sup_{\substack{x \in K \\ ||y|| \le 1}} |Ux - LIx + \varepsilon Ly|,$$

in view of the fact that K and $||y|| \le 1$ are balanced.

It remains now to show that for any $z \in \mathbb{C}$,

$$\sup_{||y||\le 1} |z + \varepsilon Ly| = |z| + \varepsilon||L|| ,$$

a fact which is easily established. Thus we conclude that Theorem 6 continues to hold in the complex case.

In summary, then, in this section we have shown that under rather weak hypotheses a linear functional can be recovered optimally by means of a linear functional of the information. The next section is devoted to examples illustrating this fact.

3. EXAMPLES OF OPTIMAL RECOVERY OF LINEAR FUNCTIONALS BY LINEAR METHODS

Unless otherwise specified we suppose K to be a balanced convex subset of X, a linear space over $\mathcal{R}$, and that $Z = \mathcal{R}$. Also we suppose that K is such that the additional hypotheses in Theorems 6 and 8 are satisfied.

Example 3.1. Let $Y = \mathcal{R}^n$ normed with the sup norm, and suppose that for $x \in X$

$$Ix = (I_1 x, \ldots, I_n x)$$

where I_j, $j = 1,\ldots,n$ are linear functionals on X. Suppose also that the feasibility condition holds, then Theorem 6 implies that

$$\sup_{\substack{x\in K \\ |I_j x|\le\varepsilon}} Ux = \min_{c_i} \{\sup_{x\in K} |Ux - \sum_{i=1}^{n} c_i I_i x| + \varepsilon \sum_{i=1}^{n} |c_i|\}. \tag{19}$$

3.1.1. For example. Take $X = K = P_{n-1}$, polynomials of degree at most n-1. Take $I_i x = x(t_i)$, $i = 1,\ldots,n$ where t_i are distinct points of the real line. If $x(t) = a_o + a_1 t + \ldots + a_{n-1} t^{n-1}$ suppose $Ux = a_o u_o + \ldots + a_{n-1} u_{n-1}$. Then

$$Ux - LIx = \sum_{i=0}^{n-1} a_i u_i - \sum_{j=1}^{n} c_j x(t_j)$$

$$= \sum_{i=0}^{n-1} a_i (u_i - \sum_{j=1}^{n} c_j t_j^i),$$

from which it is clear that the feasibility condition is satisfied if and only if

$$u_i = \sum_{j=1}^{n} c_j t_j^i , \quad i = 0,\ldots,n-1.$$

These equations determine $c_1,\ldots,c_n$ uniquely, and if we let

$\ell_i(t)\epsilon P_{n-1}$ satisfy $\ell_i(t_j) = \delta_{ij}$, then $c_i = U\ell_i$. Thus (19) becomes, for $\varepsilon>0$,

$$\max_{|x(t_i)|\le\varepsilon} Ux = \varepsilon\sum_{i=1}^{n} |U\ell_i|.$$

A linear optimal algorithm is seen to consist of Lagrange interpolation followed by U, as is trivially the case when $\varepsilon = 0$.

3.1.2. Suppose W is a normed linear space and $w_1,\dots,w_n\epsilon W$ are linearly independent. Choose $X = W^*$,

$$K = \{T\epsilon X : ||T|| \le 1\},$$

$I_iT = Tw_i$, $i=1,\dots,n$ and $UT = Tw$ for some $w\epsilon W$. Now the feasibility condition is clearly satisfied, therefore (19) yields

$$\begin{aligned}\max_{\substack{||T||\le 1\\ |Tw_i|\le\varepsilon}} Tw &= \min_{c_j} \{\sup_{||T||\le 1} |Tw - \sum_{j=1}^{n} c_jTw_j| + \varepsilon\sum_{i=1}^{n}|c_j|\}.\\ &= \min_{c_j} \{||w-\sum_{j=1}^{n} c_jw_j|| + \varepsilon\sum_{j=1}^{n} |c_j|\}.\end{aligned} \tag{20}$$

when $\varepsilon = 0$, (20) reduces to the familiar duality formula

$$\max_{\substack{||T||\le 1\\ TM = 0}} Tw = \mathrm{dist}(w,M) = \min_{c_j} ||w - \sum_{j=1}^{n} c_jw_j|| , \tag{21}$$

where M is the linear space spanned by $w_1,\dots,w_n$. Note that when $\varepsilon>0$, (20) depends on a particular choice of basis in M, which is not the case in (21).

Example 3.2 Suppose $X = (X,|\cdot|)$ and $Y=(Y,||\cdot||)$ are Hilbert spaces and $K = \{x\epsilon X:|x|\le 1\}$. Let $[x,x] = |x|^2$, $(y,y) = ||y||^2$, $Ux = [u,x]$, $u\epsilon X$, $u\ne 0$ and suppose I is a bounded linear operator from X onto Y. The feasibility condition being obviously satisfied we have, according to our theory

$$\max_{\substack{|x|\le 1\\ ||Ix||\le\epsilon}}[u,x] = \min_{y\epsilon Y} (\max_{|x|\le 1} ([u,x] - (y,Ix) + \varepsilon||y||). \tag{22}$$

The adjoint $I^* : Y \to X$ of I is defined by

$$(y,Ix) = [I^*y,x] , \quad x\epsilon X,\ y\epsilon Y$$

and (22) becomes

$$\max_{\substack{|x|\le 1\\ ||Ix||\le\epsilon}}[u,x] = \min_{y\epsilon Y} (|u-I^*y| + \varepsilon||y||). \tag{23}$$

Let $x(\varepsilon)$ be a worst function and $y(\varepsilon)$ an optimal linear algorithm (see Theorems 8 and 10 of Section 2). Clearly, $y(\varepsilon)$ is unique since (23) shows it to be the value at which a strictly convex function attains its minimum. According to Theorem 11 we have

$$\varepsilon||y(\varepsilon)|| = (y(\varepsilon), Ix(\varepsilon)) \tag{24}$$

and

$$\max_{|x|\leq 1} [u-I^* y(\varepsilon),x] = [u-I^* y(\varepsilon), x(\varepsilon)]. \tag{25}$$

Using the criteria for equality in Schwarz's inequality we see that (24) and (25) are equivalent to

$$y(\varepsilon) = 0 \text{ or } Ix(\epsilon) = \frac{\varepsilon y(\varepsilon)}{||y(\varepsilon)||} \tag{26}$$

and

$$v = u-I^* y(\varepsilon) = 0 \text{ or } x(\varepsilon) = \frac{v}{|v|}. \tag{27}$$

It is convenient to analyze these equations by considering several cases.

Case 1. $y(\varepsilon) = 0$.

In this case $v=u \neq 0$, hence $x(\varepsilon) = v/|v| = u/|u|$, and since, by definition, $||Ix(\epsilon)||\leq\varepsilon$ we conclude that

$$\varepsilon_1^2(u) = \varepsilon_1^2 = \frac{(Iu,Iu)}{[u,u]} \leq \varepsilon^2. \tag{28}$$

Conversely, if (28) holds, then $x(\varepsilon) = u/|u|$ satisfies $|x(\varepsilon)|\leq 1$, $||Ix(\varepsilon)||\leq\varepsilon$ and is a worst function. Hence $e(K,\varepsilon) = |u|$ and therefore, since the minimum problem in (23) has a unique solution it must be $y(\varepsilon) = 0$. Note that we have also proved that if $\varepsilon\geq\varepsilon_1$ then $x(\varepsilon) = u/|u|$ is the unique worst function.

Case 2. $y(\varepsilon) \neq 0$ and $v \neq 0$.

We may solve (26) and (27) for $x(\varepsilon)$ and $y(\epsilon)$. To this end we define the linear operators

$$P_\lambda = (II^* + \lambda J)^{-1} I$$

and

$$Q_\lambda = J - I^* P_\lambda$$

where J denotes the identity map. Since I maps X onto Y the mappings P_λ and Q_λ are well-defined for $\lambda\geq 0$. A simple calculation shows that

$$y(\varepsilon) = P_\mu u \quad ,$$

and

$$x(\epsilon) = \frac{Q_\mu u}{|Q_\mu u|}.$$

μ is the uniquely determined solution of the equation, $g(\lambda) = \varepsilon$

where

$$g(\lambda) = \lambda \frac{||P_\lambda u||}{|Q_\lambda u|}.$$

Again in this case $x(\varepsilon)$ is unique.

Case 3. $u \in R(I^*)$ = range of I^*.

Suppose $u = I^* t$ for $t \in Y$. We claim that $v = o$ if, and only if,

$$(29) \qquad \varepsilon^2 \le \varepsilon_o^2(t) = \varepsilon_o^2 = \frac{(t,t)}{((II^*)^{-1}t,t)}.$$

Clearly, if $v=0$ then $I^* y(\varepsilon) = u$, and since I maps X onto Y, $y(\varepsilon)=t$ and $x(\varepsilon)$ is any vector in X with $|x(\varepsilon)| \le 1$ and $Ix(\varepsilon) = \varepsilon t/||t||$. Let s be any element of Y then, if $\varepsilon > 0$,

$$(s,t) = (s, Ix(\varepsilon)) \frac{||t||}{\varepsilon} = [I^* s, x(\varepsilon)] \frac{||t||}{\varepsilon},$$

so that, for $\varepsilon \ge 0$

$$(30) \qquad \varepsilon \le \frac{|I^* s|\,||t||}{|(s,t)|}.$$

If we choose $s = (II^*)^{-1}t$ we obtain (29).

Conversely, suppose (29) holds.
Since, for any $s \in Y$

$$|I^* s| \ge \frac{[I^* s, I^*(II^*)^{-1}t]}{|I^*(II^*)^{-1}t|} = \frac{(s,t)}{((II^*)^{-1}t,t)^{1/2}}$$

so that

$$\min_{s \in Y} \frac{|I^* s|}{|(s,t)|} = \frac{\varepsilon_o}{||t||},$$

then, (29) implies that (30) holds. Now

$$||t|| - ||t-s|| \le \frac{|(s,t)|}{||t||},$$

and we conclude that for $s \in Y$

$$\varepsilon ||t|| \le |I^* s| + \varepsilon ||t-s||,$$

or, equivalently,

$$\min_{y \in Y} (|u - I^* y| + \varepsilon ||y||) = \varepsilon ||t||,$$

hence $y(\varepsilon) = t$ and $v = 0$.

Note that we also have the inequality

$$\varepsilon_o(t) \le \varepsilon_1(u).$$

For if $\varepsilon_1(u) \le \varepsilon < \varepsilon_o(t)$ then $v = y(\varepsilon) = 0$ which implies $u = 0$, contrary to our assumption.

In summary, we see that if $u \in R(I^*)$ we have the following classification according to the size of ε.

Case 3.1. $\varepsilon \leq \varepsilon_o$

$y(\varepsilon) = t$ and $x(\varepsilon)$ is any vector in X such that

$$Ix(\varepsilon) = \epsilon \frac{t}{||t||} \text{ and } |x(\varepsilon)| \leq 1.$$

Case 3.2. $\varepsilon_o < \varepsilon < \varepsilon_1$.

$x(\varepsilon)$ and $y(\varepsilon)$ are as described in case 2.

Case 3.3. $\varepsilon_1 \leq \varepsilon$.

$x(\varepsilon)$ and $y(\varepsilon)$ are as described in case 1.

Finally, let us note that

$$\lim_{\lambda\to\infty} g(\lambda) = \varepsilon_1$$

while

$$\lim_{\lambda\to 0} g(\lambda) = \begin{cases} \varepsilon_o(t), & \text{if } u = I^*t \\ 0, & \text{if } u \notin R(I^*). \end{cases}$$

Since $y(\varepsilon)$ is unique we also conclude that $g(\lambda)$ is strictly increasing for $\lambda \geq 0$, a fact that we used in our discussion of Case 2.

Observe that in Case 3, when $u = I^*t$, $t \in Y$, then $Ux = [u,x] = (t,Ix)$. Thus Ux is a function of the exact information. However, the choice $y(\varepsilon) = t$, continues to be optimal only for $\varepsilon \leq \varepsilon_o$.

3.2.1. As an example consider the problem of predicting the value of a function f at time $T > 0$ from its values in the past, $\{f(t), t<0\}$, which we can compute with error at most $\varepsilon>0$. The case $\varepsilon=0$ is implicit in the discussion of an example given by Wiener [48, p. 65].

Let $X = \{f: f, f' \in L^2(R)\}$ and suppose

$$[f,h] = \int_{-\infty}^{\infty} f(t)h(t)dt + \int_{-\infty}^{\infty} f'(t)h'(t)dt.$$

Take $Y = L^2(-\infty,0)$ and suppose $(f,h) = \int_{-\infty}^{0} f(t)h(t)dt$.

Let I be the identity mapping of X onto Y given by $(If)(t) = f(t)$, $t<0$.

The space X has a reproducing kernel $K(x,t) = \frac{1}{2}e^{-|x-t|}$ so the adjoint of I is seen to be

$$(I^*h)(t) = \frac{1}{2} \int_{-\infty}^{0} e^{-|x-t|} h(x)dx.$$

The linear functional $Uf = f(T)$ can be represented as $[u,f]$ where

$u(t) = \frac{1}{2}e^{-|T-t|}$, $t\in\mathcal{R}$. Since $f(T)\notin Y^*$ we conclude that $u\notin R(I^*)$. Thus, according to cases 1 and 2 of Example 3.2, an optimal algorithm is obtained by estimating $f(t)$ by <u>zero</u> if $\varepsilon\geq\varepsilon_1=(e^{-T})/2$, while if $\varepsilon<\varepsilon_1$ we must examine the integral equation

$$\frac{1}{2}\int_{-\infty}^{0} e^{-|t-x|}y(x)dx + \lambda y(t) = \frac{1}{2}e^{-|T-t|}, \quad t<0, \tag{31}$$

where $y\in L^2(-\infty,0)$ and $\lambda>0$. The solution of (31) is

$$y\lambda(t) = ((1+\lambda^{-1})^{1/2}-1)e^{-T}e^{(1+\lambda^{-1})^{1/2}t}$$

and

$$g(\lambda)=\frac{\lambda||y_\lambda||}{|u-I^*y_\lambda|}=\frac{1}{((1+\lambda^{-1})^{1/2}+1)(((1+\lambda^{-1})^{1/2}-1)(e^{2T}-1)+e^{2T})^{1/2}} e^{(1+\lambda^{-1})^{1/2}t}.$$

Thus when $\varepsilon<(e^{-T})/2$ the optimal recovery of f satisfying $|f|\leq 1$, from the information, $\{f(t),\ t<0\}$, is given by

$$(\beta-1)\int_{-\infty}^{0} e^{-T}e^{\beta x}f(x)dx,$$

where β is uniquely determined by the equation

$$(\beta+1)^2\ [(\beta-1)\ (e^{2T}-1) + e^{2T}] = \varepsilon^{-2}\ .$$

<u>Example 3.3</u> We now consider optimal recovery of the derivative of a smooth function. Let
$W^{2m}(\mathcal{R}) = \{f: f^{(2m-1)}$ abs. cont. on every finite interval, $f^{(2m)}\in L^\infty(\mathcal{R})\}$, for $m\geq 1$.
Given real numbers $0\leq t_1\leq\ldots\leq t_m$, define

$$(Lf)(x) = \prod_{j=1}^{m}\left(\frac{d^2}{dx^2} - t_j^2\right)f(x).$$

Let, $X = W^{2m}(\mathcal{R})\cap L^\infty(\mathcal{R})$ equipped with the ess sup norm,

$$K = \{f\in X : ||Lf||_\infty\leq 1\},$$

$Uf= f'(0)$, $Y = L^\infty(\mathcal{R})$ and I be the identity operator. The intrinsic error in this case is

$$\sup_{\substack{||f||_\infty\leq\varepsilon \\ ||Lf||_\infty\leq 1}}|f'(0)|\ .$$

Micchelli [25] gives a detailed discussion of this extremal problem, as well as determining a best linear algorithm, which, in general, requires sampling the information at a denumerably infinite number of points. In the special case that $m=1$ and $t_1=0$ his results, in agreement with those of Newman [37] and Schoenberg [45] for this simple situation, yield

$$Ag = \frac{g(t)-g(-t)}{2t}$$

where $t = (2\varepsilon)^{1/2}$ and $g(t)$ is the available data.

An optimal algorithm for the differentiation operator is easily obtained from Micchelli's result by replacing $f(x)$ by $f(x+\tau)$.

Example 3.4 Next we consider optimal recovery of the value of a function at a given point from a finite sampling of the function, which is assumed to be "smooth" on an interval. This leads to optimal recovery of the function globally. To be precise, suppose $0 \le x_1 \le x_2 \le \ldots \le x_{n+r} \le 1$ with $x_i < x_{i+n}$, $n \ge 1$ and $i = 1,\ldots,r$. Let $X = W^n[0,1] = \{f \in C^{n-1}[0,1] : f^{(n-1)}$ abs. cont.; $f^{(n)} \in L^\infty[0,1]\}$, equipped with the sup norm, and take $K = \{f \in X : ||f^{(n)}|| \le 1\}$. Choose Y to be R^{n+r} and let I be the mapping that assigns to $f \in X$ the vector $(f(x_1),\ldots,f(x_{n+r}))$ with the convention that if

$x_{i-1} < x_i = x_{i+1} = \ldots = x_{i+s} < x_{i+s+1}$ then $f(x_{i+j}) = f^{(j)}(x_i)$, $j = 1,\ldots,s$. Finally, suppose $Uf = f(\tau)$ for some fixed τ in $[0,1]$ and $\epsilon = 0$.

The general theory of Sections 1 and 2 provides us with several approaches to this problem. The first that we discuss is based on finding the midpoint of the interval

$$H(y) = \{f(\tau) : ||f^{(n)}|| \le 1,\ f(x_i) = y_i,\ i = 1,\ldots,n+r\}.$$

This problem may be solved by appealing to a result of Micchelli and Miranker [29] on envelopes of smooth functions.

Let m, M, $m<M$, be two given constants. A function, P, is called an (m,M) perfect spline of degree n with knots $\xi_1,\ldots,\xi_k$ satisfying $-\infty = \xi_0 < \xi_1 \ldots < \xi_k < \xi_{k+1} = \infty$ if: i) P is a polynomial of degree at most n in each interval (ξ_i,ξ_{i+1}), $i = 0,\ldots,k$. ii) $P \in C^{n-1}(-\infty,\infty)$. and iii) $P^{(n)}(t) = m$, $\xi_0 < t \le \xi_1$, $P^{(n)}(t) = M$, $\xi_1 < t \le \xi_2$ and so on alternately. Thus the n-th derivative of an (m,M) perfect spline begins with m. If P is either an (m,M) or an (M,m) perfect spline we say it is an $\{m,M\}$ perfect spline. An $\{m,-m\}$ perfect spline is called a perfect spline.

The following result was proved in [29] where applications to finding the zero of real valued functions on an interval were given.

Theorem E. Given $0\le x_1\le\ldots\le x_{n+r}\le 1$, $y:(y_1,\ldots,y_{n+r})$ and $m<M$. If there exists an $f\in W^n[0,1]$ with

$$(32)\qquad f(x_i)=y_i, i=1,\ldots,n+r \;;\; m\le f^{(n)}(t)\le M,\; 0\le t\le 1,$$

then there exist two $\{m,M\}$ perfect splines of degree n, P_m and P_M, with $\le r$ knots in (x_1,x_{n+r}) such that

$$P_m(x_i) = P_M(x_i) = y_i,\; i=1,2,\ldots,n+r,$$

and for every f satisfying (32)

$$\begin{aligned}(-1)^{n+r+i}(P_M(x)-f(x)) &\ge 0 ,\\ (-1)^{n+r+i}(P_m(x)-f(x)) &\le 0\end{aligned}\qquad x_i<x<x_{i+1}$$

$(x_o = 0,\ x_{n+r+1} = 1)$.

Before we comment on the use of this theorem we will digress from Example 3.4 to describe the proof given in [29].

The first observation to be made is the fact that the interpolation conditions $f(x_i) = y_i$, $i = 1,2,\ldots,n+r$ are easily seen to be equivalent to the following moment conditions on $f^{(n)}$. Let $x_1=a$, $x_{n+r} = b$, then

$$\int_a^b M(x_i,\ldots,x_{i+n};t)f^{(n)}(t)dt = [y_i,\ldots,y_{i+n}],\; i=1,\ldots,r,$$

where $z_i = [y_i,\ldots,y_{i+n}]$ is the divided difference of $y_i,\ldots,y_{i+n}$ at $x_i,\ldots,x_{i+n}$ and $u_i(t) = M(x_i,\ldots,x_{i+n};t)$ is similarly the divided difference of $\frac{1}{(n-1)!}(x-t)_+^{n-1}$ at $x = x_i,\ldots,x_{i+n}$. This reduction to moment conditions on $f^{(n)}$ is also used by H. Burchard [6] who examined the analog Theorem E corresponding to m=0 and $M=\infty$ (of course, by this we mean that $df^{(n-1)}(x)$ is to be interpreted as a nonnegative measure). For further references concerning this problem see [6].

Now the main fact we require about the functions $u_1(t),\ldots,$ $u_r(t)$ is that they form a weak Chebyshev system. This means that $u_1,\ldots,u_r$ are continuous (we may assume that $x_i<x_{i+n-1}$ since otherwise the problem splits into independent subproblems on each of which $u_i(x)$ is continuous) linearly independent functions on [0,1] and

$$\begin{vmatrix} u_1(t_1) & \cdots & u_1(t_r) \\ \vdots & & \vdots \\ u_r(t_1) & \cdots & u_r(t_r) \end{vmatrix} \geq 0$$

for all $0<t_1<\ldots<t_r<1$ (the results in [6] also depend on this property). A Chebyshev system has the additional property that strict inequality always holds above.

It is the case that every weak Chebyshev system may be "smoothed" into a Chebyshev system, Cf. Karlin and Studden [16]. Thus for $\delta>0$ there exists a sequence $u_1(x;\delta),\ldots,u_r(x;\delta)$ of functions which is a Chebyshev system and $\lim_{\delta\to 0^+} u_i(x;\delta) = u_i(x)$, $x\in(0,1)$, boundedly in δ and uniformly on every compact subinterval of (0,1). Now, the proof of the existence of P_m and P_M follows easily from the following theorem of Krein, Cf. [18], [19] or [16]

The (restricted) moment space generated by $u_1,\ldots,u_r$ is defined as

$$F_u(m,M)=\{(\int_a^b u_1(t)h(t)dt,\ldots,\int_a^b u_r(t)h(t)dt): m\leq h(t)\leq M, \text{ a.e., } t\in[a,b]\}.$$

We will say $h\in L^\infty$, $m\leq h(t)\leq M$, a.e. $t\in[a,b]$ represents $z\in F_u(m,M)$, if $z_i = \int_a^b u_i(t)h(t)dt$. Equivalently h represents $z=(z_1,\ldots,z_r)$ $z_i = [y_i,\ldots,y_{i+n}]$ iff there is an $f\in W^n[0,1]$ with $f(x_i)=y_i$, $i=1,2,\ldots,n+r$ and $f^{(n)} = h$ on $[a,b]$.

F is a convex compact subset of $\mathbb{R}^r$ with nonempty interior.

Theorem F. Let $v_1,\ldots,v_r$ be a Chebyshev system on [0,1]. Then given any $w = (w_1,\ldots,w_r)$ in the interior of $F_v(m,M)$ there exists two distinct step functions h_m,h_M, which represent w, and have exactly r jumps; each function satisfies $(h(x)-m)(h(x)-M) = 0$, a.e., $x\in[0,1]$, and $h_m(x) = m$, $h_M(x) = M$ in some interval beginning at $x=0$.

The procedure to prove Theorem E is clear; we apply Theorem F to the smoothed system $\{u_1(x;\delta),\ldots,u_r(x;\delta)\}$ and then pass to the limit as $\delta\to 0^+$.

It is interesting to note here that the case $m=0$, $M=\infty$ treated by Burchard encounters at the stage "$\delta\to 0^+$" some difficulties which do not enter into our analysis. In particular, Burchard is dealing with measures and for general data $(z_1,\ldots,z_r)$, as $\delta\to 0^+$, infinite mass at either $x = 0$ or $x = 1$ may occur. However, this difficulty does not always occur, see Micchelli and Pinkus, [31].

Proof of Theorem E: Let $z = (z_1,\ldots,z_r)$, $z_i = [y_i,\ldots,y_{i+n}]$, $i = 1,\ldots,r$. Then our hypothesis states that $z\in F_u(m,M)$. Clearly then for any $m'<m<M<M'$, z is in the interior of $F_u(m',M')$ and for fixed m',M' z is likewise in the interior of $F_{u_\delta}(m',M')$ (the moment space determined by the smoothed system) for δ sufficiently small. Now, according to Theorem F there are two step functions $h_{m'}(t;\delta)$, $h_{M'}(t;\delta)$ which represents z. We define $h_{m'}(t)=\lim_{\delta\to 0^+} h_{m'}(t;\delta)$ (through a convergent subsequence), and $P^{(n)}_{m'}(t)=h_{m'}(t)$ and similarly for $P_{M'}(t)$. Then (by adding on the appropriate polynomials to $P_{m'}$ and $P_{M'}$) $P_{m'}(x_i) = P_{M'}(x_i) = y_i$, $i=1,2,\ldots,n+r$, and $P_{m'}$, $P_{M'}$ have $\leq r$ knots.

Let f be any function in $W^n[0,1]$ with $f(x_i)=y_i$, $i=1,2,\ldots,n+r$ $m\leq f^{(n)}(t)\leq M$, a.e. $t\in[0,1]$, and define $g=P_{M'}-f$. Then g vanishes at each x_i and hence $g^{(n-1)}$ has at least $r+1$ zeros. However, $g^{(n-1)}$ is clearly strictly monotonic between the knots of $P_{M'}$. Thus $P_{M'}$ must have exactly r knots and strictly cross f only at each x_i. Since $g^{(n-1)}(x)$ is increasing beyond its last zero we have $(-1)^{n+r+i}$ $(P_{M'}(x)-f(x))>0$, for $x_i<x<x_{i+1}$. A similar analysis shows that $(-1)^{n+r+i}$ $(P_{m'}(x)-f(x))<0$, for $x_i<x<x_{i+1}$. Letting $m'\to m$, $M'\to M$ proves the theorem.

Remark. The extremal properties of P_m and P_M are not surprising. It is well known that the principal representation described in Theorem F extremizes the "next moment". This fact is one form of the Markoff-Krein inequality (Cf., [14], Chapter 4, §8) and, of course, in our context, the next moment is $f(x)$. Our proof that P_m, P_M are envelopes uses Rolle's theorem which is simple and direct. This type of argument is, of course, the basis for the general form of the Markoff-Krein inequalities discussed in [16], [19].

Let us note some further important properties of P_m and P_M.

First we suppose that $z = (z_1,\ldots,z_r)$, $z_i = [y_i,\ldots,y_{i+n}]$, $i = 1,\ldots,r$ is an <u>interior</u> point of $F_u(m,M)$. Then clearly there is a function f_o such that

$$f_o(x_i) = y_i,\ i = 1,2,\ldots,n+r,\ m<f_o^{(n)}(t)<M, \text{a.e.}, t\in[0,1].$$

Since $f_o^{(n)}(t)$ a.e. $t\in[0,1]$ lies <u>strictly</u> in the closed interval $[m,M]$ we may use Rolle's theorem as before, comparing f_o to P_m, P_M and conclude that P_m and P_M have exactly r knots, intersect only at x_i and are respectively (m,M), (M,m) perfect splines. Now, this same argument also shows the following additional fact.

Let $\xi_1<\ldots<\xi_r$ be the knots of P_M (a similar argument holds for P_m) and suppose $s(x)$ is any spline function of degree $n-1$ with knots at $\xi_1,\ldots,\xi_r$,

$$s(t) = \sum_{j=o}^{n-1} a_j t^j + \sum_{j=1}^{r} c_j (t-\xi_j)_+^{n-1}$$

which vanishes at $x=x_1,\ldots,x_{n+r}$. Then for <u>any</u> λ, $f_o+\lambda s$ is in $W^n(\xi_i,\xi_{i+1})$, $i = 0,.,\ldots,r$ and hence we conclude by the previous Rolle's theorem argument that the function $f_o+\lambda s-P_M$ has a constant sign in each of the intervals, $[0,x_1]$, $[x_1,x_2],\ldots,[x_{n+r},1]$. Thus letting $\lambda\to\pm\infty$ we conclude that $s\equiv 0$.

We have thus shown that the determinant

$$K\begin{pmatrix}0,1,\ldots,n-1,\xi_1,\ldots,\xi_r\\ x_1,x_2,\ \ldots\ , x_{n+r}\end{pmatrix} = \begin{vmatrix} 1 & 1 & \cdots & 1 \\ x_1 & x_2 & \cdots & x_{n+r} \\ \vdots & & & \vdots \\ x_1^{n-1} & & \cdots & x_{n+r}^{n-1} \\ (x_1-\xi_1)_+^{n-1} & & & (x_{n+r}-\xi_1)_+^{n-1} \\ \vdots & & & \vdots \\ (x_1-\xi_r)_+^{n-1} & & \cdots & (x_{n+r}-\xi_r)_+^{n-1} \end{vmatrix}$$

is non zero. It is known that for <u>any choice of</u> ξ_i, x_i, $\xi_1<\ldots<\xi_r$, $x_1<\ldots<x_{n+r}$ the above determinant is nonnegative. Using these facts, we will now show that P_M (as well as P_m) is <u>unique</u>. This will be shown by demonstrating that any (M,m) perfect spline interpolating the data and having r knots is alternately the upper and lower envelope for all functions satisfying (32).

Let $x_i<x<x_{i+1}$ and define

$$u(t) = \frac{K\begin{pmatrix} 0,1,\ldots,n-1,\xi_1,\ldots,\xi_r,t \\ x_1,x_2,\ldots,x_{n+r},x \end{pmatrix}}{K\begin{pmatrix} 0,1,\ldots,n-1,\xi_1,\ldots,\xi_r \\ x_1,x_2,\ldots,x_{n+r} \end{pmatrix}}$$

where $\xi_1,\ldots,\xi_r$ are the knots for an (M,m) perfect spline with $P_M(x_i) = y_i$, $i=1,\ldots,n+r$. Then $(-1)^{n+i+\ell}u(t)\geq 0$, $\xi_\ell<t<\xi_{\ell+1}$ and $u(t) = 0$, $t\notin(x_1,x_{n+r})$. Thus by Taylor's theorem with remainder (or integration by parts)

$$\begin{aligned} P_M(x)-f(x) &= \int_{x_1}^{x_{n+r}} u(t)\ (P_M^{(n)}(t)-f^{(n)}(t))dt \\ &= (-1)^{n+r+i} \int_{x_1}^{x_{n+r}} |u(t)|\,|P_M^{(n)}(t)-f^{(n)}(t)|dt. \end{aligned}$$

Hence $(-1)^{n+r+i}(P_M(x)-f(x))\geq 0$, $x\in(x_i,x_{i+1})$. This proves the uniqueness of P_M and a similar argument holds for P_m.

We end this discussion by making some observations about representations of boundary points of $F_u(m,M)$. First we recall some general facts, due to Krein, about restricted moment spaces.

A point $z=(z_1,\ldots,z_r)$ is in $F_u(m,M)$ if and only if for every $a_1,\ldots,a_r$

$$\sum_{i=1}^{r} a_iz_i\leq M \int_a^b \Big(\sum_{i=1}^{r} a_iu_i(t)\Big)_+dt+m\int_a^b \Big(\sum_{i=1}^{r} a_iu_i(t)\Big)_-dt \tag{33}$$

$((f(t))_+ = \max\{f(t),0\}$, $(f(t))_- = \min\{f(t),0\})$. z is a boundary point of $F_u(m,M)$ if and only if there is some nontrivial function $u^o(t) = \sum_{i=1}^{r} a_i^ou_i(t)$ which yields equality above. Moreover, if $u^o(t)$ is any such function and h, $m\leq h(t)\leq M$, a.e. $t\in[a,b]$ is any representer of z then

(34) $$h(t) = \begin{cases} M, & \text{if } u^o(t) > 0 \\ m, & \text{if } u^o(t) < 0 \end{cases} .$$

It is these facts that prove see [16 p. 235] that a boundary point of a restricted moment space corresponding to a <u>Chebyshev system</u> is <u>uniquely</u> represented by a step function h, $(m-h(x))(M-h(x)) = 0$, a.e. $x \in [0,1]$ with $\leq r-1$ discontinuities. Conversely, any z represented by such a step function is necessarily a boundary point. For weak Chebyshev systems, in particular for the functions $u_1,\ldots,u_r$, the following facts are valid. If $z_i = \int_a^b u_i(t)h(t)dt$ where $(m-h(x))(M-h(x)) = 0$, a.e. $x \in [a,b]$ and h has $\leq r-1$ discontinuities then z is a boundary point.

<u>Proof</u>. Since for every $\delta>0$, $u_1(x;\delta),\ldots,u_r(x;\delta)$ is a Chebyshev system, there is a function,

$$u_\delta(x) = \sum_{i=1}^{r} a_i(\delta)u_i(x;\delta) \text{ such that}$$

$$h(t) = \begin{cases} M, & u_\delta(t) > 0 \\ m, & \text{otherwise} \end{cases}$$

Cf. [16]. We normalize the coefficients of u_δ so that $\sum_{i=1}^{r} a_i^2(\delta)=1$ and letting $\delta\to 0^+$ through a convergent subsequence we produce a non-trivial function $u^o(t) = \sum_{i=1}^{r} a_i^o u_i(t)$ with

$$h(t) = \begin{cases} M, & \text{if } u^o(t) > 0 \\ m, & \text{if } u^o(t) < 0 \end{cases} .$$

Hence, u^o gives equality in (33) and so z is a boundary point of $F_u(m,M)$.

Now, let us demonstrate that every boundary point may be so represented (but not uniquely).

The proof of this fact uses Lemma 6 of Micchelli [26]. There exists points $\mu_o = a<\mu_1<\ldots<\mu_k<b = \mu_{k+1}$, $k\leq r-1$ such that

$$\sum_{j=o}^{k} (-1)^j \int_{\mu_j}^{\mu_{j+1}} u_i(t)dt = \lambda\, w_i, \; i = 1,2,\ldots,r$$

$$w_i = z_i - \frac{M+m}{2} \int_o^1 u_i(t)dt$$

$$\lambda = \min_{\sum_{i=1}^r a_i w_i = 1} \int_a^b \left| \sum_{i=1}^r a_i u_i(t) \right| dt \quad .$$

Since z is a boundary point, (33) implies that $w=(w_1,\dots,w_r) \neq 0$ and $\lambda = \frac{2}{M-m}$. Hence

$$h(t) = \frac{M-m}{2}(-1)^j + \frac{M+m}{2}, \; \mu_i < t < \mu_{i+1}$$

is the desired representer of z.

Finally, let us observe that, since every nontrivial function $u^o(t) = \sum_{i=1}^r a_i^o u_i(t)$ must be nonzero on an interval of the form (x_ℓ, x_m), $m-\ell \geq n$, $1 \leq \ell < m \leq n+r$ we conclude from (34) that every representer of z agrees on some "core" subinterval (x_ℓ, x_m).

We summarize these facts in

Theorem G.

a) If $z = (z_1,\dots,z_r)$, $z_i = [y_i,\dots,y_{i+n}]$ is a boundary point of $F_u(m,M)$ then there is an $\{m,M\}$ perfect spline P of degree n, with $\leq r-1$ such that $P(x_i) = y_i$, $i=1,2,\dots,r$.

b) If z is a boundary point then there exists a "core" interval (x_ℓ, x_m), $m-\ell \geq M$ such that if $f(x_i) = y_i$, $i=1,2,\dots,n+r$ and $m \leq f^{(n)}(x) \leq M$, a.e. $x \in (x_1, x_{n+r})$ then $f(x) = P(x)$, $x \in (x_\ell, x_m)$.

c) If there is an $\{m,M\}$ perfect spline P with $\leq r-1$ knots such that $P(x_i) = y_i$, $i=1,2,\dots,n+r$ then z is on the boundary of $F_u(m,M)$.

This theorem is a statement, in the terminology of moment theory of some results of Fisher and Jerome [10] and Karlin [15] (see also de Boor [4]). To explain this further let $M = -m = L$ and suppose z is a boundary of $F_u(m,M)$. Then

$$L^{-1} = \min_{\sum_{i=1}^r a_i z_i = 1} \int_o^1 \left| \sum_{i=1}^r a_i u_i(t) \right| dt.$$

Now, according to Lemma 6 of[26]

$$L = \min\{\|h\|: \int_0^1 h(t)u_i(t)dt = z_i,\ i=1,\dots,r\}$$

(35)

$$= \min\{\|f^{(n)}\|: f(x_i) = y_i,\ i=1,2,\dots,n+r\},$$

$(\|\cdot\| = \text{ess sup on } (x_1, x_{n+r}))$.

Thus part a) is Karlin's result which states that the extremal problem (35) has a perfect spline solution with $\leq r-1$ knots. c) is also Karlin's observation that any perfect spline with $\leq r-1$ knots which interpolates the data solves the extremal problem. Part b) of Theorem G is Fisher and Jerome's result which shows that any two solutions of (35) agree on some core subinterval of (x_1, x_{n+r}). The Envelope Theorem gives sharp pointwise bounds on all solutions to the extremal problem. The envelope is necessarily squeezed together on the core subinterval and in general neither the upper or lower envelope is a Karlin solution to (35).

Returning to Example 3.4, we use Theorem E as follows:

Specializing Theorem E to $M = -m = 1$ we see that the midpoint of the interval $H(y)$ is $c(y) = (P_{-1}(\tau) + P_1(\tau)/2$, and its length is $|P_{-1}(\tau) - P_1(\tau)|/2$. Thus $c(y)$ is an optimal algorithm for recovering $f(\tau)$.

Gaffney and Powell [11] discovered Theorem E independently and used it as follows: As pointed out in Section 1, for any $\lambda, 0<\lambda\leq 1$,

$$\lim_{\lambda\to 0^+} \frac{P_{-1}(t;\lambda y)+P_1(t;\lambda y)}{2} = s(t;y)$$

is also an optimal algorithm. Gaffney and Powell proved that $s(t;y)$ is a spline interpolant of degree $n-1$ to the data. The knots of s are determined by the perfect spline of degree $n, q(t)$, defined by

$$\lim_{\lambda\to 0^+} \frac{P_1(t;\lambda y)-P_{-1}(t;\lambda y)}{2} = q(t).$$

q has r knots and is zero at each x_i, $i=1,\dots,n+r$. This determines q uniquely (see Micchelli, Winograd and Rivlin [35]).

The approach in Micchelli, Winograd and Rivlin [35] begins with $q(t)$, defines a linear recovery which is the spline interpolant $s(t;y)$ and proves directly that it is optimal. Neither approach uses the duality formula proved in Theorem 6, which reads in this case ,

$$\max_{\substack{f(x_i)=0,\, i=1,\ldots,n+r \\ ||f^{(n)}||\le 1}} |f(\tau)| = \min_{c_1,\ldots,c_{n+r}} \sup_{||f^{(n)}||\le 1} \left|f(\tau) - \sum_{i=1}^{n+r} c_i f(x_i)\right|.$$

The minimum problem can be written, using Taylor's theorem with remainder, as

$$\min \int_0^1 |S(t)|\,dt,$$

where the minimum is taken over all (spline) functions

$$S(t) = \frac{1}{(n-1)!}\{(\tau-t)_+^{n-1} - \sum_{j=1}^{n+r} c_j (x_j - t)_+^{n-1}\}\ .$$

subject to the boundary conditions $S^{(i)}(0) = 0$, $i=0,\ldots,n-1$. The remarks in [26] imply that this minimum problem has a unique solution. The perfect spline, $q(t)$, can then be obtained directly from the minimum problem. Also, since in this case the minimum problem has a unique solution

$$\phi(y) = \max_{\substack{f(x_i) = y_i \\ ||f^{(n)}||\le 1}} f(\tau)$$

is differentiable at zero and its derivative in the direction $w=(w_1,\ldots,w_{n+r})$ is $s(t;w)$.

Remarks. 1. Suppose that U is the identity operator in $W^n[0,1]$. Since the perfect spline q mentioned above is independent of τ it is clear that

$$\max_{\substack{f(x_i) = 0,\, i=1,\ldots,n+r \\ ||f^{(n)}||\le 1}} ||f|| = ||q||\ .$$

Moreover, the above-mentioned interpolating spline approximates f globally with error not exceeding $||q||$. It fails to be an optimal algorithm because it may lack one derivative of being in $W^n[0,1]$. However, by smoothing it appropriately we can produce a bona-fide algorithm as close to optimal as we desire. This proves that for the identity operator, U, we still have $e(K,0) = E(K,0)$.

These facts also hold if we measure the error in recovery in any monotone norm. Thus if $U:X\to L^p[0,1]$, $1\le p\le\infty$ the recovery error is $||q||_p$, the L^p norm of q.

2. Other linear functionals such as $f'(\tau)$ where $\tau = x_i$ or $\tau \in [0,1]-(x_1,x_{n+r})$ are served by the same optimal algorithm, as is clear from the proof of Micchelli, Rivlin, Winograd [35].

3. There is a set of nodes, x^*, with respect to which E(K,0) is minimal. These nodes are the zeros of the Chebyshev perfect spline of degree n with r knots in [0,1] when the error is measured in L^∞, see Tihomirov [46]. For the L^p case, $p \geq 1$, see Micchelli and Pinkus [30], [34].

4. There are also results known where X is

$$W_p^{(n)}[0,1] = \{f \in C^{n-1}[0,1]: f^{(n-1)} \text{abs. cont.}; f^{(n)} \in L^p[0,1]\},$$

$1 \leq p$. For instance, Bojanov [2] studies this problem for some restricted classes of nodes, **x**.

5. If X is taken to be the set of n-fold integrals of functions of bounded variation on [0,1] the L^1 norm, and K is the subset of functions stemming from those of total variation not exceeding 1, then Micchelli and Pinkus [33] obtain what may be considered the L^1 analog of the result mentioned in Remark 1. They exhibit an optimal algorithm, which is spline interpolation of the information and show that e(K,0) = E(K,0). They also determine the set of points at which to sample in order to minimize the intrinsic error. ([32])

6. A periodic version of this problem is studied by Bojanov, [3].

7. The optimal recovery of $f(\tau)$ from $f(x_1),\ldots,f(x_{n+r})$ when $\epsilon > 0$ is discussed in [28]. The approach is based on Theorem 6.

We now turn to some problems of optimal recovery of analytic functions in the plane.

<u>Example 3.5</u> Recall that H^∞ is the set of bounded analytic function in the unit disc $D: |z| < 1$. If $f \in H^\infty$, $||f|| = \sup|f(z)|$, $z \in D$. Put $X = H^\infty$ and let $K = \{f \in X: ||f|| \leq 1\}$. $z_1,\ldots,z_n$ are given points of D, $Y = \mathbb{C}^n$ and for each $f \in K$, $If = (f(z_1),\ldots,f(z_n))$, with the convention that if some of the points $z_1,\ldots,z_n$ coincide the corresponding functions values are replaced by consecutive derivatives in the obvious way. Suppose that $\zeta \in D$ is given, $Uf = f(\zeta)$ and $\varepsilon = 0$.

Let

$$B_n(z) = \prod_{j=1}^{n} \frac{z-z_j}{1-\bar{z}_j z},$$

then the elegant theory of extremal problems in H^∞ for bounded linear functionals which stem from rational kernels (Cf. Duren [9, Ch. 8]) tells us that the coefficients, $\alpha_1,\ldots,\alpha_n$ of the optimal recovery are determined uniquely from

$$(36) \quad Lf = f(\zeta)+\sum_{j=1}^{n} \alpha_j f(z_j) = \frac{1}{2\pi i}\int_{|z|=1} \frac{B_n(\zeta)}{B_n(z)} \frac{1-|\zeta|^2}{1-z\bar{\zeta}} \frac{1}{z-\zeta} f(z)dz$$

and $B_n(z)$ is the unique (up to rotation) worst function and $|B_n(\zeta)|$ is the intrinsic error. In making the proper choice of the kernel in (36) we use Theorem 6 of Section 2 in this problem. The calculation of the parameters $\alpha_1,\ldots,\alpha_n$ of an optimal algorithm from (36) is easy. For example, if $n=1$ and $z_1=0$ we obtain $\alpha_1 = -(1-|\zeta|^2)$ and $E(K,0) = |\zeta|$. But note that (Cf. Golusin [13, p. 287])

$$\left|f(\zeta)-f(0)\frac{1-|\zeta|^2}{1-|f(0)|^2|\zeta|^2}\right| \le |\zeta| \frac{1-|f(0)|^2}{1-|f(0)|^2|\zeta|^2} \le |\zeta|,$$

which informs us that in addition to the linear optimal algorithm $(1-|\zeta|^2)f(0)$, there is a non-linear optimal algorithm, namely,

$$\frac{1-|\zeta|^2}{1-|f(0)|^2|\zeta|^2} f(0),$$

and indeed there are uncountably many non-linear optimal algorithms.

Remark. Osipenko [38] solves the optimal recovery problem when in our notation K is the set of functions analytic and bounded by a fixed constant in a simply connected domain symmetric with respect to the real axis and containing an interval J, of the real axis. He assumes that ζ as well as the sampling points are situated in J. He also studies the question of the optimal choice of sampling points so as to minimize the maximum intrinsic error for all $\zeta\in J$.

Example 3.6. Assumptions are all the same as in Example 3.5 except that $Uf = f'(\zeta)$. Our main result and the general theory lead to the conclusion that if $\alpha_1,\ldots,\alpha_n$ are the parameters of an optimal algorithm then

$$(37) \qquad Lf = f'(\zeta) + \sum_{j=1}^{n} \alpha_j f(z_j) = \frac{1}{2\pi i}\int_{|z|=1} K(z)f(z)dz$$

where either

(38) $$K(z) = Ce^{-i\gamma} \frac{(1-\bar{a}z)^2}{(z-\zeta)^2(1-\bar{\zeta}z)^2} \frac{1}{B_n(z)}$$

or

(39) $$K(z) = Ce^{-i\gamma} \frac{(z-a)}{(z-\zeta)^2} \frac{(1-\bar{a}z)}{(1-\bar{\zeta}\, z)^2} \frac{1}{B_n(z)} ,$$

for approiately chosen $C(>0)$, a satisfying $|a| \leq 1$ and a real γ, all depending on ζ. In other words for each ζ in D there is a C, γ and $\underline{a}$ determined. The corresponding (unique) worst functions are

(40) $$F(z) = e^{i\gamma} \frac{z-a}{1-\bar{a}z} B_n(z)$$

and

(41) $$F(z) = e^{i\gamma} B_n(z)$$

respectively.

In order for (37) to hold we must have

(42) $$[K(z)(z-\zeta)^2]'_{z=\zeta} = 0$$

which enables us to determine $\underline{a}$ in terms of ζ. The complex number $Ce^{i\gamma}$ is then determined by the fact that the coefficient of $f'(\zeta)$ in (37) is 1, so that

$$[K(z)(z-\zeta)^2]_{z=\zeta} = 1.$$

Once K is thus specified the calculation of the parameters $\alpha_1, \ldots, \alpha_n$ of the optimal algorithm is straightforward.

Let us now carry out the computation indicated by (42) when K is given by (38) and then (39).

a) $$\left[\frac{(1-\bar{a}z)^2}{(1-\bar{\zeta}z)^2} \frac{1}{B_n(z)}\right]'_{z=\zeta} = 0$$

implies that

(43) $$\bar{a} = \frac{(1-|\zeta|^2)B_n'(\zeta) - 2\bar{\zeta}B_n(\zeta)}{\zeta(1-|\zeta|^2)B_n'(\zeta) - 2B_n(\zeta)} .$$

The condition that $\underline{a}$ as given by (43) is satisfactory is $|\bar{a}| \leq 1$, or equivalently, $|\bar{a}|^2 \leq 1$, which is satisfied, if, and only if

(44) $$\left|\frac{B_n'(\zeta)}{B_n(\zeta)}\right| \leq \frac{2}{1-|\zeta|^2} .$$

Thus if $\zeta \in D$ satisfies (44) then $\underline{a}$ is given by (43) and with this $\underline{a}$

the problem is solved as indicated above. The set of $\zeta\epsilon D$ for which (44) is violated must then lead to the kernel given in (39). Let us see next how this works out explicitly.

b) $$\left[\frac{(z-a)(1-\bar{a}z)}{(1-\bar{\zeta}z)^2 B_n(z)}\right]'_{z=\zeta} = 0$$

implies that

$$(45) \qquad \frac{1}{2}\left\{\frac{\bar{a}-\bar{\zeta}}{1-\zeta\bar{a}} + \frac{1-\bar{\zeta}a}{a-\zeta}\right\} = \frac{B_n'(\zeta)}{B_n(\zeta)} \frac{1-|\zeta|^2}{2} .$$

Now

$$w = \frac{\bar{a}-\bar{\zeta}}{1-\zeta\bar{a}}$$

gives a 1-1 map of $|\bar{a}|\leq 1$ onto $|w|\leq 1$ and

$$\tau = \frac{1}{2}(w+\bar{w}^{-1})$$

gives a 1-1 map of $|w|\leq 1$ onto $|\tau|\leq 1$. Hence

$$(46) \qquad \left|\frac{B_n'(\zeta)}{B_n(\zeta)}\right| \geq \frac{2}{1-|\zeta|^2}$$

if, and only if, (45) holds for $\underline{a}$ satisfying $|\underline{a}|\leq 1$, and if $\zeta\epsilon D$ satisfies (46) the $\underline{a}$ given by (45) solves the problem, as we have seen.

Thus D is partitioned into two subsets D_a and D_b which meet along the curve

$$\left|\frac{B_n'(\zeta)}{B_n(\zeta)}\right| = \frac{2}{1-|\zeta|^2} .$$

These subsets can sometimes be simply described. For example, suppose $z_1=z_2= \dots = z_n = 0$, so that If = (f(0), f'(0),..., $f^{(n-1)}(0)$). In this case $B_n(z) = z^n$ and so the curve becomes

$$|\zeta|^2 + \frac{2}{n}|\zeta| - 1 = 0$$

or

$$|\zeta| = \sqrt{1+\frac{1}{n^2}} - \frac{1}{n} = r_n .$$

Thus if $|\zeta|\leq r_n$ we are in case b), $e^{i\gamma}z^n$ is a worst function and the intrinsic error is n. If $|\zeta|\geq r_n$ we are in case a). We obtain

$$\bar{a} = \frac{(1-|\zeta|^2)n-2|\zeta|^2}{\zeta[n(1-|\zeta|^2)-2]}$$

and then use (38), (40) and (37) to find all the required parameters.

When n = 1 we recover a theorem of Dieudonné (Cf. Caratheodory [7, p. 19].).

Example 3.6 The following example was communicated to us by Donald J. Newman. Again $X = H^\infty$, but now $K = \{f \in X: ||f'|| \le 1\}$, $z_1 = z_2 = \ldots = z_n = 0$ and $If = (f(0),\ldots,f^{(n-1)}(0))$, while $Uf = f(\zeta)$, $|\zeta|<1$ and $\varepsilon = 0$. Put $\zeta = re^{i\phi}$, and

$$f(z) = \sum_{j=0}^{\infty} a_j z^j.$$

Clearly, $g(z) = z^n/n$ is in K, $Ig = 0$ and $|g(\zeta)| = |\zeta|^n/n$. Thus $e(K,0) \ge |\zeta|^n/n$. But consider the linear functional

$$Lf = \frac{1}{2\pi i}\int_{|z|=1} \{\frac{\zeta^n}{nz^n} + \sum_{k=1}^{\infty} \frac{1}{n+k}((\frac{\zeta}{z})^{n+k} + r^{2k}(\frac{\zeta}{z})^{n-k})\} f'(z)dz$$

$$= a_n\zeta^n + \sum_{k=1}^{\infty} a_{n+k}\zeta^{n+k} + \sum_{k=1}^{n-1} \frac{n-k}{n+k} r^{2k} a_{n-k}\zeta^{n-k}$$

$$= f(\zeta) + \sum_{j=0}^{n-1} (\frac{j}{2n-j} r^{2(n-j)} - 1)\zeta^j a_j$$

$$= f(\zeta) + \sum_{j=0}^{n-1} d_j f^{(j)}(0) = f(\zeta) - A(If)$$

we have

$$|Lf| \le |\zeta|^n \frac{1}{2\pi}\int_0^{2\pi} |\frac{1}{n} + 2\sum_{k=1}^{\infty} \frac{r^k}{n+k}\cos k(\phi-\Theta)|\, d\Theta. \tag{47}$$

The sequence $\{r^k/(n+k)\}$ is convex, hence the quantity whose absolute value is the integrand in (47) is non-negative, as summation by parts twice reveals. The absolute value signs can be erased and we obtain $|Lf| \le |\zeta|^n/n$; which shows that A is optimal.

4. EXAMPLE OF OPTIMAL RECOVERY OF A FUNCTION

This section is devoted to a specific example of the optimal recovery of a function (i.e. U is the identity operator) which we believe has some interesting features. It is a generalization of Example 1.2.

Let $X = \ell^p(\mathbb{C})$ for $p \geq 1$, that is $f = \{f_j\}$, $f_j \in \mathbb{C}$, $j=0,\pm1,\pm2,\ldots$ is contained in X if, and only if,

$$(\Sigma|f_j|^p)^{\frac{1}{p}} < \infty.$$

(Throughout this section all sums are from $-\infty$ to ∞ unless otherwise indicated.) Let $a = \{a_j\}$, $a_j \in \mathbb{C}$, $j=0,\pm1,\pm2,\ldots$ be a given sequence. Put

$$K = K(a,p) = \{f \in X : (\Sigma|a_j f_j|^p)^{\frac{1}{p}} \leq 1\} .$$

K is a convex (in view of Minkowski's inequality) balanced subset of X. Let $J = \{k_1,\ldots,k_m\}$ be a given set of m distinct integers, and let J' denote its complement in the set of all integers. Take $Y = \mathbb{C}^m$, equipped with the p-norm, define I by

$$If = (f_{k_1},\ldots,f_{k_m}) ,$$

and suppose $Uf = f$, all $f \in X$, so that $Z = X$.
Put

$$A = \inf_{j \in J'} |a_j| ; \quad B = \min_{j \in J} |a_j|$$

and if $A > 0$

$$\frac{B}{A} = C .$$

Recall that, in the present setting

$$e(K,\varepsilon) = \sup_{\substack{f \in K(a,p) \\ \sum_{j \in J} |f_j|^p \leq \varepsilon^p}} (\Sigma|f_j|^p)^{1/p} .$$

Then we have

<u>Lemma 4</u>. i) If $A > 0$, $C \leq 1$ and $\varepsilon B \leq 1$ then

$$e(K,\varepsilon) = (A^{-p} + (1-C^p)\,\varepsilon^p)^{1/p} . \tag{48}$$

ii) If $A > 0$, $C \leq 1$ and $\varepsilon B \geq 1$ then

$$e(K,\varepsilon) = \frac{1}{B} . \tag{49}$$

(50) iii) If $A > 0$ and $C > 1$ then

$$e(K,\varepsilon) = \frac{1}{A}.$$

iv) If $A = 0$ than $e(K,\varepsilon) = \infty$.

Proof. Suppose $A > 0$, then

$$A^p \sum_{j\in J'} |f_j|^p \le \sum_{j\in J'} |a_j f_j|^p \le 1 - \sum_{j\in J} |a_j f_j|^p .$$

Therefore,

$$\sum_{j\in J'} |f_j|^p \le A^{-p} - \sum_{j\in J} \left|\frac{a_j}{A}\right|^p |f_j|^p$$

and

$$\Sigma |f_j|^p \le A^{-p} + \sum_{j\in J} \left(1-\left|\frac{a_j}{A}\right|^p\right) |f_j|^p$$

$$\le A^{-p} + (1-C^p) \sum_{j\in J} |f_j|^p .$$

Note that

$$\sum_{j\in J} |f_j|^p \le B^{-p}$$

since

$$\sum_{j\in J} |a_j|^p |f_j|^p \le 1.$$

Thus: i) if $C \le 1$ and $\varepsilon B \le 1$ then

(51) $$e(K,\varepsilon) \le (A^{-p} + (1-C^p)\, \varepsilon^p)^{\frac{1}{p}} .$$

ii) if $C \le 1$ and $\varepsilon B \ge 1$ then

(52) $$e(K,\varepsilon) \le \frac{1}{B} .$$

iii) If $C > 1$ then

(53) $$e(K,\varepsilon) \le \frac{1}{A} .$$

We show next that equality holds in (51)-(53) by exhibiting extremal functions. Choose any $\delta > 0$. Consider f defined by $f_j=0$. $j \ne k,q$, where $k\in J$, $q\in J'$, $B = |a_k|$, $|a_q| < A + \delta$ and

$$f_k = \lambda \ ; \quad f_q = \left(\frac{1-\lambda^p B^p}{|a_q|^p}\right)^{1/p}$$

with

$$\lambda \le \min(\varepsilon, \tfrac{1}{B}).$$

For this choice of f we have

$$\Sigma|a_j f_j|^p = 1, \quad \sum_{j\in J} |f_j|^p = \lambda^p \le \varepsilon^p$$

and

$$\Sigma|f_j|^p = \lambda^p + \frac{1-\lambda^p B^p}{|a_q|^p} > \lambda^p + \frac{1-\lambda^p B^p}{(A+\delta)^p} .$$

Now choose $\lambda = \min(\varepsilon, B^{-1})$ and then $\lambda = 0$, and since δ was arbitrary (48)-(50) hold, and (iv) is proved as well.

We turn now to optimal algorithms. If $y\in Y$ define $y'\in\ell^p$ by $y'_j = y_j$ for $j\in J$, $y'_j = 0$ for $j\in J'$. We now define the following three algorithms.

(54) $\quad \iota : y \to y'$

(55) $\quad \zeta : y \to 0$

(56) $\quad \beta : y \to (1-C^p)y'$

Since there is no optimal algorithm if $A = 0$ we suppose henceforth that $A > 0$. Then we have

Theorem 12. i) If $C = 0$ or $\varepsilon = 0$, ι is an optimal algorithm with $E(K,\varepsilon) = (A^{-p} + \varepsilon^{-p})^{1/p}$.

ii) If $C \ge 1$ or $C < 1$ and $\varepsilon B \ge 1$, ζ is an optimal algorithm with $E(K,\varepsilon) = \max(A^{-1}, B^{-1})$.

iii) If $0 < C < 1$ and $\varepsilon B \le 1$, β is an optimal algorithm with $E(K,\varepsilon) = (A^{-p}+\varepsilon^p(1-C^p))^{1/p}$.

Proof. Let $\tilde{f} : (\tilde{f}_{k_1},\ldots,\tilde{f}_{k_m})$ denote an element of Y.

i) $$E^p_\iota(K,\varepsilon) = \sup_{\substack{f\in K(a,p) \\ \sum_{j\in J}|f_j-\tilde{f}_j|^p\le\varepsilon^p}} \Sigma|f_j-(\iota\tilde{f})_j|^p \le \varepsilon^p + \sum_{j\in J'}|f_j|^p .$$

Since

$$\sum_{j\in J'}|f_j|^p \le A^{-p} - C^p \sum_{j\in J}|f_j|^p \le A^{-p}$$

i) follows from (48) and (50).

ii) $$E^p_\zeta(K,\varepsilon) = \sup_{\sum_{j\in J}|f_j-\tilde{f}_j|^p\le\varepsilon^p} \Sigma|f_j|^p$$

$$\leq A^{-p} + (1-C^p) \sup_{\substack{f\in K(a,p) \\ \sum_{j\in J} |f_j-\tilde{f}_j|^p \leq \epsilon^p}} \sum_{j\in J} |f_j|^p$$

Since we always have

$$\sum_{j\in J} |f_j|^p \leq B^{-p}$$

ii) follows from Lemma 4.

iii) $$E_\beta^p(K,\varepsilon) = \sup_{\substack{f\in K(a,p) \\ \sum_{j\in J} |f_j-\tilde{f}_j|^p \leq \varepsilon^p}} \{\sum_{j\in J} |f_j-(1-C^p)\tilde{f}_j|^p + \sum_{j\in J'} |f_j|^p\}.$$

As usual we have

$$\sum_{j\in J'} |f_j|^p \leq \frac{1}{A^p} - C^p \sum_{j\in J} |f_j|^p . \tag{57}$$

Also, an application of Minkowski's inequality yields

$$\sum_{j\in J} |f_j - (1-C^p)\tilde{f}_j|^p = \sum_{j\in J} |(1-C^p)(f_j-\tilde{f}_j) + C^p f_j|^p$$

$$\leq \left[(\sum_{j\in J} |(1-C^p)(f_j-\tilde{f}_j)|^p)^{1/p} + (\sum_{j\in J} |C^p f_j|^p)^{1/p} \right]^p$$

$$\leq \left[(1-C^p) (\sum_{j\in J} |f_j-\tilde{f}_j|^p)^{1/p} + C^p (\sum_{j\in J} |f_j|^p)^{1/p} \right]^p$$

$$\leq [(1-C^p)u + C^p v]^p ,$$

where we put

$$u = (\sum_{j\in J} |f_j-\tilde{f}_j|^p)^{1/p} \; ; \; v = (\sum_{j\in J} |f_j|^p)^{1/p} .$$

Since the function t^p, $p>1$ is convex for $0<t$ we have for $p\geq 1$

$$[(1-C^p)u + C^p v]^p \leq (1-C^p)u^p + C^p v^p$$

so that,

$$\sum_{j\in J} |f_j-(1-C^p)\tilde{f}_j|^p \leq (1-C^p) \sum_{j\in J} |f_j-\tilde{f}_j|^p + C^p \sum_{j\in J} |f_j|^p$$

$$\leq (1-C^p)\varepsilon^p + C^p \sum_{j\in J} |f_j|^p \quad ,$$

which taken with (57) and (48) completes the proof of (iii) and the theorem.

Remark 1. Theorem 12 is summarized in Figure 3.

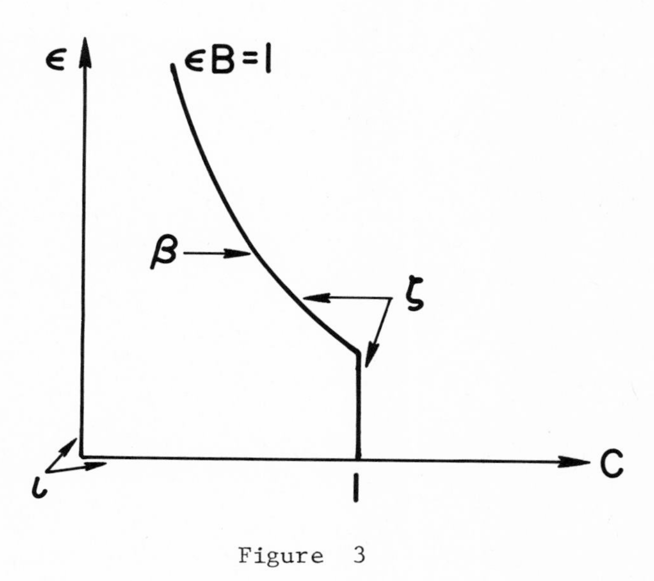

Figure 3

Remark 2. Suppose $X = L^2[0,1]$,

(58) $$K = \{f\in X : \int_0^1 (f'(t))^2 dt \leq 1\},$$

$Y = \mathbb{C}^{2k+1}$ with the L^2-norm, $If = (f_{-k},\ldots,f_k)$ where

$$f_j = \int_0^1 f(t)e^{-2\pi ijt}dt,$$

and $\varepsilon\geq 0$. Thus we consider the problem of recovering a function satisfying (58) from, possibly noisy, values of its Fourier coefficients of order at most k.

If $f \sim \Sigma\, c_j e^{2\pi ijt}$ then (58) is equivalent to

$$4\pi^2 \,\Sigma\, j^2 |c_j|^2 \leq 1,$$

and by identifying X with ℓ^2 ($\mathbb{C}$) and $2\pi j$ with a_j, Theorem 12 can be invoked. Since C = 0, Figure 3 tells us that an optimal algorithm is obtained by simply forming the trigonometric polynomial of degree k whose coefficients are the observed Fourier coefficients $\tilde{f}_{-k},\ldots,\tilde{f}_k$.

5. OPTIMAL RECOVERY BY RESTRICTED ALGORITHMS

The material in this section is suggested by some recent results of H. F. Weinberger [47].

Referring back to Example 3.4 we see that the dimension of the optimal recovery spline is equal to the number of data values. If this number is excessively large it may be desirable to have an algorithm which requires fewer parameters. For instance, we may desire a method of recovering $f(\tau)$ which is a polynomial of low degree, e.g., a linear function of τ,

$$A(f(x_1),\ldots,f(x_{n+r})) = \alpha_o(f(x_1),\ldots,f(x_{n+r}))+\tau\alpha_1(f(x_1),\ldots,f(x_{n+r})).$$

In [47] Weinberger treats such questions in Hilbert spaces.

Following Weinberger we consider the following problem of restricted recovery from exact data. Let U,X,Y,Z,K and I be as before, except that X is a Hilbert space, K is its unit ball (actually K may be the unit ball determined by a Hilbert space semi-norm), $Y = R^n$, $Ix = (I_1x,\ldots,I_nx)$, $I_j \in X^*$ and $\varepsilon = 0$. However, instead of allowing any transformation of R^n into Z as an admissible algorithm we require the transformation to have the form MA(Ix) where M is some fixed mapping from R^m into Z and A is any transformation from R^n into R^m. Figure 1 now becomes

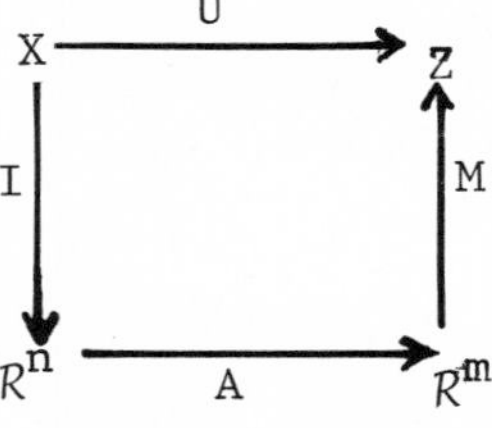

Figure 4

We suppose that the range of M is a subspace of Z of dimension m, which we also denote by M. Thus algorithms are restricted to map into $M \subset Z$. As before we assume that I maps X onto R^n, and the intrinsic error is defined to be

$$E(M) = \inf_{A} \sup_{||x|| \le 1} ||Ux - MA\, Ix||.$$

The analog of Theorem 1 in the present context is easily seen to be

$$e(M) = \max \Big(\sup_{\substack{Ix=0 \\ ||x|| \le 1}} ||Ux||, \quad \sup_{||x|| \le 1} \inf_{z \in M} ||Ux-z|| \Big) \le E(M).$$

Weinberger proved that the intrinsic error in approximating U by MAI, where A is an n×m matrix is precisely e(M), when Z is a Hilbert space.

Theorem H (Weinberger) If X and Z are Hilbert spaces and A(n,m) is the set of n×m real matrices then

$$\min_{A \in A(n,m)} \sup_{||x|| \le 1} ||Ux - MAIx|| = e(M).$$

Hence there is always a linear optimal recovery for U of the form MA, $A \in A(m,n)$.

This theorem of Weinberger has some interesting consequences for unrestricted recovery of U using the information I. We showed in Example 1.1 that the (unrestricted) intrinsic error satisfied

$$E(K,0) = \sup_{\substack{Ix=0 \\ ||x|| \le 1}} ||Ux||.$$

Hence, if M is any finite dimensional subspace of Z such that

$$\sup_{\substack{Ix=0 \\ ||x|| \le 1}} ||Ux|| \ge \sup_{||x|| \le 1} \inf_{z \in M} ||Ux - z||$$

then there is an (unrestricted) optimal recovery of the form MAIx where A is an n×m matrix. In many cases this inequality is easily demonstrated.

For example, if U is a compact operator and $\lambda_1 \ge \lambda_2 \ge \ldots$ are the eigenvalues of U^*U then by the min-max characterization of the eigenvalues of U^*U we have

$$\max_{\substack{Ix=0 \\ ||x|| \le 1}} ||Ux||^2 \ge \lambda_{n+1}.$$

Hence if M is any n dimensional subspace of Z such that

$$\sup_{||x|| \le 1} \inf_{z \in M} ||Ux-z||^2 = \lambda_{n+1}$$

then we obtain an optimal algorithm of the form MA. Since the n-width of the set $U = \{Ux: ||x|| \le 1\}$

$$d_n = \inf_{\dim M \le n} \sup_{||x|| \le 1} \inf_{z \in M} ||Ux - z||$$

equals λ_{n+1} this requirement on M means that it is an optimal subspace for approximating the set U. There are many subspaces which have this desired property (Cf. Karlovitz [17], Melkman and Micchelli [23]). If the associated eigenfunctions are given by $U^*U\phi_i = \lambda_i\phi_i, (\phi_i,\phi_j) = \delta_{ij}$, then a classical choice for such a subspace is the one spanned by $U\phi_1,\ldots,U\phi_n$.

For the special Example 1.3 these remarks imply that in addition to natural spline interpolation to the data we may also choose an optimal recovery which is an "approximation" by natural splines with knots at $\eta_1,\ldots,\eta_{n+r}$. Thus in the notation of Example 1.3 we have

$$\sup_{\substack{f(x_i)=0 \\ ||f^{(n)}||_2 \le 1}} ||f||_2 = \sup_{||f^{(n)}||_2 \le 1} ||f - Sf||_2 \ge d_{n+r}(K) = \frac{1}{\mu_{n+r+1}} = \sup_{||f^{(n)}||_2 \le 1} \inf_{g \in S_\eta} ||f-g||_2$$

where S_η = the space of natural splines of degree 2n-1 with knots at $\eta_1,\ldots,\eta_{n+r}$ and $K = \{f: ||f^{(n)}||_2 \le 1\}$.

Thus we have a trade-off between two desirable properties. The first method of recovery, natural spline interpolation Sf, has the advantage of being interpolatory. However the knots of the splines, being at $x_1,\ldots,x_{n+r}$ change if we alter even one sampling point. On the other hand, the second method uses the same set of knots regardless of the sampling points. However, the method of approximation is more complicated. We must find an $n+r \times n+r$ matrix, B, such that

$$\min_{A_{ij}} \max_{||f^{(n)}||_2 \le 1} ||f - \sum_{i=1}^{n+r} g_i A_{ij} f(x_j)||_2 = \max_{||f^{(n)}||_2 \le 1} ||f - \sum_{i=1}^{n+r} g_i B_{ij} f(x_j)||_2,$$

where $(g_1,\ldots,g_{n+r})$ is a basis for the natural splines of degree 2n-1 with knots at $\eta_1,\ldots,\eta_{n+r}$.

Acknowledgement: We are grateful to Dr. A. Melkman for reading an earlier version of this manuscript and suggesting many improvements.

REFERENCES

1. Bakhvalov, N. S., On the optimality of linear methods for operator approximation in convex classes of functions, USSR Computational Mathematics and Mathematical Physics 11 (1971), 244-249.

2. Bojanov, B.D., Optimal methods of interpolation in $W^{(r)}L_q(M;a,b)$, Comptes Rendus de l'Acadamie Bulgare des Sciences 27 (1974), 885-888.

3. Bojanov, B.D., Best methods of interpolation for certain classes of differentiable functions, Matematicheskie Zametki 17 (1975), 511-524.

4. Bojanov, B.D., Favard's interpolation and best approximation of periodic functions, preprint.

5. deBoor, C., A remark concerning perfect splines, Bull. Amer. Math. Soc. 80(1974), 724-727.

6. Burchard, H., Interpolation and approximation by generalized convex functions, Ph.D., Dissertation, Purdue University, Lafayette, Indiana, 1968.

7. Caratheodory, C., Theory of Functions of a Complex Variable, Volume Two, 2nd English Edition, Chelsea, New York, 1960.

8. Danskin, John M., The theory of max-min, with applications. SIAM Journal, 14 (1966), 641-664.

9. Duren, Peter L., Theory of H^p spaces, Academic Press, New York 1970.

10. Fisher, S.D. and J.W. Jerome, The existence, characterization and essential uniqueness of solutions of L^∞ extremal problems Trans. Amer. Math. Soc. 187(1974), 391-404.

11. Gaffney, P.W. and M.J.D. Powell, Optimal Interpolation, C.S.S. 16 Computer Science and Systems Division, A.E.R.E., Harwell, Oxfordshire, England 1975.

12. Golomb, M., and H.F. Weinberger, Optimal approximation and error bounds, in On Numerical Approximation, R.E. Langer ed. The University of Wisconsin Press, Madison (1959), 117-190.

13. Golusin, G.M. Geometrische Funktionentheorie, VEB, Berlin 1957.

14. Holmes, R.B., A Course on Optimization and Best Approximation Lecture Notes Series 257, Springer-Verlag, Berlin 1972.

15. Karlin, S., Interpolation properties of generalized perfect splines and the solution of certain extremal problems I, Trans. Amer. Math. Soc. 206 (1975), 25-66.

16. Karlin, S. and W.J. Studden, Tchebycheff Systems with Applications in Analysis and Statistics, Interscience Publishers, New York 1966.

17. Karlovitz, L.A., Remarks on variational characterization of Eigenvalues and n-widths problems, J. Math. Anal. Appl. 53 (1976), 99-110.

18. Krein, M.G., The L-problem in abstract linear normed space in some questions in the theory of moments (N.I. Ahiezer, M.G. Krein, Eds.), Translations of Mathematical Monographs, Vol. 2 Amer. Math. Soc. Providence, R.I., 1962.

19. Krein, M.G., The ideas of P.L. Chebyshev and A.A. Markov in the theory of limiting values of integrals and their further developments, Amer. Math. Soc. Transl. 12 (1951), 1-122.

20. Marchuk, A.G., K. Yu Osipenko, Best approximation of functions specified with an error at a finite number of points, Matematicheskie Zametki 17 (1975), 359-368.

21. Meinguet, J., Optimal approximation and error bounds in seminormed spaces, Numer. Math. 10 (1967) 370-388.

22. Melkman, A.A., n-widths and optimal interpolation of time- and band-limited functions , these proceedings.

23. Melkman, A.A. and C.A. Micchelli, On nonuniqueness of optimal subspaces for L^2 n-width, IBM Research Report 6113 (1976).

24. Micchelli, C.A., Saturation classes and iterates of operators, Ph.D. Dissertation, Stanford University, 1969.

25. Micchelli, C.A., On an optimal method for the numerical differentiation of smooth functions, J. Approx. Theory 18(1976)189-204.

26. Micchelli, C.A., Best L^1-approximation by weak Chebyshev systems and the uniqueness of interpolating perfect splines, to appear in J. Approx. Theory.

27. Micchelli, C.A., Optimal estimation of linear functionals, IBM Research Report 5729 (1975).

28. Micchelli, C.A., Optimal estimation of smooth functions from inaccurate data, in preparation.

29. Micchelli, C.A. and W. Miranker, High order search methods for finding roots, J. of Assoc. of Comp. Mach. 22(1975), 51-60.

30. Micchelli, C.A. and A. Pinkus, On n-widths in L^{∞}, to appear Trans. Amer. Math. Soc.

31. Micchelli, C.A. and A. Pinkus, Moment theory for weak Chebyshev systems with applications to monosplines, quadrature formulae and best one-sided L^1-approximation by spline functions with fixed knots, to appear in SIAM J. of Math. Anal.

32. Micchelli, C.A. and A. Pinkus, Total positivity and the exact n-width of certain sets in L^1, to appear in Pacific Journal of Mathematics.

33. Micchelli, C.A. and A. Pinkus, On a best estimator for the class M^r using only function values, Math. Research Center, Univ. of Wisconsin, Report 1621 (1976).

34. Micchelli, C.A., and A. Pinkus, Some problems in the approximation of functions of two variables and the n-widths of integral operators, to appear as Math. Research Center Report, University of Wisconsin.

35. Micchelli, C.A., T.J. Rivlin, S. Winograd, Optimal recovery of smooth functions, Numer. Math. 260 (1976), 191-200.

36. Morozov, V.A. and A.L. Grebennikov, On optimal approximation of operators, Soviet Math. Dokl. 16(1975), 1084-1088.

37. Newman, D., Numerical differentiation of smooth data, preprint.

38. Osipenko, K. Yu, Optimal interpolation of analytic functions, Mathematicheskie Zametki 12(1972), 465-476.

39. Osipenko, K. Yu, Best approximation of analytic functions from information about their values at a finite number of points, Matematischeski Zametki 19(1976), 29-40.

40. Royden, H.L., Real Analysis, MacMillan Company, New York 1963.

41. Rudin, Walter, Functional Analysis, McGraw-Hill, New York 1973.

42. Sard, A., Optimal approximation, J. Funct. Anal. 1 (1967), 222-244; Addendum 2(1968), 368-369.

43. Schultz, M.H., Complexity and differential equations, in Analytic Computational Complexity, Ed. J.F. Traub, Academic Press 1976.

44. Smolyak, S.A., On an optimal restoration of functions and functionals of them, Candidate Dissertation, Moscow State University 1965.

45. Schoenberg, I.J., The elementary cases of Landau's problems of inequalities between derivatives, Amer. Math. Monthly 80(1973), 121-158.

46. Tihomirov, V.M., Best methods of approximation and interpolation of differentiable functions in the space C[-1,+1], Math. USSR Sbornik, 9(1969), 275-289.

47. Weinberger, H.F., On optimal numerical solution of partial differential equations, SIAM J. Numer. Anal. 9(1972), 182-198.

48. Wiener, N., Extrapolation, Interpolation, and Smoothing of Stationary Time Series, the Technology Press of MIT and J. Wiley New York, 1950.

49. Winograd, S., Some remarks on proof techniques in analytic complexity, in Analytic Computational Complexity, Ed., J.F. Traub, Academic Press, 1976.

n-WIDTHS AND OPTIMAL INTERPOLATION OF TIME- AND BAND-LIMITED FUNCTIONS

Avraham A. Melkman

IBM Thomas J. Watson Research Center

Yorktown Heights, New York 10598

1. INTRODUCTION

Let $\mathscr{L}^2(R)$ denote the space of complex valued square integrable functions on the real line. Define

$$\mathscr{G} = \{f \in \mathscr{L}^2(R) \Big| \int_{|t|>T} |f(t)|^2 \, dt \le \epsilon_T^2, \int_{|\omega|>\sigma} |\hat{f}(\omega)|^2 d\omega \le \eta_\sigma^2\} \quad (1)$$

with $\hat{f}$ the Fourier transform of f. Following Slepian [7] $\mathscr{G}$ may be regarded as the class of functions time-limited to $(-T,T)$, at level ϵ_T, and bandlimited to $(-\sigma,\sigma)$, at level η_σ (only $f \equiv 0$ is strictly time- and band-limited). For this class we consider the n-widths and the optimal recovery of a function from its sampled values as well as the interrelation between them.

Section 2 is devoted to n-widths. It is based on the classical papers of Landau, Pollak and Slepian [3,4,8] and Slepian [7]. We have attempted to make the proofs more geometrical in nature and to provide a unified framework. The main result for the n-widths is

$$d_n^2(\mathscr{G};\mathscr{L}^2(R)) = \frac{1}{2}\frac{(\epsilon_T+\eta_\sigma)^2}{1-\sqrt{\lambda_n}} + \frac{1}{2}\frac{(\epsilon_T-\eta_\sigma)^2}{1+\sqrt{\lambda_n}} \quad (2)$$

where λ_n are decreasing numbers dependent only on σT. This result can be put to two uses of particular interest. By the definition of the zero-width as the radius of $\mathscr{G}$ it follows that for all $f \in \mathscr{G}$, $||f||^2 \le (\epsilon_T^2+\eta_\sigma^2 + 2\sqrt{\lambda_0}\,\epsilon_T\,\eta_\sigma)/(1-\lambda_0)$. This inequality may then be turned around to provide, for any f, a relationship between the fraction of its norm outside (-T,T) and outside $(-\sigma,\sigma)$, the uncertainty relationship of Landau and Pollak [3]. Another use of formula (2) rests upon an examination of λ_n for σT large. This shows that for any $\epsilon^2 > \epsilon_T^2 + \eta_\sigma^2$ and $\delta > 0$. σT can be chosen large enough so that for $n = (1+\delta)4\sigma T$ $d_n^2 < \epsilon^2$ while for $n = (1-\delta)4\sigma T$ $d_n^2 > \epsilon^2$. Slepian [7], in rephrasing Landau and Pollak's [4] dimension theorem, interpreted this fact to mean that the approximate dimension of the set of time- and band-limited signals is asymptotically $4\sigma T$ as σ or T becomes large.

In section 3 we turn to the optimal recovery of a function from its values at a fixed sampling set $\{s_i\}_1^n$. To make the problem meaningful $\eta_\sigma=0$ is required which turns $\mathcal{G}$ into an ellipsoidal set. Consequently it follows from general arguments, e.g., Micchelli and Rivlin in these Proceedings, that the optimal procedure is to interpolate with the set of functions $\{K(t,s_i)\}_1^n$, $K(t,t')$ a suitable kernel (if the ϵ_T^2 bound is replaced by $\int_{-\infty}^{\infty}|f(t)|^2\,dt \le 1$ the resulting kernel is simply $\sin 2\pi\sigma(t-t')/\pi(t-t')$). Since the procedure is a particular kind of linear approximation the worst error that can be encountered in recovery of the class $\mathcal{G}$ cannot be less than the n-width d_n. A natural question is therefore whether this bound can be achieved. This is indeed the case, the corresponding points being the zeros of the "worst" function for the n-width problem. Thus as a fringe benefit this approach identifies an (additional) optimal approximating subspace, which moreover achieves the n-width by the simple device of interpolation. Similar results have been described by Melkman and Micchelli [6] for totally positive kernels. The present result shows that the condition of total positivity is not necessary.

2. n-WIDTHS

We briefly review some basics of n-widths; for more details consult Lorentz [5]. The n-width of a subset $\mathcal{A}$ of a normed linear space X is obtained by considering its distance from an arbitrary n-dimensional subspace $X_n \subset X$, and then looking for the optimal approximating subspace X_n^* which achieves the least distance, the n-width with respect to X,

$$d_n(\mathcal{A};X) = \inf_{X_n \subset X} \sup_{f \in \mathcal{A}} \inf_{g \in X_n} ||f-g|| \qquad (3)$$

When X is a Hilbert space $\mathcal{A}$ is typically an ellipsoidal set for which n-widths results are known in great generality. Although the present set $\mathcal{G}$ is not ellipsoidal it is illuminating to consider a simple example of this kind:

$$X = R^3 \text{ and } \mathcal{A} = \{(x_0,x_1,x_2) \mid \sum_{i=o}^{2} x_i^2/\lambda_i \leq 1\}$$

with $\lambda_0 \geq \lambda_1 \geq \lambda_2 > 0$. Then the zero width is the radius of the smallest ball containing $\mathcal{A}$, $d_0(\mathcal{A};R^3) = \sqrt{\lambda_0}$. For the 1-width one has to look for the optimal line through the origin. The major axis is clearly such a line, $d_1(\mathcal{A};R^3) = \sqrt{\lambda_1}$, but any other line in the x_0-x_2 plane with slope not exceeding $\sqrt{(\lambda_1-\lambda_2)/(\lambda_0-\lambda_1)}$ is optimal too.

The same line of reasoning shows that non-uniqueness of the optimal subspace prevails for general ellipsoidal sets, e.g., Karlovitz [1], and even in the case at hand which is not quite ellipsoidal. But more on that in the next section.

The above example intimates that a convenient representation for $\mathcal{G}$, defined in (1), will greatly facilitate the calculation of its n-widths. To that end let us introduce the following notation.

Denote by $\mathcal{D}$ the space of functions $f \in \mathcal{L}^2(R)$ vanishing outside $(-T,T)$ (strictly time-limited), by $\mathcal{B}$ those with Fourier transform vanishing outside $(-\sigma,\sigma)$ (strictly band-limited). Let D and B be the projections on these spaces, i.e.,

$$Df(t) = \begin{cases} f(t) & |t| \leq T \\ 0 & |t| > T \end{cases} \qquad (4)$$

$$Bf(t) = \int_{-\infty}^{\infty} f(t') \frac{\sin 2\pi\sigma(t-t')}{\pi(t-t')} dt' \qquad (5)$$

corresponding to time-limiting and band-limiting. Landau and Pollak [3] made the crucial observation that the minimum angle formed between $\mathcal{D}$ and $\mathcal{B}$ is non-zero, meaning

$$\sup_{\substack{d \in \mathcal{D} \\ b \in \mathcal{B}}} \frac{\text{Re}(d,b)}{||d|| \; ||b||} < 1$$

with the usual inner product

$$(d,b) = \int_{-\infty}^{\infty} d(t)\,\overline{b}(t)dt, \quad (f,f) = ||f||^2.$$

Using this property they prove

Lemma 1.

The space $\mathscr{D} \dotplus \mathscr{B}$ is closed. Consequently any $f \in \mathscr{L}^2(R)$ may be decomposed as

$$f = d+b+g \text{ with } d \in \mathscr{D},\ b \in \mathscr{B},\ Dg = Bg = 0.$$

There is no convenient representation for the multitude of functions g with $Dg = Bg = 0$, but fortunately they play no essential role here. The spaces $\mathscr{D}$ and $\mathscr{B}$ on the other hand possess bases with remarkable properties as shown by Slepian and Pollak [8], Landau [2].

Let ψ_i and λ_i, $i = 0,1,\ldots$ be the eigenfunctions and eigenvalues of the integral equation

$$\int_{-T}^{T} \frac{\sin 2\pi\sigma(t-t')}{\pi(t-t')}\psi(t') = \lambda\psi(t) \tag{6}$$

Noting that it has a completely continuous, positive definite, symmetric kernel the λ_i may be assumed nonnegative decreasing to zero and the ψ_i real and normalized $(\psi_i,\psi_j) = \delta_{ij}$. Denoting

$$\phi_i(t) = \frac{1}{\sqrt{\lambda_i}} D\psi_i(t) \tag{7}$$

the integral equation implies

$$B\phi_i = \sqrt{\lambda_i}\,\psi_i\,,\ (\phi_i,\phi_j) = \delta_{ij}.$$

Clearly $\psi_i \in \mathscr{B}$, $\phi_i \in \mathscr{D}$. Out of the many other properties of these functions to be found in the cited references we will find the following particularly useful.

1. The set $\{\psi_i\}_0^\infty$ is complete in $\mathscr{B}$, the set $\{\phi_i\}_0^\infty$ in $\mathscr{D}$.
2. ψ_i has exactly i simple zeros in $(-T,T)$.
3. $1>\lambda_0>\lambda_1> \ldots,\ \lim_{j\to\infty} \lambda_j = 0.$
4. The eigenvalues depend only on the product σT,

   $\lambda_{[4\sigma T]+1}(\sigma T) \leq .5,\quad \lambda_{[4\sigma T]-1}(\sigma T) \geq .5$,

$$\lim_{\sigma T\to\infty} \lambda_n(\sigma T) = \begin{cases} 0 & n = [(1+\eta)4\sigma T] \\ 1 & n = [(1-\eta)4\sigma T] \end{cases}$$

5. Of all functions $f \in \mathscr{B}$ orthogonal to $\psi_0,\ldots,\psi_n$ the one most concentrated in (-T,T) is ψ_{n+1}, i.e., this function achieves

$$\sup_{\substack{f \in \mathscr{B} \\ (f,\psi_i)=0 \ i=o,\ldots,n}} \frac{||Df||^2}{||f||^2} = \lambda_{n+1}$$

The last property provides the rationale behind the ψ_i : it leads directly to the integral equation $BDf = \lambda f$, i.e., (6). It may be worth mentioning that the ψ_i are also (regular) solutions of a singular Sturm-Liouville type differential equation, the prolate spheroidal wave equation, so that much is known about them.

Property 1 and Lemma 1 immediately provide the sought after representation for $\mathscr{G}$, which is then used to calculate its n-widths.

Lemma 2.

Any function $f \in \mathscr{L}^2(R)$ has the representation

$$f(t) = \sum_{i=o}^{\infty} (d_i\phi_i(t) + b_i\psi_i(t)) + g(t) \tag{8}$$

where $Dg = Bg = 0$. If in addition $f \in \mathscr{G}$, i.e., $||(1-D)f||^2 \le \epsilon_T^2$, $||(1-B)f||^2 \le \eta_\sigma^2$ then

$$\sum_{i=o}^{\infty} |b_i|^2(1-\lambda_i) + ||g||^2 \le \epsilon_T^2 \tag{9}$$

$$\sum_{i=o}^{\infty} |d_i|^2(1-\lambda_i) + ||g||^2 \le \eta_\sigma^2 \tag{10}$$

Theorem 1.

$$d_n(\mathscr{G};\mathscr{L}^2(R)) = \frac{1}{2}\,\frac{(\epsilon_T+\eta_\sigma)^2}{1-\sqrt{\lambda_n}} + \frac{1}{2}\,\frac{(\epsilon_T-\eta_\sigma)^2}{1+\sqrt{\lambda_n}} \tag{11}$$

and any set of the form $\{\delta_i\phi_i(t) + \beta_i\psi_i(t)\}_o^{n-1}$, $|\delta_i|^2+|\beta_i|^2 > 0$, Re $\delta_i\beta_i \ge 0$, spans an optimal subspace.

Proof.

We show first that the right hand side is an upper bound by calculating the distance between $\mathcal{G}$ and the particular space spanned by

$$g_i(t) = \delta_i\phi_i(t) + \beta_i\,\psi_i(t).$$

Let P_nf denote the projection of f given by (8) on $\{g_i\}_o^{n-1}$.

A short calculation using $(\phi_i,\phi_j) = (\psi_i,\psi_j) = \delta_{ij}, (\phi_i,\psi_j) = \sqrt{\lambda_i}\,\delta_{ij}$ shows

$$||d_i\phi_i + b_i\psi_i - \frac{(f,g_i)}{(g_i,g_i)} g_i||^2$$

$$= [|b_i|^2+|d_i|^2 - \frac{|\overline{\delta}_i d_i+\overline{\beta}_i b_i|^2 + 2\sqrt{\lambda_i}\,(|b_i|^2+|d_i|^2)\mathrm{Re}\,\delta_i\overline{\beta}_i}{|\delta_i|^2+|\beta_i|^2 + 2\sqrt{\lambda_i}\mathrm{Re}\,\delta_i\overline{\beta}_i}](1-\lambda_i)$$

$$\leq (|b_i|^2+|d_i|^2)(1-\lambda_i)$$

since $\mathrm{Re}\,\delta_i\overline{\beta}_i \geq 0$. Hence

$$||f-P_nf||^2$$

$$\leq \sum_{i=o}^{n-1} (|b_i|^2+|d_i|^2)(1-\lambda_i) + \sum_{i=n}^{\infty} [|b_i|^2+|d_i|^2 + 2\sqrt{\lambda_i}\,\mathrm{Re}\,d_i\,\overline{b}_i]+||g||^2$$

$$\leq \frac{\sum_{i=o}^{\infty} |b_i|^2(1-\lambda_i)+||g||^2+ \sum_{i=o}^{\infty} |d_i|^2(1-\lambda_i)+||g||^2+2\sqrt{\lambda_n}\sum_{i=o}^{\infty} |b_id_i|(1-\lambda_i)}{1-\lambda_n}$$

$$\leq \frac{\epsilon_T^2 + \eta_\sigma^2 + 2\sqrt{\lambda_n}\,\epsilon_T\eta_\sigma}{1-\lambda_n}$$

making use of $(\phi_i,g) = (\psi_i,g) = 0$, $0 < \lambda_i < \lambda_n < 1$ $i \geq n$ and Cauchy-Schwarz. Examination of the inequalities reveals that equality is attained only for $h_n = (\eta_\sigma\phi_n + \epsilon_T\,\psi_n)/\sqrt{1-\lambda_n}$.

In order to show that no n-dimensional subspace can improve this bound we use the standard technique of finding an n+1-dimensional ball of this radius contained in $\mathcal{G}$, i.e., a set $\{h_i\}_o^n$ such that

$$\|\sum_{i=o}^{n} a_i h_i\|^2 \leq (\epsilon_T^2 + \eta_\sigma^2 + 2\sqrt{\lambda_n}\,\epsilon_T\eta_\sigma)/(1-\lambda_n) \quad \text{implies}$$

$\sum_{i=o}^{n} a_i h_i \in \mathscr{G}$. Indeed, the distance of this ball from an arbitrary n-dimensional subspace equals its radius, for there always is a function on the ball which is orthogonal to the subspace and hence its best approximation is zero. Taking

$$h_i = (\eta_\sigma\phi_i + \epsilon_T\,\psi_i)/\sqrt{1-\lambda_i}$$

one has $\|(1-D)h_i\| = \epsilon_T$, $\|(1-B)h_i\| = \eta_\sigma$ and hence $\sum_{i=o}^{n} a_i h_i \in \mathscr{G}$ if $\sum_{i=o}^{n} |a_i|^2 \leq 1$. That the ball is indeed contained in $\mathscr{G}$ follows therefore from

$$\|\sum_{i=o}^{n} a_i g_i\|^2 = \sum_{i=o}^{n} |a_i|^2 \frac{\epsilon_T^2+\eta_\sigma^2+2\sqrt{\lambda_i}\,\epsilon_T\eta_\sigma}{1-\lambda_i} \geq \sum_{i=o}^{n} |a_i|^2 \frac{\epsilon_T^2+\eta_\sigma^2+2\sqrt{\lambda_n}\,\epsilon_T\eta_\sigma}{1-\lambda_n} \quad .$$

Remark. From the proof it follows that $h_n = (\eta_\sigma\phi_n + \epsilon_T\,\psi_n)/\sqrt{1-\lambda_n}$ is the unique function with the largest norm of all those in $\mathscr{G}$ orthogonal to $\{h_i\}_o^{n-1}$. In this sense the system $\{h_i\}$ is the most natural one to approximate with, though from the n-width point of view one may choose the approximating subspace independently from ϵ_T and η_σ.

In this theorem $\mathscr{G}$ is approximated on the whole real line. If one is interested in obtaining an approximation only on (-T,T) then $\mathscr{G}$ should be considered as a subset of $\mathscr{L}^2(-T,T)$ instead of $\mathscr{L}^2(R)$. This is done by Slepian [7] who, using variational arguments derived the bounds

$$\lambda_n\epsilon_T^2/(1-\lambda_n) \leq d_n(\mathscr{G};\mathscr{L}^2(-T,T)) \leq (\sqrt{\lambda_n}\,\epsilon_T+\eta_\sigma)^2/(1-\lambda_n).$$

The next theorem shows that the upper bound is sharp.

Theorem 2.

$$d_n(\mathscr{G};\mathscr{L}^2(-T,T)) = \frac{(\sqrt{\lambda_n}\,\epsilon_T+\eta_\sigma)^2}{1-\lambda_n} \tag{12}$$

and $\{\phi_i\}_o^{n-1}$ *spans an optimal subspace.*

Proof.

The proof of the previous theorem showed that, with P_n the projection on $\{\phi_i\}_o^{n-1}$,

$$||f-P_nf||^2 \le \frac{||(1-D)f||^2+||(1-B)f||^2+2\sqrt{\lambda_n}\,||(1-D)f||\cdot||(1-B)f||}{1-\lambda_n}$$

Since $D\phi_i = \phi_i$, $f-P_nf = (1-D)f+D(f-P_nf)$ and so

$$||D(f-P_nf)||^2 \le \frac{\lambda_n||(1-D)f|^2+||(1-B)f||^2+2\sqrt{\lambda_n}||(1-D)f||\cdot||(1-B)f||}{1-\lambda_n}$$

establishing the right hand of (12) as an upper bound. The ball argument again shows that equality holds.

Two consequences of these results have particularly interesting interpretations. The first one addresses the question to what extent a signal can be concentrated in the time interval $(-T,T)$ and simultaneously in the frequency interval $(-\sigma,\sigma)$. This problem was solved by Landau and Pollak [3]. We give here an equivalent version which instead shows how small the fractional concentrations outside the time and band intervals, $||(1-D)f||/||f||$ and $||(1-B)f||/||f||$, can be.

Corollary 1. For $f \in \mathscr{L}^2(R)$ let $||f|| = 1$. Then the possible values of $\epsilon_T = ||(1-D)f||$ and $\eta_\sigma = ||(1-B)f||$ fill up the unit square $[0,1]\times[0,1]$ except for the points $(0,1)$, $(1,0)$ and the region inside the ellipse

$$\frac{1}{2}\,\frac{(\epsilon_T+\eta_\sigma)^2}{1-\sqrt{\lambda_o}}+\frac{1}{2}\,\frac{(\epsilon_T-\eta_\sigma)^2}{1+\sqrt{\lambda_o}} = 1. \tag{13}$$

Proof.

The zero-width of theorem 1 establishes the ellipse as a boundary since, by definition, $||f|| \le d_o(\mathscr{G};\mathscr{L}^2(R))$ for any $f \in \mathscr{G}$. The boundary is attained by the functions $h_o = (\eta_\sigma\phi_o+\epsilon_T\psi_o)/\sqrt{1-\lambda_o}$. The point $(1,0)$ is not permitted since a strictly band-limited function cannot vanish identically in an interval. However the points $(1,\lambda_i)$ come arbitrarily close and are attained by $(1-D)\psi_i/\sqrt{1-\lambda_i}$. The point $(1,1)$ corresponds to all functions in $\mathscr{L}^2(R)-(\mathscr{D}+\mathscr{B})$. Finally all other points may be reached by the device of shifting frequency via a factor $e^{2\pi i\delta t}$. This leaves ϵ_T intact and increases η_σ to 1 as δ increases.

The other application we want to bring concerns the notion that there are $4\sigma T$ signals of duration $2T$ and bandwidth σ. It is based upon an analysis of the eigenvalues as functions of σT, as summarized in property 5. The idea is to show that $4\sigma T$ functions suffice to approximate $\mathscr{G}$ on the real line to within the order of magnitude of $\epsilon_T^2 + \eta_\sigma^2$.

Corollary 2. Let $N(\epsilon_T,\eta_\sigma;\epsilon)$ be the least integer such that $d_N(\mathscr{G};\mathscr{L}^2(R)) \le \epsilon$. Then

a. Landau and Pollak [4]: for $\epsilon^2=2(\epsilon_T^2+\eta_\sigma^2+\sqrt{2}\epsilon_T\eta_\sigma)$ and all σT

$$[4\sigma T]-1 \le N(\epsilon_T,\eta_\sigma;\epsilon) \le [4\sigma T]+1$$

b. Slepian [7]: for any $\epsilon^2 > \epsilon_T^2 + \eta_\sigma^2$ and $\delta > 0$ σT can be chosen large enough so that

$$(1-\delta)[4\sigma T] \le N(\epsilon_T,\eta_\sigma;\epsilon) \le (1+\delta)[4\sigma T]$$

3. OPTIMAL INTERPOLATION

Consider the problem of recovery of $f \in \mathscr{G}$, on the real line or $(-T,T)$, from the knowledge of its values at a fixed set of sampling points $\{s_i\}_1^n$. Since point evaluations have no meaning in the context of $\mathscr{G}$ we confine attention to the subset $\mathscr{G}_o$ for which $\eta_\sigma = 0$

$$\mathscr{G}_o = \{f \in \mathscr{L}^2(R) \mid \ ||(1-D)f|| \le \epsilon_T \ , \ Bf = f\} \tag{14}$$

Thus $\mathscr{G}_o$ is a subset of $\mathscr{B}$ the Paley Wiener class of entire functions of exponential type $\le 2\pi\sigma$. For comparison we shall also consider the recovery problem for $f \in \hat{\mathscr{B}} = \{f \in \mathscr{B} \mid \ ||f||^2 \le 1\}$.

Taking the latter problem first, recall that $\mathscr{B}$ has the reproducing kernel

$$K(t,t') = \frac{\sin 2\pi\sigma(t-t')}{\pi(t-t')} \tag{15}$$

i.e., $f(t) = (K(t,\cdot),f)$ for all $f \in \mathscr{B}$ (this may also be deduced from (5) and $Bf = f$). For this situation an optimal scheme for recovery on $(-T,T)$, or even pointwise, is well known, see, e.g., the survey of Micchelli and Rivlin, and consists of projecting f on $\{K(t,s_i)\}_1^n$. Indeed, projection obviously decreases the norm and moreover it uses only the available information because the orthogonality conditions imply it is equivalent to interpolation

$$0 = (K(\cdot,s_k),\ f - \sum_{i=1}^{n} a_i K(\cdot,s_i)) = f(s_k) - \sum_{i=1}^{n} a_i K(s_k,s_i).$$

Hence the worst error that can be encountered as f ranges over $\mathscr{B}$ occurs for zero data,

$$E(\hat{\mathscr{B}};S) = \max_{\substack{f \in \hat{\mathscr{B}} \\ f(s_i)=0 \quad i=1,\ldots,n}} ||Df||$$

The procedure of projection-interpolation with the functions $\sin 2\pi\sigma(t-s_i)/\pi(t-s_i)$ is of course widely used in practice. However this procedure is not optimal for $\mathscr{G}_o$, for in that case we should use the kernel $K_o(t,t')$ which reproduces $\mathscr{B}$ with respect to the inner product

$$\langle f,g\rangle = (f,(1-D)g) \tag{16}$$

This kernel can be defined implicitly by the integral equation

$$K_T(t,t') = K(t,t') + \int_{-T}^{T} K(t,s)\ K_T(s,t')ds \tag{17}$$

or explicitly by its expansion

$$K_T(t,t') = \sum_{i=o}^{\infty} \frac{\psi_i(t)\psi_i(t')}{1-\lambda_i}\ . \tag{18}$$

Proceeding as before one finds that an optimal recovery procedure is projection with respect to $\langle\cdot,\cdot\rangle$ on $\{K_T(t,s_i)\}_1^n$ which again is equivalent to interpolation by this set. The error in recovery of $f \in \mathscr{G}_o$ on $(-T,T)$ is therefore

$$E(\mathscr{G}_o;S) = \max_{\substack{f \in \mathscr{G}_o \\ f(s_i)=0,i=1,\ldots,n}} ||Df|| = \epsilon_T E(\hat{\mathscr{B}};S)/(1-E(\hat{\mathscr{B}};S)) \tag{19}$$

Since the optimal recovery scheme involves linear approximation from a particular n-dimensional subspace, the inequality

$$E(\mathscr{G}_o;S) \geq d_n\ (\mathscr{G}_o;\mathscr{L}^2(-T,T))$$

is immediate. We want to show next that with a propitious choice of the points equality can be achieved. A similar result will hold for $\hat{\mathscr{B}}$. The following fact will be needed.

Lemma 3.

Let $f^* \in \mathscr{B}$ achieve $E(\hat{\mathscr{B}};S)$ or $E(\mathscr{G}_o;S)$. Then the only zeros of f^* in $[-T,T]$ are those s_i lying in the interval.

Proof.

Suppose f^* does have an additional zero, s_o, in $[-T,T]$. Consider the function

$$f_1(t) = f^*(t)\,\frac{T^2 - ts_o}{T(t-s_o)} \;.$$

f_1 is an entire function of exponential type $2\pi\sigma$ and $f_1 \in \mathscr{L}^2(R)$ because the same is true for f^* and $f^*(s_o) = 0$. Hence by the Paley-Wiener theorem $f_1 \in \mathscr{B}$. However $|f_1(t)| \geq |f^*(t)|$ for $|t| \leq T$, $|f_1(t)| \leq |f^*(t)|$ for $|t| \geq T$ and therefore

$$\frac{||Df_1||}{||(1-D)f_1||} > \frac{||Df^*||}{||(1-D)f^*||}$$

Thus f^* cannot possibly attain $E(\mathscr{G}_o;S)$ nor, from (19), $E(\hat{\mathscr{B}};S)$.

Theorem 3.

Let the (simple) zeros of $\psi_n(t)$ in $(-T,T)$ be $\Xi = \{\xi_i\}_i^n$. Then the recovery error of the optimal scheme based on these points equals the n-width

$$E(\hat{\mathscr{B}};\Xi) = \sqrt{\lambda_n} \tag{20}$$

$$E(\hat{\mathscr{G}}_o;\Xi) = \epsilon_T\sqrt{\lambda_n/(1-\lambda_n)} \tag{21}$$

Thus Ξ is an optimal sampling set.

Proof.

We prove (20) since (21) follows from it via (19). What is needed is an explicit evaluation of max $||Df||^2/||f||^2$, the maximum being taken over the space $f \in \mathscr{B}$, $f(\xi_i) = 0$ $i=1,\ldots,n$. This is most easily done by considering the reproducing kernel for this space.

$$K^n(t,t') = \frac{K\begin{pmatrix} t, \xi_1, \ldots, \xi_n \\ t', \xi_1, \ldots, \xi_n \end{pmatrix}}{K\begin{pmatrix} \xi_1, \ldots, \xi_n \\ \xi_1, \ldots, \xi_n \end{pmatrix}}$$

where the numerator and denominator are appropriate Fredholm determinants of $K(t,t')$. With this notation $E(\mathscr{B};\Xi)$ is the largest eigenvalue of the integral operator $H = K^n D K^n$ with kernel

$$H(s,t) = \int_{-T}^{T} K^n(s,t')K^n(t',t)dt'.$$

The eigenfunction associated with the top eigenvalue is the "worst" function for recovery. Clearly $\psi_n(t)$ is an eigenfunction of this operator, with eigenvalue λ_n, since $(K^n(t,\cdot),\psi_n) = \psi_n(t)$, $(K^n(t,\cdot),D\psi_n) = \lambda_n\psi_n(t)$, by the choice of Ξ. The question is whether it is the largest eigenvalue. Any eigenfunction associated with a different eigenvalue is orthogonal to $\psi_n(t)$ on $(-T,T)$. Moreover, all eigenfunctions must vanish at the ξ_i. Since these are the only zeros of ψ_n in $(-T,T)$ the orthogonality implies that all other eigenfunctions (which may be taken real) must vanish in addition somewhere else in $(-T,T)$. Lemma 3 now finishes the proof of (20).

That $d_n(\hat{\mathscr{B}};\mathscr{L}^2(-T,T)) = \sqrt{\lambda_n}$ follows by a standard argument, cf. Lorentz [5], once it is observed that $\hat{\mathscr{B}}$ is an ellipsoidal set with respect to $\mathscr{L}^2(-T,T)$. Indeed, while any $f \in \mathscr{L}^2(-T,T)$ may be expanded as

$$f(t) = \sum_{i=o}^{\infty} c_i\phi_i(t)$$

the condition $f \in \hat{\mathscr{B}}$ is equivalent to $\sum_{i=o}^{\infty} |c_i|^2/\lambda_i \leq 1$.

<u>Corollary 3.</u> <u>The subspace spanned by</u> $\{DK_o(t,\xi_i)\}_1^n$ <u>is an optimal n-dimensional approximating subspace for</u> $\mathscr{G}_o$, <u>with respect to both</u> $\mathscr{L}^2(-T,T)$ <u>and</u> $\mathscr{L}^2(-\infty,\infty)$, <u>in addition to the subspaces given in theorems 1, 2.</u>

Proof.

Optimality for $\mathscr{L}^2(-T,T)$ is inherent in theorem 3. From it the optimality on the whole real line follows since, denoting by $P_n f$ the interpolant of f by $\{DK_o(t,\xi_i)\}_1^n$

$$||f-P_n f||^2=||(1-D)f+D(f-P_n f)||^2 \leq \epsilon_T^2+\epsilon_T^2 \frac{\lambda_n}{1-\lambda_n} = \frac{\epsilon_T^2}{1-\lambda_n}$$

which equals $d_n(\mathscr{G}_o;\mathscr{L}^2(-\infty,\infty))$, by theorem 1.

Remark. We have been informed that Logan (unpublished) obtained a similar theorem on the basis of a formula, valid for all $f \in \mathscr{B}$, expressing $||Df||^2 - \lambda_n||f||^2$ as a weighted sum of $\{|f(x_k)|^2\}$ where $\{x_k\}$ are the zeros of $\psi_n(t)$ in $(-\infty,\infty)$ and the weights are positive for $|x_k| < T$ and negative otherwise.

The above proof has been patterned after Melkman and Micchelli [6]. There we considered sets of the form $f = Kh$, $||h|| \leq 1$ with K a totally positive kernel, and used the total positivity to arrive at similar results. Though total positivity does not hold here, e.g., $\sin 2\pi\sigma(t-t')/\pi(t-t')$ is not TP for all t,t', an important consequence of it, the zero properties of the eigenfunctions, continues to hold. This, together with lemma 3, was the crux of the proof. An open problem is what the most general conditions are under which a conclusion such as theorem will be valid. The following example demonstrates that the zero properties in themselves are neither sufficient nor necessary.

Let $\{f_i\}_0^2$ be real orthonormal functions, $\lambda_0 \geq \lambda_1 \geq \lambda_2 > 0$, and consider the set of functions

$$f(t) = \sum_{i=o}^{2} a_i f_i(t) \qquad \sum_{i=o}^{2} a_i^2/\lambda_i \leq 1 .$$

An optimal recovery scheme when given $f(\xi)$ consists of interpolation by

$$K(t,\xi) = \sum_{i=o}^{2} \lambda_i f_i(t) f_i(\xi)$$

and the largest error E encountered occurs for zero data. Thus, taking ξ to be a zero of f_1, one has to find the maximum of Σa_i^2 under the conditions $\Sigma a_i^2/\lambda_i = 1$, $a_o f_o(\xi) + a_2 f_2(\xi) = 0$.

$$E^2 = \max_{a_2} \left[\left(\frac{f_2(\xi)}{f_o(\xi)} \right)^2 (\lambda_0 - \lambda_1) - (\lambda_1 - \lambda_2) \right] a_2^2 + \lambda_1$$

Hence $E^2 = \lambda_1$, the 1-width, if and only if

$$\left| \frac{f_2(\xi)}{f_0(\xi)} \right| \le \sqrt{\frac{\lambda_1 - \lambda_2}{\lambda_0 - \lambda_1}}$$

This condition can be made to hold, by choosing λ_i, when ξ is not the only zero of f_1, or made to fail when it is.

References

1. L. A. Karlovitz, Remarks on variational characterization of eigenvalues and n-width problems, J. Math. Anal. Appl. 53 (1976), 99-110.

2. H. J. Landau, The eigenvalue behavior of certain convolution equations, Trans. Am. Math. Soc. 115 (1965), 242-256.

3. H. J. Landau and H. O. Pollak, Prolate spherical wave functions, Fourier analysis and uncertainty. II, Bell System Tech. J. 40 (1961), 65-84.

4. ______________________________, Prolate spheroidal wave functions. Fourier analysis and uncertainty. III, Bell System Tech. J. 41 (1962), 1295-1336.

5. G. G. Lorentz, Approximation of functions. Holt, Rinehart and Winston, 1966.

6. A. A. Melkman and C. A. Micchelli, On non-uniqueness of optimal subspaces for L^2 n-width, IBM Research Report, RC 6113 (1976).

7. D. Slepian, On bandwidth, Proc. IEEE, 64 (1976), 292-300.

8. D. Slepian and H. O. Pollak, Prolate spheroidal wavefunctions, Fourier analysis and uncertainty, I, Bell System Tech. J. 40 (1961), 43-64.

COMPUTATIONAL ASPECTS OF OPTIMAL RECOVERY

Carl de Boor

Mathematics Research Center, U. Wisconsin-Madison

610 Walnut St., Madison, WI 53706 USA

1. INTRODUCTION

This paper offers a Fortran program for the calculation of the optimal recovery scheme of Micchelli, Rivlin and Winograd [15]. A short derivation of that recovery scheme is given first, in order to make the paper selfcontained and also to provide an alternative to the original derivation in [15]. For the latter reason, a derivation of the related envelope construction of Gaffney and Powell [8] is also given. From a computational point of view, these schemes are special cases of the following computational problem: to construct an extension with prescribed norm of a linear functional on some finite dimensional linear subspace to all of $\mathbb{L}_1[a,b]$.

2. THE OPTIMAL RECOVERY SCHEME OF MICCHELLI, RIVLIN AND WINOGRAD

This scheme concerns the recovery of functions from partial information about them. Let $n > k$ and let $\tau := (\tau_i)_1^n$ be nondecreasing, in some interval $[a,b]$, with $\tau_i < \tau_{i+k}$, all i. For $f \in \mathbb{L}_\infty^{(k)} := \{g \in C^{(k-1)}[a,b] : g^{(k-1)}$ abs. cont., $g^{(k)} \in \mathbb{L}_\infty\}$, denote by $f|_\tau$ the restriction of f to the data point sequence τ, i.e., $f|_\tau := (f_i)_1^n$ with

$$f_i := f^{(j)}(\tau_i),\ j := j(i) := \max\{m : \tau_{i-m} = \tau_i\} .$$

We call a map $S : \mathbb{L}_\infty^{(k)} \to \mathbb{L}_\infty$ a <u>recovery scheme</u> (<u>on</u> $\mathbb{L}_\infty^{(k)}$) <u>with respect to</u> τ provided Sf depends only on $f|_\tau$, i.e., provided $f|_\tau = g|_\tau$ implies that $Sf = Sg$. The (possibly infinite) constant

$$\text{const}_S := \sup\{\|f - Sf\| / \|f^{(k)}\| : f \in \mathbb{L}^{(k)}\}$$

measures the extent to which such a recovery scheme S may fail to recover some f since it provides the sharp error bound

$$(1) \quad \|f - Sf\| \le \text{const}_S \|f^{(k)}\|, \quad \text{all } f \in \mathbb{L}_\infty^{(k)} .$$

Here and below,

$$\|g\| := \text{ess.sup}\{|g(t)| : a \le t \le b\} .$$

A recovery scheme S is <u>optimal</u> if its const_S is as small as possible, i.e., if the worst possible error is as small as possible, as measured by (1). We write

$$\text{const}_\tau := \inf_S \text{const}_S$$

for the best possible constant. Here is a quick lower bound for that constant. We have

$$\begin{aligned} \text{const}_S &= \sup_f \sup_{g|_\tau = f|_\tau} \|g - Sf\| / \|g^{(k)}\| \\ &\ge \sup_{g|_\tau = 0} \|g - S0\| / \|g^{(k)}\| \\ &= \sup_{g|_\tau = 0} \max\{\|g - S0\|, \|-g - S0\|\} / \|g^{(k)}\| \\ &\ge \sup_{g|_\tau = 0} \|g\| / \|g^{(k)}\| , \end{aligned}$$

for every recovery scheme S, hence

$$(2) \quad \text{const}_\tau \ge c(\tau) := \sup_{g|_\tau = 0} \|g\| / \|g^{(k)}\| ~$$

But, actually, equality holds here. For the proof, we need the notion of perfect splines.

A perfect spline s of degree k with (simple) knots $\zeta_1 < \dots < \zeta_r$ in $[a,b]$ is any function s of the form

$$s(x) = P(x) + c \sum_{i=0}^{r} (-)^i \int_{\zeta_i}^{\zeta_{i+1}} (x-y)_+^{k-1} dy \qquad (\zeta_0 := a,\ \zeta_{r+1} := b)$$

with $P \in \mathbb{P}_k$:= polynomials of degree $< k$ and c some constant. In other words, such a function s is any k-th anti-derivative of an absolutely constant function with r sign changes or jumps in $[a,b]$. Their connection with the present topic stems from

Proposition 1 (Karlin [10]). If s is a perfect spline of degree k with $n-k$ knots then

$$\|s^{(k)}\| = \inf\{\|f^{(k)}\| : f \in \mathbb{L}_\infty^{(k)},\ f|_\tau = s|_\tau\} .$$

This proposition follows directly from the following lemma.

Lemma 1. If $h \in \mathbb{L}_\infty^{(k)}$ is such that, for $a = \zeta_0 < \dots < \zeta_{r+1} = b$,

$$\sigma_i h^{(k)} > 0 \text{ a.e. on } (\zeta_i, \zeta_{i+1}) \text{ for some } \sigma_i \in \{-1,1\},\ i=0,\dots,r ,$$

and $h|_\tau = 0$, then $r \geq n-k$. Further, if $r \leq n-k$, hence $r = n-k$, then

$$(3) \qquad \tau_i < \zeta_i < \tau_{i+k}, \quad \text{all } i .$$

Assuming this lemma for the moment, we see that if, in Proposition 1, we had $\|s^{(k)}\| > \|f^{(k)}\|$ for some $f \in \mathbb{L}_\infty^{(k)}$ with $f|_\tau = s|_\tau$, then $h := s - f$ would satisfy the hypotheses of the lemma for some $r < n-k$, which would contradict the conclusion of the lemma.

Proof of Lemma 1: By Rolle's theorem, there must be points

$$t_1 < \dots < t_{n+1-k}$$

at which $h^{(k-1)}$ vanishes, and for which

(4) $\tau_i \le t_i \le \tau_{i+k-1}$, all i .

Then, each interval $[\zeta_i, \zeta_{i+1}]$ contains at most one of the t_j since, on each such interval, $h^{(k-1)}$ is strictly monotone, by assumption. Therefore, if n_j denotes the number of such intervals to which t_j belongs, then

$$n + 1 - k \le \Sigma n_j \le r + 1 ,$$

or $n - k \le r$. This also shows that, if $n - k = r$, then $n_j = 1$, all j, and so

$$\zeta_{i-1} < t_i < \zeta_i, \quad \text{all } i \text{ (except that, possibly, } \zeta_0 = t_1, \; t_{n+1-k} = \zeta_{r+1})$$

which, combined with (4) then implies (3). QED.

We also need

Theorem 1 (Karlin [10]). *If* $f \in \mathbb{L}_\infty^{(k)}$, *then there exists a perfect spline of degree* k *with* $< n - k$ *knots which agrees with* f *at* τ.

A simple proof can be found in [2].

We are now fully equipped for the attack on the optimal recovery scheme. By Karlin's theorem, the set

$$Q_\tau := \{q : q \text{ is a perfect spline of degree } k \text{ with } \le n - k \text{ knots, } q|_\tau = 0\}$$

is not empty. By Lemma 1, each $q \in Q_\tau \backslash \{0\}$ has, in fact, exactly $n - k$ knots and does not vanish off τ (since otherwise it would have $n + 1$ zeros yet only $n - k$ sign changes in its k-th derivative, contrary to the lemma). Therefore, if $f \in \mathbb{L}_\infty^{(k)}$ and $f|_\tau = 0$ and $x \notin \tau$, then $(f(x)/q(x))q$ is a well defined perfect spline of degree k with $n - k$ knots which agrees with f at the n points of τ and the additional point x, hence, Proposition 1 implies that

(5) $$\|f^{(k)}\| \ge |f(x)/q(x)| \, \|q^{(k)}\| .$$

It follows that

(6) $\sup\{|f(x)|/\|f^{(k)}\| : f \in \mathbb{L}_\infty^{(k)},\ f|_\tau = 0\} = |q(x)|/\|q^{(k)}\|$.

Since the left side is independent of q, this shows that Q_τ is the span of just one function, say of $\hat{q}$, normalized to have

$$\hat{q}^{(k)}(0^+) = 1 .$$

It also shows that

(7) $c(\tau) = \|\hat{q}\|$.

This gives a means of computing $c(\tau)$, but not quite yet the equality $\text{const}_\tau = c(\tau)$ nor the optimal recovery scheme. For this, let $\xi_1,\ldots,\xi_{n-k}$ be the $n-k$ knots of $\hat{q}$ and consider

$\$$:= splines of order k with simple knots $\xi_1,\ldots,\xi_{n-k}$

$:= \mathbb{P}_{k,\xi} \cap C^{(k-2)}$

$:= \{f \in C^{(k-2)} : f|_{(\xi_i,\xi_{i+1})} \in \mathbb{P}_k, \text{ all } i\}$ $\quad(\xi_0 = a, \xi_{n+1-k} = b)$.

Theorem 2 (Micchelli, Rivlin & Winograd [15]). The rule

$(\overset{\bullet}{S}f)_\tau = f|_\tau$ and $\overset{\bullet}{S}f \in \$$, all $f \in \mathbb{L}_\infty^{(k)}$

defines a map $\overset{\bullet}{S}$ (read "ess crown(ed)") on $\mathbb{L}_\infty^{(k)}$ which is linear and an optimal recovery scheme with respect to τ.

Proof. First, we prove that $\overset{\bullet}{S}$ is well defined. Since, by the lemma, $\tau_i < \xi_i < \tau_{i+k}$, we ask for matching $f^{(k-1)}$ at some point τ_i only when such $\tau_i \notin \xi$, i.e., when $s^{(k-1)}(\tau_i)$ makes sense for each $s \in \$$. Secondly, $\dim \$ = k +$ number of polynomial pieces $- 1 = n$, therefore $\overset{\bullet}{S}$ is well defined if and only if

(8) $f|_\tau = 0$ and $f \in \$$ implies $f = 0$.

For this, we would like to use inequality (5), but, since $\$$ is not contained in $\mathbb{L}_\infty^{(k)}$ but only in the larger space

$$\mathbb{L}_{\infty,\xi}^{(k)} := \mathbb{L}_\infty^{(k-1)} \cap \bigcap_{i=0}^{n-k} \mathbb{L}_\infty^{(k)}(\xi_i,\xi_{i+1}) ,$$

we must first extend (5) to such spaces, with the convention that

$$\| f^{(k)} \| := \max_i \| f^{(k)}|_{(\xi_i,\xi_{i+1})} \|_\infty \quad \text{for} \quad f \in \mathbb{L}^{(k)}_{\infty,\xi} .$$

This requires the following slight strengthening of Lemma 1.

<u>Lemma</u> 1'. <u>If</u> $h \in \mathbb{L}^{(k)}_{\infty,\zeta}$ <u>for some</u> $a = \zeta_0 < \dots < \zeta_{r+1} = b$ <u>and, for some</u> $\sigma \in \{-1,1\}$,

$$\sigma(-)^i h^{(k)} > 0 \quad \underline{\text{a.e. on}} \quad (\zeta_i,\zeta_{i+1}), \quad i = 0,\dots,r ,$$

<u>then</u> $h|_\tau = 0$ <u>implies</u> $n - k \le r$.

Here, we mean by $h^{(k-1)}(x) = 0$ that $h^{(k-1)}(x^-)h^{(k-1)}(x^+) \le 0$, in case the number x occurs k-fold in τ.

<u>Proof</u>. Rolle's theorem gives again $n + 1 - k$ distinct zeros of $h^{(k-1)}$ which again consists of $r + 1$ strictly monotone pieces, but may fail to be continuous across the points $\zeta_1,\dots,\zeta_r$. But, since $h^{(k)}$ alternates in sign, this latter fault can easily be remedied by appropriate local linear interpolation across

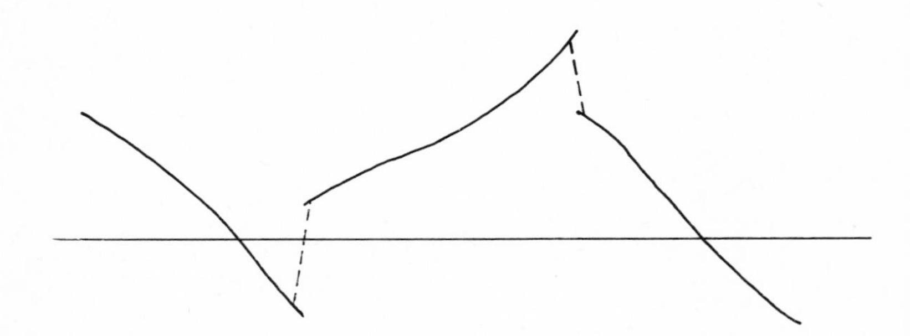

a small neighborhood of each discontinuity ζ_i without disturbing the other two properties and now $n - k \le r$ follows as before; Q.E.D.

This gives

<u>Proposition 1'</u>. <u>If</u> s <u>is a perfect spline with</u> $< n - k$ <u>knots, say with knots</u> $\zeta_1 < \dots < \zeta_r$ <u>where</u> $r < n - k$, <u>and of</u>

<u>degree</u> k, <u>then even</u>

$$\| s^{(k)} \| = \inf\{ \| f^{(k)} \| : f \in \mathbb{L}_{\infty,\zeta}^{(k)},\ f|_\tau = s|_\tau \} .$$

From this we conclude, with $\| \hat{q}^{(k)} \| = 1$, that

(5') $\| f^{(k)} \| \geq |f(x)/\hat{q}(x)|$ <u>for</u> $x \notin \tau,\ f \in \mathbb{L}_{\infty,\xi}^{(k)},\ f|_\tau = 0$.

But this implies (8) since $\| f^{(k)} \| = 0$ for $f \in \$$, and so shows that $\overset{*}{S}$ is well defined. Finally, (5') also implies that, for $f \in \mathbb{L}_\infty^{(k)}$,

$$\| f^{(k)} \| = \| (f - \overset{*}{S}f)^{(k)} \| \geq |f(x) - \overset{*}{S}f(x)| / |\hat{q}(x)|$$

or

(9) $|f(x) - \overset{*}{S}f(x)| \leq |\hat{q}(x)| \, \| f^{(k)} \|$

showing, with (2) and (7), that $\overset{*}{S}$ is an optimal recovery scheme. This proves Theorem 2.

MRW actually insist that a recovery scheme S map into $\mathbb{L}_\infty^{(k)}$, hence they are not quite done at this point, since $\overset{*}{S}$ only maps into $\mathbb{L}_\infty^{(k-1)}$. But, since $\$$ can be viewed as spline functions of degree k with <u>double</u> knots at the ξ_i, we can produce an element of $\mathbb{L}_\infty^{(k)}$ arbitrarily close to $\overset{*}{S}f$ merely by pulling all these double knots apart ever so slightly. This shows that $\inf \text{const}_S = c(\tau)$ even if the inf is restricted to S mapping into $\mathbb{L}_\infty^{(k)}$ but now the inf is not attained apparently for $k > 1$.

3. THE ENVELOPE CONSTRUCTION

The preceding discussion allows a simple derivation of sharp estimates for the value of a function $f \in \mathbb{L}_\infty^{(k)}$ at some point x, given the vector $f|_\tau$ and a bound L on its k-th derivative on [a,b], as follows.

We are to construct the set

$$I_x := \{f(x) : f \in F\}$$

with

$$F := \{f \in \mathbb{L}_\infty^{(k)} : f\big|_\tau = \alpha,\ \|f^{(k)}\| \le L\}$$

for some given α and L. If F is not empty, then I_x is a closed interval,

$$I_x =: [a_x, b_x]$$

say, since F is closed, convex and bounded and $[x] : f \mapsto f(x)$ is a continuous linear functional on $\mathbb{L}_\infty^{(k)}$. Assume that

$$F^o := \{f \in F : \|f^{(k)}\| < L\} \ne \phi .$$

Then

(10) $(a_x, b_x) = [x]F^o$.

Karlin's theorem then implies that, for $x \notin \tau$,

$$Q_x := \{q : q \text{ is perfect spline of degree } k \text{ with } \le n-k \text{ knots, } q\big|_\tau = \alpha,\ q(x) = a_x\}$$

is not empty. Further, $Q_x \subseteq F$, since, by definition of a_x, there exists $f \in F$ with $f(x) = a_x$ while each $q \in Q_x$ agrees with such an f at the $n+k+1$ points τ and x, therefore $\|q^{(k)}\| \le \|f^{(k)}\| \le L$, by the proposition. On the other hand, $Q_x \cap F^o = \phi$, since, if $q \in Q_x \cap F^o$, then $a_x = q(x) \in (a_x, b_x)$, by (10), a contradiction.

It follows that

$$\text{for all } y,\ a_y \le q(y) \le b_y .$$

But, for any $g \in F^o$, $h := g - q$ has $\le n-k$ sign changes in its k-th derivative and vanishes at τ, therefore q has exactly $n-k$ knots and h does not vanish off τ, by Lemma 1. Hence, if $\tau_{i+1}, \ldots, \tau_{i+j}$ are all the points of τ between x and $y \notin \tau$, then

$$(-)^j (g(y) - q(y)) > 0 .$$

This shows that

$$q(y) = \begin{Bmatrix} a_y \\ b_y \end{Bmatrix} \text{ if the interval between x and y contains an } \begin{Bmatrix} \text{even} \\ \text{odd} \end{Bmatrix}$$

number of points of τ. It follows that Q_x contains exactly one function and that this function supplies half of the entire boundary of

$$(11)\quad \{(x,f(x)) : x \in [a,b],\ f \in F\} ,$$

the other half being supplied by the perfect spline p of degree k with $n-k$ knots for which $p|_\tau = \alpha$ and $p(x) = b_x$.

We note the curious corollary that the perfect spline s of Karlin's theorem is unique if there exists g which agrees with f at all points of τ but one and for which $\| g^{(k)} \| < \| s^{(k)} \|$, i.e., if at least one of the interpolation constraints is active. Put differently, it says that if there are two different perfect splines s, $\hat{s}$ of degree k with $\leq n-k$ knots for which $s|_\tau = \hat{s}|_\tau = \alpha$ and which agree at some $x \notin \tau$, then

$$\| s^{(k)} \| = \| \hat{s}^{(k)} \| = L_\alpha := \min\{ \| f^{(k)} \| : f \in \mathbb{L}_\infty^{(k)},\ f|_\tau = \alpha\} .$$

Let now q be the half of the envelope of (11) with $q^{(k)}(0^+) = L$, and let p be the one with $p^{(k)}(0^+) = -L$. Gaffney and Powell [8] choose

$$S_L\alpha := (p + q)/2$$

as a good interpolant, its graph being clearly the center of (11). Since p and q are uniquely defined for $L > L_\alpha$ by the requirement that they are perfect splines of degree k with $n - k$ knots, equal to α at τ and to a_x, resp. b_x at x, they are necessarily continuous functions of L and α in that range. In particular, with $q =: q_{L,\alpha}$, we have $q_{L,\alpha} = Lq_{1,\alpha/L}$ and $q_{1,\alpha/L} \to q_{1,0} = \hat{q}$ as $L \to \infty$, and, similarly, $p_{L,\alpha/L} \to -\hat{q}$ as $L \to \infty$. This shows that, with $u_1 < \dots < u_{n-k}$ the knots of q and $v_1 < \dots < v_{n-k}$ the knots of p,

$$\lim_{L\to\infty} u_i = \lim_{L\to\infty} v_i = \xi_i, \quad i = 1,\dots,n-k .$$

In particular, for large enough L, the sequence

$$0 = \xi_{0+}, \xi_{1-}, \xi_{1+}, \dots, \xi_{n-k-}, \xi_{n-k+}, \xi_{n-k+1-} = 1$$

with

$$\xi_{i-} := \min\{u_i, v_i\},\ \xi_{i+} := \max\{u_i, v_i\}$$

is nondecreasing and

$$(S_L\alpha)^{(k)} = \begin{cases} 0 & \text{on } (\xi_{i+},\xi_{i+1-}) \\ \pm 2L & \text{on } (\xi_{i-},\xi_{i+}) \end{cases}$$

hence

$$\| (S_L\alpha)^{(k)} \|_1 \le 2L\| u - v \|_1 \xrightarrow[L\to\infty]{} 0 .$$

This shows that $S_L\alpha$ converges to an element of \$, while

$$S_L\alpha\big|_\tau = \alpha \quad \text{for all} \quad L ,$$

therefore, $S_L\alpha$ converges to the optimal interpolant for the data.

It was in this way that Gaffney and Powell [8] constructed, quite independently from Micchelli, Rivlin and Winograd [15], the same optimal recovery or interpolation scheme $\overset{*}{S}$.

The problem of constructing the set I_x was posed originally in the basic paper by Golomb and Weinberger [9], although they gave detailed attention to such problems only when the (semi)norm involved comes from an inner product. Micchelli and Miranker [14] solved the problem posed at the beginning of this section in the sense that they correctly described the boundary of (11) as being given by just two perfect splines of degree k, each with n - k knots, and with their k-th derivative equal to L in absolute value. In fact, Micchelli and Miranker consider the slightly more general situation where $f^{(k)}$ is only known to map [a,b] into some interval [m,M]. They state that these matters could be proved along the lines used by Burchard [6] to solve a related restricted moment problem and refer specifically to Karlin and Studden [11, VIII, §3] for requisite facts concerning principal representations of interior points of moment spaces. Of course, these facts go back to Krein [12]. Quite independently, Gaffney and Powell [8] also solved this problem, with the proofs in Gaffney's thesis based on Chebyshev type inequalities as found in Karlin and Studden [11, VIII, §8] and adapted by him to weak Chebyshev systems.

4. THE CONSTRUCTION OF NORM PRESERVING EXTENSIONS TO ALL OF $\mathbb{L}_1$

Both the optimal recovery scheme $\overset{*}{S}$ of Section 2 and the envelope of Section 3 require the construction of an absolutely constant function h with no more than a specified number of jumps which provides an integral representation of a linear functional given on a linear space of spline functions.

In the optimal recovery, we are to construct a perfect spline $\hat{s}$ of degree k with n - k knots and with $\| \hat{s}^{(k)} \| = 1$ which vanishes at the given n-point sequence τ. Let $(M_i)_1^n$ be the sequence of B-splines of order k with knot sequence τ, each normalized to have unit integral, i.e.,

(12) $M_i(x) := M_{i,k,\tau}(x) := k[\tau_i,\ldots,\tau_{i+k}] (\cdot - x)_+^{k-1}$,

with $[\tau_i,\ldots,\tau_{i+k}]f$ the k-th divided difference of the function f at the points $\tau_i,\ldots,\tau_{i+k}$. Then, from Taylor's expansion with integral remainder,

$$[\tau_i,\ldots,\tau_{i+k}]f = \int_{\tau_i}^{\tau_{i+k}} M_{i,k,\tau}(x) f^{(k)}(x)dx/k! \text{ for all } f \in \mathbb{L}_\infty^{(k)} .$$

The points $\xi = (\xi_i)_1^{n-k}$ are therefore characterized by the requirement that the function

$$h_\xi(x) := \text{sign} \prod_{i=1}^{n-k} (x - \xi_i) = \pm \hat{s}^{(k)}(x)$$

be orthogonal to each of the n - k functions $M_1,\ldots,M_{n-k}$.

Before considering the computational details of determining ξ from this orthogonality condition, I want to comment on the fact that this is a problem of representing or extending a linear functional on some subspace of $\mathbb{L}_1$ and is therefore closely related to the problem of computing the norm of a linear functional on some subspace of $\mathbb{L}_1$. This is also explored in a forthcoming paper by Micchelli [13].

Indeed, if T is a linear subspace of $\mathbb{L}_1$ of dimension n + 1 - k, and λ is a linear functional on T, we might ask for ξ and α so that

(13) $\alpha \int h_\xi g = \lambda g$ for all $g \in T$.

But then, in particular, h_ξ is orthogonal to ker λ, a subspace of dimension n - k. Conversely, if we have already found h_ξ orthogonal to ker λ then there will be exactly one α so that αh_ξ

represents λ on T in the sense of (13), unless h_ξ is even orthogonal to all of T. But this latter event cannot happen in case T is weak Chebyshev since h_ξ has only n - k jumps.

It is clear that any such representation h_ξ for λ produces an upper bound for the norm of λ. In fact, $\|\lambda\| = |\alpha|$ in case T is weak Chebyshev. This is actually how I became interested two years ago in the numerical construction of representers of linear functionals [3], [4]. I was interested in computing, or at least closely estimating, the norm of certain linear functionals on certain spline subspaces in $\mathbb{L}_1$. In a way, this is a trivial problem, viz. the maximization of a linear function over a finite dimensional compact convex set, and there was the feeling that there ought to be special methods available. Perhaps some reader can steer me towards such methods. I found, for the particular cases of concern to me in which T was always weak Chebyshev, nothing more effective for calculating $\|\lambda\|$ than to construct such a representation (13).

Finally, the envelope construction corresponds to the slightly different situation where dim T = n - k, $\lambda \in T'$ and α with $|\alpha| > \|\lambda\|$ is prescribed and one seeks ξ so that again (13) holds. We have now one less condition to satisfy but also one less parameter to do it with.

5. CONSTRUCTION OF THE KNOTS FOR THE OPTIMAL RECOVERY SCHEME

As we saw in the preceding section, the knots $\xi = (\xi_i)_1^{n-k}$ for the optimal recovery scheme are the solution to the following problem. We are given $(\tau_i)_1^n$ nondecreasing, with $\tau_i < \tau_{i+k}$, all i, and n > k, and wish to construct $\xi_0 < \ldots < \xi_r$ with

$$r := n - k + 1$$

and with $\xi_0 = a := \tau_1$, $\xi_r = b := \tau_n$ so that

$$(14) \quad \sum_{j=1}^{r} \beta_j \int_{\xi_{j-1}}^{\xi_j} M_{i,k} = 0, \quad i = 1,\ldots,n - k$$

while also

$$(15) \quad \beta_j + \beta_{j+1} = 0, \quad j = 1,\ldots,n - k .$$

Extend τ by

$$\tau_n =: \tau_{n+1} =: \tau_{n+2} =: \dots =: \tau_{n+k} .$$

Then one verifies easily that

$$\int_a^x M_{i,k} = \sum_{m \geq i} N_{m,k+1}(x) \quad \text{for} \quad x \geq a$$

with

$$N_{m,k+1} := \frac{\tau_{m+k+1} - \tau_m}{k+1} M_{m,k+1}, \quad \text{all} \quad m .$$

Therefore, (14) is equivalent to

$$\sum_{j=1}^{r} \beta_j \sum_{m \geq i} (N_{m,k+1}(\xi_j) - N_{m,k+1}(\xi_{j-1})) = 0, \quad i = 1,\dots,n-k ,$$

or, on subtracting equation i from equation i - 1 for i = 2,...,n-k,

$$\sum_{j=1}^{r} \beta_j (N_{i,k+1}(\xi_j) - N_{i,k+1}(\xi_{j-1})) = 0, \; i = 1,\dots,n-k-1$$

$$\sum_{j=1}^{r} \beta_j \sum_{m \geq n-k} (N_{m,k+1}(\xi_j) - N_{m,k+1}(\xi_{j-1})) = 0 .$$

Using the fact that $N_{m,k+1}(\xi_0) = 0$ for $m \geq 1$, this can also be written

$$\sum_{j=1}^{n-k} (\beta_j - \beta_{j+1}) N_{i,k+1}(\xi_j) = -\beta_r N_{i,k+1}(\xi_r), \; i = 1,\dots,n-k-1$$

$$\sum_{j=1}^{n-k} (\beta_j - \beta_{j+1}) \sum_{m \geq n-k} N_{m,k+1}(\xi_j) = -\beta_r \sum_{m \geq n-k} N_{m,k+1}(\xi_r) .$$

Since $N_{i,k+1}(\xi_r) = 0$ for $i < n-k$, while $\sum_{m \geq n-k} N_{m,k+1}(\xi_r) = 1$, the right side becomes simply $(0,\dots,0,-\beta_r)$.

Choose now $\beta_r = -1$ to make things definite. Then $\beta_j = (-)^{r-j-1} = (-)^{n-k-j}$ by (15), and (19) and (15) are seen to be equivalent to

(16) $F(\xi) = 0$

with $F : \mathbb{R}^{n-k} \to \mathbb{R}^{n-k}$ given by

$$(17)\quad F(\xi)_i := \begin{cases} a_i & , \ i < n-k \\ \sum_{m=n-k}^{n} a_m, & i = n-k \end{cases}$$

where

$$(18)\quad a_i := \begin{cases} \sum_{j=1}^{n-k} (-)^{n-k-j} N_{i,k+1}(\xi_j), & i = 1,\ldots,n-1 \\ -1/2 & , \ i = n\,. \end{cases}$$

We solve (16) by Newton's method. From the current guess ξ, we compute a new guess $\xi^* = \xi + \delta\xi$, with $\delta\xi$ the solution to the linear system

$$(19)\quad F'(\xi)\delta\xi = -F(\xi)\ .$$

Since $N'_{i,k+1} = M_{i,k} - M_{i+1,k}$, addition of equation i to equation i-1, $i = n-k,\ldots,2$, in (19) produces the equivalent linear system

$$\sum_{j=1}^{n-k} \sum_{m=i}^{n-1} (M_{m,k} - M_{m+1,k})(\xi_j)\ (-)^{n-k-j}\delta\xi_j = -\sum_{m=i}^{n} a_m,\ i=1,\ldots,n-k.$$

But, since $\sum_{i}^{n-1} (M_{m,k} - M_{m+1,k})(t) = M_{i,k}(t) - M_{n,k}(t) = M_{i,k}(t)$ for $t \le b$, this shows that (19) is equivalent to the linear system

$$(20)\quad Cx = d$$

with

$$(21)\quad (-)^{n-k-j}\delta\xi_j = x_j,\quad d_i = \Big(\sum_{m=i}^{n} -a_m\Big)(\tau_{i+k} - \tau_i)/k,\quad c_{ij} = N_{i,k}(\xi_j)$$

$$i,j = 1,\ldots,n-k.$$

The matrix C is totally positive and (2k - 1)-banded, hence can be stored in 2k - 1 bands of length n - k each, and can be factored cheaply and reliably within these bands by Gauss elimination _without_ pivoting (see [5]).

The iteration is carried out in the program SPLOPT below, starting with the initial guess

$$(22)\quad \xi_i = (\tau_{i+1} + \ldots + \tau_{i+k-1})/(k-1),\ i = 1,\ldots,n-k\ .$$

A first version of the program was equipped to carry out Modified Newton iteration: ξ^* is computed as the first vector in the sequence

$$\xi + 2^{-h}\delta\xi, \quad h = 0,1,2,\ldots$$

for which $\| F(\xi + 2^{-h}\delta\xi) \|_2 < \| F(\xi) \|_2$. But, in all examples tried, the initial guess (22) was apparently sufficiently close to the solution to have always $h = 0$, i.e., $\| F(\xi + \delta\xi) \|_2 < \| F(\xi) \|_2$. In fact, the termination criterion

$$\| \delta\xi \|_\infty \leq 10^{-6}(\tau_n - \tau_1)/(n - k)$$

was usually reached in three or four clearly quadratically converging iterations. For this reason, the program SPLOPT below carries out simple Newton iteration. It would be nice to prove that Newton iteration, starting from (22), necessarily converges. But, such a proof would necessarily have to be in control of the norm of the inverse of the matrix $F'(\xi)$, hence in control of the norm of the inverse of $C = (N_i(\xi_j))$ as a function of ξ. Good estimates for $\| (N_i(\xi_j))^{-1} \|$ have been searched for in the past by some who were interested in bounding (the error in) spline interpolation, but without much success. E.g., the simple conjecture that, for the initial guess (22), $\| (N_i(\xi_j))^{-1} \|_1 \leq \text{const}_k$ independent of ξ has been proved so far only for $k \leq 4$.

```
      SUBROUTINE SPLOPT ( TAU, N, K, SCRTCH, T, IFLAG )
COMPUTES THE KNOTS  T  FOR THE OPTIMAL RECOVERY SCHEME OF ORDER  K
C  FOR DATA AT  TAU (I),I=1,...,  N .  TAU  MUST BE STRICTLY INCREASING.
C  SEE TEXT FOR DEFINITION OF VARIABLES AND METHOD USED.
C    IFLAG  =  1 OR 2 DEPENDING ON WHETHER OR NOT  T  WAS CONSTRUCTED.
C       DIMENSION SCRTCH((N-K)*(2*K+3)+5*K+3), T(N+K)
        DIMENSION TAU(N),SCRTCH(1),T(1)
        DATA NEWTMX,TOLRTE / 10,.000001/
        NMK = N-K
        IF (NMK)                                    1,56,2
    1   PRINT 601,N,K
  601   FORMAT(13H ARGUMENT N =,I4,29H IN  SPLOPT  IS LESS THAN K =,I3)
                                                    GO TO 999
    2   IF (K .GT. 2)                               GO TO 3
        PRINT 602,K
  602   FORMAT(13H ARGUMENT K =,I3,27H IN  SPLOPT  IS LESS THAN 3)
                                                    GO TO 999
    3   NMKM1 = NMK - 1
        FLOATK = K
        KPK = K+K
        KP1 = K+1
        KPKP1 = K+KP1
        KM1 = K-1
        KPKM1 = K+KM1
        KPN = K+N
        SIGNST = -1.
        IF (NMK .GT. (NMK/2)*2)  SIGNST = 1.
```

```
C   SCRTCH(I) = TAU-EXTENDED(I), I=1,...,N+K+K
      NX = N+KPKP1
C   SCRTCH(I+NX) = XI(I),I=0,...,N-K+1
      NA = NX + NMK + 1
C   SCRTCH(I+NA) = -A(I), I=1,...,N
      ND = NA + N
C   SCRTCH(I+ND) = X(I) OR D(I), I=1,...,N-K
      NV = ND + NMK
C   SCRTCH(I+NV) = VNIKX(I),I=1,...,K+1
      NC = NV + KP1
C   SCRTCH((J-1)*(N-K)+I + NC) = CHAT(I,J),I=1,...,N-K,J=1,...,2*K-1
      LENGCH = NMK*KPKM1
C   EXTEND  TAU  TO A KNOT SEQUENCE AND STORE IN SCRTCH.
      DO 5 J=1,K
         SCRTCH(J) = TAU(1)
    5    SCRTCH(KPN+J) = TAU(N)
      DO 6 J=1,N
    6    SCRTCH(K+J) = TAU(J)
C   FIRST GUESS FOR  SCRTCH (.+NX)  =  XI .
      SCRTCH(NX) = TAU(1)
      SCRTCH(NMK+1+NX) = TAU(N)
      DO 10 J=1,NMK
         SUM = 0.
         DO 9 L=1,KM1
    9       SUM = SUM + TAU(J+L)
   10    SCRTCH(J+NX) = SUM/KM1
C   LAST ENTRY OF  SCRTCH (.+NA)  =  - A  IS ALWAYS ...
      SCRTCH(N+NA) = .5
C   START NEWTON ITERATION.
      NEWTON = 1
      TOL = TOLRTE*(TAU(N) - TAU(1))/NMK
C   START NEWTON STEP
COMPUTE THE 2K-1 BANDS OF THE MATRIX C AND STORE IN SCRTCH(.+NC),
C   AND COMPUTE THE VECTOR  SCRTCH(.+NA) = -A.
   20 DO 21 I=1,LENGCH
   21    SCRTCH(I+NC) = 0.
      DO 22 I=2,N
   22    SCRTCH(I-1+NA) = 0.
      SIGN = SIGNST
      ILEFT = KP1
      DO 28 J=1,NMK
         XIJ = SCRTCH(J+NX)
   23       IF (XIJ .LT. SCRTCH(ILEFT+1))    GO TO 25
            ILEFT = ILEFT + 1
            IF (ILEFT .LT. KPN)          GO TO 23
            ILEFT = ILEFT - 1
   25    CALL BSPLVN(SCRTCH,K,1,XIJ,ILEFT,SCRTCH(1+NV))
         ID = MAX0(0,ILEFT-KPK)
         INDEX = NC+(J-ID+KM1)*NMK+ID
         LLMAX = MIN0(K,NMK-ID)
         LLMIN = 1 - MIN0(0,ILEFT-KPK)
         DO 26 LL=LLMIN,LLMAX
            INDEX = INDEX - NMKM1
   26       SCRTCH(INDEX) = SCRTCH(LL+NV)
         CALL BSPLVN(SCRTCH,KP1,2,XIJ,ILEFT,SCRTCH(1+NV))
         ID = MAX0(0,ILEFT-KPKP1)
         LLMIN = 1 - MIN0(0,ILEFT-KPKP1)
         DO 27 LL=LLMIN,KP1
            ID = ID + 1
   27       SCRTCH(ID+NA) = SCRTCH(ID+NA) - SIGN*SCRTCH(LL+NV)
   28    SIGN = -SIGN
```

```
      CALL BANFAC(SCRTCH(1+NC),NMK,NMK,KPKM1,K,IFLAG)
                                                 GO TO (45,44),IFLAG
   44 PRINT 644
  644 FORMAT(32H C IN  SPLOPT  IS NOT INVERTIBLE)
                                                 RETURN
COMPUTE  SCRTCH (.+ND) =  D  FROM  SCRTCH (.+NA) = - A .
   45 DO 46 I=N,2,-1
   46    SCRTCH(I-1+NA) = SCRTCH(I-1+NA) + SCRTCH(I+NA)
      DO 49 I=1,NMK
   49    SCRTCH(I+ND) = SCRTCH(I+NA)*(TAU(I+K)-TAU(I))/FLOATK
COMPUTE  SCRTCH (.+ND) =  X .
      CALL BANSUB(SCRTCH(1+NC),NMK,NMK,KPKM1,K,SCRTCH(1+ND))
COMPUTE  SCRTCH (.+ND) = CHANGE IN  XI . MODIFY, IF NECESSARY, TO
C  PREVENT NEW  XI  FROM MOVING MORE THAN 1/3 OF THE WAY TO ITS
C  NEIGHBORS. THEN ADD TO  XI  TO OBTAIN NEW  XI  IN SCRTCH(.+NX).
      DELMAX = 0.
      SIGN = SIGNST
      DO 53 I=1,NMK
         DEL = SIGN*SCRTCH(I+ND)
         DELMAX = AMAX1(DELMAX,ABS(DEL))
         IF (DEL .GT. 0.)                        GO TO 51
         DEL = AMAX1(DEL,(SCRTCH(I-1+NX)-SCRTCH(I+NX))/3.)
                                                 GO TO 52
   51    DEL = AMIN1(DEL,(SCRTCH(I+1+NX)-SCRTCH(I+NX))/3.)
   52    SIGN = -SIGN
   53    SCRTCH(I+NX) = SCRTCH(I+NX) + DEL
CALL IT A DAY IN CASE CHANGE IN  XI  WAS SMALL ENOUGH OR TOO MANY
C  STEPS WERE TAKEN.
      IF (DELMAX .LT. TOL)                       GO TO 54
      NEWTON = NEWTON + 1
      IF (NEWTON .LE. NEWTMX)                    GO TO 20
      PRINT 653,NEWTMX
  653 FORMAT(33H NO CONVERGENCE IN  SPLOPT  AFTER,I3,14H NEWTON STEPS.)
   54 DO 55 I=1,NMK
   55    T(K+I) = SCRTCH(I+NX)
   56 DO 57 I=1,K
         T(I) = TAU(1)
   57    T(N+I) = TAU(N)
                                                 RETURN
  999 IFLAG = 2
                                                 RETURN
      END
```

The subroutine SPLOPT has <u>input</u> TAU(i) = τ_i, i = 1,...,n, assumed to be nondecreasing and to satisfy $\tau_i < \tau_{i+k}$, all i, the integer N = n and the desired order k in K. The routine needs a work array SCRTCH, of size $\geq$ (n - k) (2k + 3) + 5k + 3; n(2k + 3) is more than enough. The routine has <u>output</u> T(i) = t_i, i = 1,...,n + k, the knot sequence for the optimal recovery scheme, in case IFLAG = 1. For IFLAG = 2, something went wrong.

The routine uses the subroutine BSPLVN for the evaluation of all B-splines of a given order on a given knot sequence which do not vanish at a given point. This routine, and others for dealing computationally with splines and B-splines, can be found in [1]. For completeness, we also list here the subroutines BANFAC and BANSUB, used in SPLOPT for the solution of the banded system (20).

```
      SUBROUTINE BANFAC ( A, NROW, N, NDIAG, MIDDLE, IFLAG )
      DIMENSION A(NROW,NDIAG)
      IFLAG = 1
      ILO = MIDDLE - 1
      IF (ILO)                                      999,10,19
   10 DO 11 I=1,N
         IF (A(I,1))                                11,999,11
   11    CONTINUE
                                                    RETURN
   19 IHI = NDIAG - MIDDLE
      IF (IHI)                                      999,20,29
   20 DO 25 I=1,N
         IF (A(I,MIDDLE))                           21,999,21
   21    JMAX = MINO(ILO,N-I)
         IF (JMAX)                                  25,25,22
   22    DO 23 J=1,JMAX
   23       A(I+J,MIDDLE-J) = A(I+J,MIDDLE-J)/A(I,MIDDLE)
   25    CONTINUE
                                                    RETURN
   29 DO 50 I=1,N
         DIAG = A(I,MIDDLE)
         IF (DIAG)                                  31,999,31
   31    JMAX = MINO(ILO,N-I)
         IF (JMAX)                                  50,50,32
   32    KMAX = MINO(IHI,N-I)
         DO 33 J=1,JMAX
            IPJ = I+J
            MMJ = MIDDLE-J
            A(IPJ,MMJ) = A(IPJ,MMJ)/DIAG
            DO 33 K=1,KMAX
   33          A(IPJ,MMJ+K) = A(IPJ,MMJ+K) - A(IPJ,MMJ)*A(I,MIDDLE+K)
   50    CONTINUE
                                                    RETURN
  999 IFLAG = 2
                                                    RETURN
      END
```

BANFAC factors an N × N band matrix C whose NDIAG bands are contained in the columns of the NROW × NDIAG array A, with the MIDDLE column containing the main diagonal of C. It uses Gauss elimination without pivoting and stores the factors in A.

```
      SUBROUTINE BANSUB ( A, NROW, N, NDIAG, MIDDLE, B )
      DIMENSION A(NROW,NDIAG),B(N)
      IF (N .EQ. 1)                            GO TO 21
      ILO = MIDDLE - 1
      IF (ILO)                                 21,21,11
   11 DO 19 I=2,N
         JMAX = MINO(I-1,ILO)
         DO 19 J=1,JMAX
   19       B(I) = B(I) - B(I-J)*A(I,MIDDLE-J)
C
   21 I = N
      IHI = NDIAG-MIDDLE
      DO 29 II=1,N
         JMAX = MINO(N-I,IHI)
         IF (JMAX)                             28,28,22
   22    DO 25 J=1,JMAX
   25       B(I) = B(I) - B(I+J)*A(I,MIDDLE+J)
   28    B(I) = B(I)/A(I,MIDDLE)
   29    I = I - 1
      END
```

BANSUB then uses the factorization of C into the product of a lower and an upper triangular matrix computed in BANFAC to solve the equation $Cx = b$ for given b (input in B) by forward and back substitution. The solution x is contained in B, on output.

6. CONSTRUCTION OF THE OPTIMAL INTERPOLANT

With the break points $\xi_1 < \ldots < \xi_{n-k}$ for the optimal interpolant $\mathring{S}f$ determined from τ in SPLOPT, it remains to compute $\mathring{S}f$. This we propose to do by determining its B-spline coefficients. SPLOPT has produced the knot sequence $\underline{\underline{t}} = (t_i)_1^{n+k}$, with

$$t_1 = \ldots = t_k = \tau_1,\ t_{k+i} = \xi_i,\ i = 1,\ldots,n-k,$$

$$t_{n+1} = \ldots t_{n+k} = \tau_n .$$

Let $(N_i)_1^n$ be the corresponding sequence of normalized B-splines of order k, i.e.,

$$N_i(x) := N_{i,k,\underline{\underline{t}}}(x) := (t_{i+k}-t_i)[t_i,\ldots,t_{i+k}](\cdot-x)_+^{k-1},\ \text{all } i.$$

Then, according to Curry & Schoenberg [7], every piecewise polynomial function of order k on $[a,b] := [\tau_1,\tau_n]$, with $k - 2$ continuous derivatives and break points $\xi_1,\ldots,\xi_{n-k}$, i.e., every spline of order k on [a,b] with knot sequence $\underline{\underline{t}}$, has a unique representation as a linear combination of the n functions $N_1,\ldots,N_n$. Therefore

$$\hat{S}f = \sum_{i=1}^{n} a_i N_i$$

with $a = (a_i)_1^n$ the solution of the linear system

$$(23) \quad \sum_{j=1}^{n} N_j(\tau_i)a_j = f(\tau_i), \quad i = 1,\ldots,n .$$

In case τ is strictly increasing, - the only case considered here,- Lemma 1 implies that $t_i < \tau_i < t_{i+k}$, all i, which, together with the fact that N_j vanishes outside the interval $[t_j,t_{j+k}]$, all j, shows that the coefficient matrix of (23) is $2k - 1$ banded. Since the coefficient matrix is also totally positive, we can therefore (see [5]) solve (23) by Gauss elimination without pivoting and within the $2k - 1$ bands required for the storage of the nonzero entries of the matrix. The following subroutine SPLINT generates the linear system (23), given on input the arrays TAU(i) = τ_i, FTAU(i) = $f(\tau_i)$, i = 1,...,N, T(i) = t_i, i = 1,...,N + K and the numbers N and K. The system is then solved, using BANFAC and BANSUB given in Section 5, and using a working array Q, of size N(2K - 1). The output consists of the B-coefficients a_i = BCOEF(i), i = 1,...,N, in case IFLAG = 1. If IFLAG = 2, then the linear system (23) was not invertible.

```
      SUBROUTINE SPLINT ( TAU, FTAU, T, N, K, Q, BCOEF, IFLAG )
C     SPLINT  PRODUCES THE B-SPLINE COEFF.S  BCOEF  OF THE SPLINE OF ORDER
C     K  WITH KNOTS  T (I), I=1,..., N + K , WHICH TAKES ON THE VALUE
C     FTAU (I) AT  TAU (I), I=1,..., N .
C     TAU  IS ASSUMED TO BE STRICTLY INCREASING.
C    SEE TEXT FOR DESCRIPTION OF VARIABLES AND METHOD.
C    DIMENSION T(N+K),Q(N,2*K-1)
      DIMENSION TAU(N), FTAU(N), T(1), Q(N,1), BCOEF(N)
      NP1 = N + 1
      KPKM1 = 2*K - 1
      ILEFT = K
      DO 30 I=1,N
         TAUI = TAU(I)
         ILP1MX = MIN0(I+K,NP1)
         DO 13 J=1,KPKM1
   13       Q(I,J) = 0.
         ILEFT = MAX0(ILEFT,I)
         IF (TAUI .LT. T(ILEFT))            GO TO 998
   15       IF (TAUI .LT. T(ILEFT+1))       GO TO 16
            ILEFT = ILEFT + 1
            IF (ILEFT .LT. ILP1MX)          GO TO 15
         ILEFT = ILEFT - 1
         IF (TAUI .GT. T(ILEFT+1))          GO TO 998
   16    CALL BSPLVN ( T, K, 1, TAUI, ILEFT, BCOEF )
C        NOTE THAT  BCOEF  IS USED HERE FOR TEMP.STORAGE.
         L = ILEFT - I
         DO 30 J=1,K
            L = L+1
   30       Q(I,L) = BCOEF(J)
      NP2MK = N+2-K
      CALL BANFAC ( Q, N, N, KPKM1, K, IFLAG )
                                            GO TO (40,999), IFLAG
   40 DO 41 I=1,N
   41    BCOEF(I) = FTAU(I)
      CALL BANSUB ( Q, N, N, KPKM1, K, BCOEF )
                                            RETURN
  998 IFLAG = 2
  999 PRINT 699
  699 FORMAT(41H LINEAR SYSTEM IN  SPLINT  NOT INVERTIBLE)
                                            RETURN
      END
```

Note that [1] contains programs which might facilitate subsequent use of the optimal interpolant determined in this way via SPLOPT and SPLINT.

Finally, the linear system (23) can be generated in $O(nk^2)$ operations and, because of the band structure, can be solved in $O(nk)$ operations. The linear system (20), to be generated and solved at each Newton step for finding ξ, is of similar nature (with $n - k$ rather than n equations and a coefficient matrix which is the transposed of the kind of matrix appearing in (23)) hence requires a similar effort for its construction and solution. Therefore, if it takes indeed only three to four Newton iterations to find ξ to sufficient accuracy, then it takes only four to five times as much work to construct the optimal interpolant rather than any spline interpolant to the same data. Also, the total effort is only $O(nk^2)$ which, for large n, compares very favorably with such interpolation schemes as polynomial interpolation which takes $O(n^2)$ operations, or more general schemes which take as much as $O(n^3)$ operations.

ACKNOWLEDGMENT

This work was sponsored by the United States Army under Contract No. DAAG29-75-C-0024.

REFERENCES

1. C. de Boor, Package for calculating with B-splines, MRC TSR 1333 (1973); SIAM J. Numer. Anal. (to appear).

2. C. de Boor, A remark concerning perfect splines, Bull. Amer. Math. Soc. 80 (1974) 724-727.

3. C. de Boor, A smooth and local interpolant with 'small' k-th derivative, in "Numerical solutions of boundary value problems for ordinary differential equations", A. K. Aziz ed., Academic Press, New York, 1975, 177-197.

4. C. de Boor, On local linear functionals which vanish at all B-splines but one, in "Theory of Approximation with Applications", A. G. Law & B. N. Sahney eds., Academic Press, New York, 1976, 120-145.

5. C. de Boor & A. Pinkus, Backward error analysis for totally positive linear systems, MRC TSR 1620, Jan. 1976; Numer. Math. (to appear).

6. H. G. Burchard, Interpolation and approximation by generalized convex functions, Ph.D. Thesis, Purdue U., Lafayette, IN., 1968.

7. H. B. Curry and I. J. Schoenberg, On Pólya frequency functions. IV: The fundamental spline functions and their limits, J. d'Analyse Math. 17 (1966) 71-107.

8. P. W. Gaffney & M. J. D. Powell, Optimal interpolation, in "Numerical Analysis", G. A. Watson ed., Lecture Notes in Math. Vol. 506, Springer, Heidelberg, 1976.

9. M. Golomb and H. F. Weinberger, Optimal approximation and

error bounds, in "On numerical approximation", R. E. Langer ed., Univ. Wisconsin Press, Madison, WI., 1959, 117-190.

10. S. Karlin, Some variational problems on certain Sobolev spaces and perfect splines, Bull. Amer. Math. Soc. 79 (1973) 124-128.

11. S. Karlin and W. J. Studden, Tchebycheff systems: with applications in analysis and statistics, Interscience Publishers, John Wiley, New York, 1966.

12. M. G. Krein, The ideas of P. L. Chebyshev and A. A. Markov in the theory of limiting values of integrals and their further development, Uspekhi Mat. Nauk (n.S.) 6 (1951), 3-120; Engl. Transl.: Amer. Math. Soc. Transl. Series 2, 12 (1959) 1-122.

13. C. A. Micchelli, Best L^1-approximation by weak Chebyshev systems and the uniqueness of interpolating perfect splines, J. Approximation Theory (to appear).

14. C. A. Micchelli & W. L. Miranker, High order search methods for finding roots, J. ACM 22 (1975) 51-60.

15. C. A. Micchelli, T. J. Rivlin and S. Winograd, The optimal recovery of smooth functions, Numer. Math. 26 (1976) 279-285.

INTERPOLATION OPERATORS AS OPTIMAL RECOVERY SCHEMES FOR CLASSES OF ANALYTIC FUNCTIONS

Michael Golomb

Purdue University
West Lafayette, Indiana 47907

1. INTRODUCTION

Suppose that the only information we have about a function f is that it belongs to a certain class $\mathscr{B}$ (usually a ball in a normed space) and that it takes on given values at some finitely many points $x_1,\ldots,x_n$ of its domain (or that some other finitely many linear functionals $\ell_1,\ldots,\ell_n$ have given values at f). Suppose we are to assign a value to $f(x)$ where x does not belong to the set $\{x_1,\ldots,x_n\}$ (or to $\ell(f)$ where ℓ is not in the span of $\{\ell_1,\ldots,\ell_n\}$). The value α_* for $f(x)$ is considered optimal if for any other assignment α there is some $f_\alpha \in \mathscr{B}$ with $f_\alpha(x_i) = f(x_i)$ $(i = 1,\ldots,n)$ for which the error $|\alpha - f_\alpha(x)|$ is at least as large as $|\alpha_* - f(x)|$. If moreover a function $s_* \in \mathscr{B}$ can be found such that s_* evaluated at x gives the optimal value α_* and this is so for every x in the domain of the functions f then s_* is considered an <u>optimal interpolant</u> (or extrapolant) for these functions.

If $\mathscr{B}$ is a closed ball in a Hilbert space (and in some other cases) it is known that an optimal interpolant s_* exists (see [1], [2]). The interpolant s_* has several other extremal properties, and is in particular characterized uniquely by the property that among all functions in $\mathscr{B}$ which have the given values at $x_1,\ldots,x_n$, s_* has minimum norm. This allows s_* to be calculated from the variational equation and the interpolation conditions. Most of these interpolants are called <u>splines</u> of one kind or another.

More recently the problem of reconstructing in an optimal way the functions f of a class $\mathscr{B}$ from data $f(x_1),\ldots,f(x_n)$ has been given a wider formulation. If F_n denotes the data vector $\{f(x_1),\ldots,f(x_n)\}$ one asks for a mapping S (linear or non-linear) from the space of data vectors to the class $\mathscr{B}$, so that SF_n is a good substitute for the only partially known function f. Each such mapping is called a recovery scheme and S_* is said to be an optimal recovery scheme (with respect to a chosen norm or seminorm) if the norm of the largest error $SF_n - f$ that occurs in $\mathscr{B}$ is minimized by S_* (for this concept and results based on it see [3], [4]).

It is not too surprising that in the case where $\mathscr{B}$ is a ball in a Hilbert space the two problems lead to the same solution: $S_*F_n = s_*$. This is true, no matter what the norm or seminorm is that is used for measuring the error $Ef = SF_n - f$. Those used most commonly in this paper are $|Ef(x_o)|$ for some fixed x_o in the domain of $f \in \mathscr{B}$, and $\sup_{x\in D}|Ef(x)|$ for some subset D of the domain. The limitation imposed by the Hilbert space setting is compensated for by the simplicity of the recovery algorithm. The mapping $S_*:F_n \mapsto s_*$ is, of course, linear, moreover the mapping $f \mapsto s_*$ is an orthogonal projection onto the n-dimensional subspace of spline interpolants with nodes at $x_1,\ldots,x_n$. These and other general results, which are hardly new, are presented in Sections 2-3 below. Practically all the Hilbert spaces of functions of use for numerical analysis have a reproducing kernel. This must be so because the place value functional $x \mapsto f(x)$ is necessarily continuous. This is, in particular, true of the spaces of analytic functions in one complex variable that are treated in detail in this paper. In Section 4 formulas for the optimal recovery operator and the optimal error in terms of the reproducing kernel are presented.

The "splines" of our Hilbert spaces of analytic functions are themselves analytic functions. The higher order discontinuities - the "knots" of the usual polynomial splines - are replaced by poles outside of the domain of analyticity, which have a simple geometric relationship to the interpolation nodes. The spaces considered in Sections 5-8 have been selected so that they yield simple spline bases, which can be obtained without computation. This is not to imply that the methods are restricted to these examples. Simplicity is achieved by considering only simple domains (disk, annulus, strip, ellipse) and specially chosen norms. These norms, although not the most natural (or traditional) ones from the point of view of function theory, are topologically equivalent to them, and the metric difference is rather irrelevant to the numerical analyst. If, for example, in the case of the strip (with periodicity) the area integral of the square of the absolute value had been chosen for the norm, as was done in some recent papers for similar problems (see [5], [6]), the resulting splines would be elliptic functions rather than the elementary ones obtained by our choice of norm (this will be shown in a forthcoming note).

After a basis for the splines has been found, it remains to solve the interpolation conditions. For the simplest of spaces these conditions are solved explicitly and simple expressions for the error $\sup_{f\in\mathcal{B}}|Ef(x_o)|$ are established, which yield precise results on the asymptotic behavior of this error as n (= number of data) or R (= radius of the disk of analyticity tends to infinity (Section 5). The error $\sup_{f\in\mathcal{B}}|\lambda Ef|$ for an arbitrary continuous linear functional λ is also evaluated explicitly and is then minimized with respect to the location of the interpolation nodes. This results in "hyperoptimal λ-rules", which are generalizations of the quadrature formulas of Gauss and Wilf (Section 6).

In the less simple spaces the interpolation conditions require, in general, numerical matrix inversion. We choose specially simple arrays for the interpolation nodes so that the interpolation conditions can be solved analytically (for a similar method applied to periodic functions of a real variable see [7]). This results in simple explicit formulas for the spline interpolants (optimal recovery schemes), which are well suited for programmed computation. They make it also possible to obtain rather precise bounds and explicit asymptotic limits for the error. In Section 9 the error in optimal recovery from n data is compared to the n-width of the class $\mathcal{B}$ with respect to a coarser than the initial Hilbert space norm, for which $\mathcal{B}$ is compact. In most cases the asymptotic unity of the error and the n-width are the same. The upper bounds for the recovery error also provide some new upper bounds for n-widths.

From the error bounds given it follows that the splines s_n (n = number of data) converge to f uniformly in any compact part of the domain $\mathcal{D}$ (at an exponential rate). The sequence $\{s_n(z)\}$ allows analytic continuation of f from the arc containing the data points to the arbitrary $z \in \mathcal{D}$. Several numerical analysts have pointed out that analytic continuation of functions only given by their values in some infinite compact set is not a well-posed problem since small errors in the data can produce arbitrarily large errors of the continuation. Methods have been proposed to cope with this problem (see, e.g., [11], [12]). Some of our examples, reported in the Appendix, show that even the values of a simple elementary function, computed with ordinary precision, are "noisy" enough to render useless the higher order approximations. In Section 10 a simple modification of the splines is given which achieves stability of the error. The device makes use of two parameters: a known upper bound ε for the magnitude of the noise and a known upper bound μ for the function f. The procedure is similar to but not identical with the one used by Keith Miller in [12].

2. INTERPOLATING $\mathcal{H}$- SPLINES

Suppose $\mathcal{H}$ is a Hilbert space of complex-valued functions which has a reproducing kernel k. Thus if $\mathcal{D}$ is the domain of the functions, $(\cdot,\cdot)_{\mathcal{H}}$ is the inner product in $\mathcal{H}$, then

$$(2.1) \qquad f(x) = (f,\ k(\ ,x))_{\mathcal{H}} , \qquad \forall f \in \mathcal{H},\ \forall x \in \mathcal{D} .$$

Clearly $\|k(\cdot,x)\|^2_{\mathcal{H}} = k(x,x)$ and $k(\cdot,x_o)/k(x_o,x_o)$ is the unique function in $\mathcal{H}$ that minimizes $\|f\|_{\mathcal{H}}$ among all the functions $f \in \mathcal{H}$ for which $f(x_o) = 1$. In fact if $g \in \mathcal{H}$ and $g(x_o) = 0$ then

$$(2.2) \qquad (g,k(\cdot,x_o))_{\mathcal{H}} = 0, \quad \|k(\cdot,x_o)/k(x_o,x_o) + g\|^2_{\mathcal{H}} =$$

$$= \|k(\cdot,x_o)/k(x_o,x_o\|^2_{\mathcal{H}} + \|g\|^2_{\mathcal{H}} .$$

More generally, given $f_o \in \mathcal{H}$ and n distinct points $x_1,\ldots,x_n$ in $\mathcal{D}$, the unique function in $\mathcal{H}$ that minimizes $\|f\| = \|f\|_{\mathcal{H}}$ among all the functions $f \in \mathcal{H}$ for which

$$(2.3) \qquad f(x_i) = f_o(x_i) , \quad i = 1,\ldots,n$$

is

$$(2.4a) \qquad s = \sum_{j=1}^{n} \gamma_j k(\cdot,x_j) .$$

with the γ_j so chosen that the interpolation conditions

$$(2.4b) \qquad s(x_i) = f_o(x_i) , \quad i = 1,\ldots,n$$

are satisfied. That there is a unique such interpolant follows from the fact that the matrix $(k(x_i,x_j))$ is the Gramian of the linearly independent functions $k(\cdot,x_1),\ldots,k(\cdot,x_n)$. If $g \in \mathcal{H}$, $g(x_i) = 0$ $(i = 1,\ldots,n)$ then $(g,s)_{\mathcal{H}} = 0$, $\|s + g\|^2 = \|s\|^2 + \|g\|^2$.

Since the above minimizing property is characteristic for many classes of splines that have been defined in the past, it is justified to call s, as defined by (2.4a,b), an <u>interpolating $\mathcal{H}$-spline</u>; $x_1,\ldots,x_n$ are the <u>interpolation nodes</u> of the spline. The n-dimensional subspace $\mathcal{S}$ of $\mathcal{H}$ spanned by the splines $k(\cdot,x_1),\ldots,k(\cdot,x_n)$ is the <u>spline space</u> of $\mathcal{H}$ <u>based on</u> $(x_1,\ldots,x_n)$.

Let the set of distinct interpolation nodes $x_1,\ldots,x_n$ be fixed. Then if f is any function $\mathcal{D} \to \mathbb{C}$ we write F for the n-tuple $(f(x_1),\ldots,f(x_n))$ (F_o for $(f_o(x_1),\ldots,f_o(x_n))$, G for $(g(x_1),\ldots,g(x_n))$, etc.) We denote the spline S defined by

(2.4a,b) by either Sf_o or SF_o. Clearly $F \mapsto SF$ is a linear operator from $\mathbb{C}^n$ to $\mathcal{S}$.

The interpolation conditions (2.4b) may also be interpreted in another way. Let $P_{\mathcal{S}}$ denote the orthogonal projector onto $\mathcal{S}$. Since $f_o - Sf_o$ vanishes at the points $x_1,\ldots,x_n$, $f_o - Sf_o \in \mathcal{S}^{\perp}$, hence $P_{\mathcal{S}} f_o = Sf_o$. We formulate this well known result as

Theorem 2.1. The $\mathcal{A}$-spline Sf that interpolates $f \in \mathcal{A}$ at the points $x_1,\ldots,x_n$ is the same as the orthogonal projection of f onto the spline space based on $x_1,\ldots,x_n$.

There is another minimizing property of the interpolating spline Sf, which is of crucial importance for its use in optimal recovery. Let $\mathcal{B}$ denote the unit ball in $\mathcal{A}$:

$$\mathcal{B} = \{f \in \mathcal{A} : \|f\| \le 1\}. \tag{2.5}$$

We assert:

Lemma 2.1. For any $x_o \in \mathcal{D}$ $(x_o \neq x_1,\ldots,x_n)$, $\eta \in \mathbb{C}$, $f_o \in \mathcal{B}$

$$\sup_{f\in\mathcal{B},F=F_o} |f(x_o) - \eta| \ge \sup_{f\in\mathcal{B},F=F_o} |f(x_o) - Sf_o(x_o)|$$

Proof. Let h_{x_o} be an element of $\mathcal{A}$ of norm 1 which vanishes at $x_1,\ldots,x_n$ and for which

$$\sup_{f\in\mathcal{B},F=0} |f(x_o)| = h_{x_o}(x_o) \quad . \tag{2.6}$$

Put for $\alpha \in \mathbb{R}$

$$f_\alpha = Sf_o + (1 - \|Sf_o\|)^{\frac{1}{2}} e^{i\alpha} h_{x_o} \quad . \tag{2.7}$$

Then since $(SF_o, h_{x_o})_{\mathcal{A}} = 0$ we have

$$\|f_\alpha\| = 1, \quad F_\alpha = F_o \ , \tag{2.8}$$

thus f_α is in the competing class. By definition $(f_\alpha - Sf_o)(x_o) = (1-\|Sf_o\|^2)^{\frac{1}{2}} e^{i\alpha} h_{x_o}(x_o)$, hence for any $\eta \in \mathbb{C}$

$$\max_{0\le\alpha\le 2\pi} |f_\alpha(x_o) - \eta| \ge (1 - \|Sf_o\|^2)^{\frac{1}{2}} h_{x_o}(x_o)$$

and, a fortiori,

$$\begin{aligned}\sup_{f\in\mathcal{B},F=F_o} |f(x_o) - \eta| &\ge (1 - \|Sf_o\|^2)^{\frac{1}{2}} h_{x_o}(x_o) \\ &= (1 - \|Sf_o\|^2)^{\frac{1}{2}} \sup_{f\in\mathcal{B},F=0} |f(x_o)| \\ &\ge \sup_{f\in\mathcal{B},F=F_o} |f(x_o) - Sf_o(x_o)|.\end{aligned}$$

Q.E.D.

The content of Lemma 2.1 is that $Sf_o(x_o)$ is the "center" of the set $\{f(x_o)\}_{f\in\mathscr{B}, F=F_o} \subset \mathbb{C}$. The proof also gives us the radius and extremal elements of this set.

Lemma 2.2. Let c_{x_o} be the $\mathscr{A}$-spline with interpolating nodes $x_o, x_1, \ldots, x_n$, for which $c_{x_o}(x_o) = 1$, $c_{x_o}(x_i) = 0$ $(i = 1,\ldots,n)$. Then

$$E_{\mathscr{B}}(x_o,F_o) := \inf_{\eta\in\mathbb{C}} \sup_{f\in\mathscr{B}, F=F_o} |f(x_o) - \eta|$$

$$= \|c_{x_o}\|^{-1}(1 - \|SF_o\|^2)^{\frac{1}{2}} \le \|c_{x_o}\|^{-1} .$$

The only functions $f_* \in \mathscr{B}$ for which $|f_*(x_o) - Sf_*(x_o)| = \|c_{x_o}\|^{-1}$ are the functions $e^{i\alpha}c_{x_o}/\|c_{x_o}\|$, $\alpha \in \mathbb{R}$.

Proof. From the minimizing property of the spline c_{x_o} and by (2.5) we have

$$(2.10) \qquad \|c_{x_o}\| = \inf_{f\in\mathscr{A}, F=0} \frac{\|f\|}{|f(x_o)|} = 1/\sup_{f\in\mathscr{B}, F=0} |f(x_o)| = 1/h_{x_o}(x_o).$$

For every $f \in \mathscr{B}$, $(1 - \|Sf\|^2)^{-\frac{1}{2}}(f - Sf)$ is in $\mathscr{B}$ and vanishes at $x_i (i = 1,\ldots,n)$, hence by (2.10),

$$(2.11) \qquad |f(x_o) - Sf(x_o)| \le \|c_{x_o}\|^{-1}(1 - \|Sf\|^2)^{\frac{1}{2}} .$$

On the other hand, by (2.9)

$$(2.12) \qquad \sup_{f\in\mathscr{B}, F=F_o} |f(x_o) - Sf(x_o)| \ge \|c_{x_o}\|^{-1}(1 - \|Sf_o\|^2)^{\frac{1}{2}}$$

(2.11) and (2.12) prove the first assertion of the Lemma. Now $f_* = e^{i\alpha}c_{x_o}/\|c_{x_o}\|$ $(\alpha \in \mathbb{R})$ is in $\mathscr{B}$ and satisfies $|f_*(x_o) - Sf_*(x_o)| = \|c_{x_o}\|^{-1}$, and if g is any other function in $\mathscr{B}$ then $(1 - \|Sg\|^2)^{-\frac{1}{2}}(g - Sg)$ is in $\mathscr{B}$ and vanishes at x_i $(i = 1,\ldots,n)$, hence

$$(1 - \|Sg\|^2)^{-\frac{1}{2}}|g(x_o) - Sg(x_o)| < \sup_{f\in\mathscr{B}, F=0} |f(x_o)| = \|c_{x_o}\|^{-1} ,$$

so that the second part of the Lemma is also proved.

Lemmas 2.1 and 2.2 deal with the set in $\mathbb{C}$ that is obtained by applying the functional $\delta_{x_o}: \delta_{x_o}(f) = f(x_o)$ to $\mathscr{B}$. They are easily extended to the case of a general linear functional.

Lemma 2.3. For every continuous linear functional λ on $\mathscr{A}$ which is not in the span of $\delta_{x_1},\ldots,\delta_{x_n}$, and for each $\eta \in \mathbb{C}$, $f_o \in \mathscr{B}$

$$\sup_{f\in\mathscr{B},F=F_o} |\lambda f - \eta| \geq \sup_{f\in\mathscr{B},F=F_o} |\lambda(f) - \lambda(Sf_o)| .$$

If $c_\lambda \in \mathscr{A}$ minimizes $\|f\|$ among all $f \in \mathscr{A}$ for which $f(x_1) = \cdots = f(x_n) = 0$, $\lambda(f) = 1$, then

$$\inf_{\eta\in\mathbb{C}} \sup_{f\in\mathscr{B},F=F_o} |\lambda f - \eta| = \|c_\lambda\|^{-1}(1 - \|SF_o\|^2)^{\frac{1}{2}} \leq \|c_\lambda\|^{-1} .$$

The only functions $f_* \in \mathscr{B}$ for which $|\lambda f_* - \lambda Sf_*| = \|c_\lambda\|^{-1}$ are the functions $e^{i\alpha}c_\lambda/\|c_\lambda\|$, $\alpha \in \mathbb{R}$.

Proof. The proof is identical with that of Lemmas 2.1 and 2.2 except that h_{x_o} is replaced by $h_\lambda \in \mathscr{A}$ of norm 1 which vanishes at $x_1,\ldots,x_n$ and satisfies $\sup_{f\in\mathscr{B},F=0} |\lambda f| = \lambda h_\lambda$.

3. OPTIMAL RECOVERY SCHEMES FOR A BALL IN $\mathscr{A}$.

Let $\mathscr{R} = \mathscr{R}_{X,\mathscr{B}}$ denote the class of mappings (not necessarily linear) from $\mathbb{C}^n$ to $\mathscr{B}$, called recovery schemes. $R \in \mathscr{R}$ is to associate an element $RF \in \mathscr{B}$ with certain elements $F \in \mathbb{C}^n$. F is interpreted as the n-tuple $(f(x_1),\ldots,f(x_n))$, where $f \in \mathscr{B}$ is the unknown function to be recovered. The subset $(f(x_1),\ldots,f(x_n))_{f\in\mathscr{B}} \subset \mathbb{C}_n$ is the domain of each $R \in \mathscr{R}_{X,\mathscr{B}}$. X denotes the set $(x_1,\ldots,x_n)$, the domain of the data.

Definition. $R_{x_o} \in \mathscr{R}_{X,\mathscr{B}}$ is an-optimal recovery scheme of $\mathscr{B}$ at $x_o \in \mathscr{D}$ if

$$E_{x_o}(\mathscr{B}) = \inf_{R\in\mathscr{R}} \sup_{f\in\mathscr{B}} |f(x_o) - RF(x_o)| = \sup_{f\in\mathscr{B}} |f(x_o) - R_{x_o}F(x_o)|.$$

The function $x \mapsto E_x(\mathscr{B})$ is the minimal error function for recovery. An element $f_{x_o} \in \mathscr{B}$ such that

$$E_{x_o}(\mathscr{B}) = |f_{x_o}(x_o) - R_{x_o}F_{x_o}(x_o)|$$

is an extremal for recovery at x_o. $R_* \in \mathscr{R}_{X,\mathscr{B}}$ is an optimal recovery scheme of $\mathscr{B}$ if

$$E_*(\mathscr{B}) = \sup_{x\in\mathscr{D}} E_x(\mathscr{B}) = \sup_{x\in\mathscr{D}} \sup_{f\in\mathscr{B}} |f(x) - R_*F(x)|$$
$$= \sup_{f\in\mathscr{B}} \|f - R_*F\|_\infty ,$$

where $\|\ \|_\infty$ denotes the sup-norm for maps $\mathscr{D} \to \mathbb{C}$. $E_*(\mathscr{B})$ is the minimal error for recovery of $\mathscr{B}$ from data on X.

With these definitions and the results of Section 2 we have the following:

Theorem 3.1. The linear spline projector S which associates with the n-tuple F (or the function f) the interpolating $\mathscr{A}$-spline SF (or Sf), with set of interpolation nodes $(x_1,\ldots,x_n)$, is the unique optimal recovery scheme of $\mathscr{B}$ at every $x \in \mathscr{D}$, $x \neq x_1,\ldots,x_n$. The minimal error function for recovery is $x \mapsto \|c_x\|^{-1}$, where c_x is the interpolating $\mathscr{A}$-spline for which $c_x(x) = 1$, $c_x(x_i) = 0$ $(i = 1,\ldots,n)$. The only extremals for recovery at x are the functions $e^{i\alpha}c_x/\|c_x\|$, $(\alpha \in \mathbb{R})$. The minimal error in recovery of $\mathscr{B}$ is

$$E_*(\mathscr{B}) = \sup_{x \in \mathscr{D}} \|c_x\|^{-1} \quad .$$

Proof. The proof is a direct consequence of Lemmas 2.1, 2.2 and the definitions.

In place of recovery of $\mathscr{B}$ at x_o we may ask for recovery at λ where λ is any continuous linear functional. Using Lemma 2.3 we have

Corollary 3.1. The unique optimal recovery scheme of $\mathscr{B}$ at the continuous linear functional λ not in the span of $\delta_{x_1},\ldots,\delta_{x_n}$ is $\lambda(SF)$. The minimal error in recovery at λ is $\|c_\lambda\|^{-1}$, where c_λ is as defined in Lemma 2.3. The only extremals for recovery at λ are the functions $e^{i\alpha}c_\lambda/\|c_\lambda\|$, $\alpha \in \mathbb{R}$.

There is another expression for the minimal error function that we find useful in many cases. In the formulation of this result we use the notation k_{x_o} for $k(\cdot,x_o)$ to avoid possible confusion.

Theorem 3.2. Suppose $\mathscr{A}$ has the reproducing kernel $k(x,x_o) = k_{x_o}(x)$. The minimal error for recovery of the ball $\mathscr{B} \subset \mathscr{A}$ at x_o from data on $(x_1,\ldots,x_n)$ is

$$E_{x_o}(\mathscr{B}) = \|k_{x_o} - Sk_{x_o}\| = [k_{x_o}(x_o) - Sk_{x_o}(x_o)]^{\frac{1}{2}}$$

where S is the projector onto the $\mathscr{A}$-spline space $\mathscr{S}$ based on $(x_1,\ldots,x_n)$.

Proof. By Theorem 3.1

$$\begin{aligned} E_{x_o}(\mathscr{B}) &= \sup_{f \in \mathscr{B}} |f(x_o) - Sf(x_o)| \\ &= \sup_{f \in \mathscr{B}} |(f,k_{x_o}) - (Sf,k_{x_o})| \\ &= \sup_{f \in \mathscr{B}} |(f,k_{x_o} - Sk_{x_o})| \\ &= \|k_{x_o} - Sk_{x_o}\| \ . \end{aligned} \tag{3.1}$$

Since S is an orthogonal projector, we also have

$$[E_{x_o}(\mathscr{B})]^2 = \|k_{x_o}\|^2 - \|Sk_{x_o}\|^2 = (k_{x_o},k_{x_o}) - (Sk_{x_o},k_{x_o})$$
$$= k_{x_o}(x_o) - Sk_{x_o}(x_o).$$

Remark 3.1. It is well known that $P_{\mathscr{M}}k$ is the reproducing kernel for the subspace $\mathscr{M} \subset \mathscr{A}$ if $P_{\mathscr{M}}$ is the orthogonal projector onto $\mathscr{M}$. Therefore, Theorem 3.2 says that $E_{x_o}(\mathscr{B})$ is the norm of the reproducing kernel at x_o of the space $\mathscr{S}^{\perp} \subset \mathscr{A}$.

4. SPACES OF ANALYTIC FUNCTIONS

In the following we consider Hilbert spaces of functions that are analytic in some region $\mathscr{D} \subset \mathbb{C}$ and have a reproducing kernel $k(z,z_o)$ with the property that the function $z \mapsto k(z,z)$ is bounded on every compact subset of $\mathscr{D}$. If $\mathscr{A}$ is such a space and $\{\phi_\nu\}_{\nu=1,2,...}$ is a complete orthogonal system in $\mathscr{A}$ then $k(\cdot,z_o) = \sum_\nu \phi_\nu(z_o)\phi_\nu$ for every $z_o \in \mathscr{D}$, the series being convergent in the sense of $\mathscr{A}$. But it is also true that

$$(4.1) \qquad k(z,z_o) = \sum_{\nu=1}^{\infty} \overline{\phi_\nu(z_o)}\, \phi_\nu(z), \qquad \forall z,\ \bar{z}_o \in \mathscr{D}$$

in the sense of pointwise convergence. Moreover the convergence is uniform in any compact subset of $\mathscr{D} \times \mathscr{D}$. In particular

$$(4.2) \qquad k(z,z) = \sum_{\nu=1}^{\infty} |\phi_\nu(z)|^2 , \qquad \forall z \in \mathscr{D}$$

and the sum is uniformly bounded in every compact subset of $\mathscr{D}$. Conversely if $\mathscr{A}$ is a Hilbert space of functions analytic in a domain $\mathscr{D} \subset \mathbb{C}$ which has a complete orthogonal system $\{\phi_\nu\}_{\nu=1,2,...}$ such that $\Sigma|\phi_\nu(z)|^2$ is uniformly bounded on every compact subset of $\mathscr{D}$ then $\mathscr{A}$ has a reproducing kernel of the above kind. Indeed the kernel is given by the series (4.1). We observe that if $f \in \mathscr{A}$ then

$$(4.3) \qquad f^{(m)}(z_o) = \overline{\frac{\partial^m}{\partial z_o^m}(k(\cdot,z_o),f)_{\mathscr{A}}}\ , \quad \forall z_o \in \mathscr{D};\ m = 0,1,...$$

In general, if λ is any element of the dual space $\mathscr{A}'$ then for any $f \in \mathscr{A}$

$$(4.4) \qquad \lambda f = \overline{(\lambda_z k(\cdot,z),f)_{\mathscr{A}}}\ .$$

Let $Z = (z_1,...,z_n)$ be a fixed n-tuple of distinct points in $\mathscr{D}$. As pointed out in Section 2, an interpolating $\mathscr{A}$-spline with nodes $(z_1,\dots,z_n)$ is of the form

$$(4.5)\qquad s = \sum_{i=1}^{n} \gamma_i k(\cdot,z_i)$$

with $\gamma_i \in \mathbb{C}$. By Section 3, the optimal recovery scheme of the unit ball $\mathscr{B} \subset \mathscr{A}$ from data $F = (f(z_1),\dots,f(z_n))$ is given by the spline

$$(4.6)\qquad SF = \sum_{i=1}^{n} \gamma_i k(\cdot,z_i)$$

where the γ_i are determined from the system

$$(4.7)\qquad \sum_{i=1}^{n} \gamma_i k(z_j,z_i) = f(z_j),\quad j = 1,\dots,n.$$

The minimal error $E_{z_o}(\mathscr{B})$ in recovery at z_o can also be obtained in terms of the kernel k. Let the numbers $\eta_o, \eta_1,\dots,\eta_n$ be determined from the system

$$(4.8)\qquad \sum_{i=0}^{n} \eta_i k(z_j\; z_i) = \delta_{0j}\;,\quad j = 0, 1,\dots,n\;.$$

Then by Theorem 3.1

$$(4.9)\qquad E_{z_o}(\mathscr{B}) = \Big\|\sum_{i=0}^{n} \eta_i k(\cdot,z_i)\Big\|_{\mathscr{A}}^{-1} = \left(\sum_{i,j=0}^{n} k(z_j,z_i)\eta_j\bar{\eta}_i\right)^{-\frac{1}{2}}\;.$$

From this expression we obtain bounds on the error. They are valid for any Hilbert space of functions with reproducing kernel.

Theorem 4.1. Suppose $\mathscr{A}$ is a Hilbert space of functions $\mathscr{D} \to \mathbb{C}$, with the reproducing kernel k, and $\mathscr{B}$ is the unit ball in $\mathscr{A}$. Then the minimal error $E_{z_o}(\mathscr{B})$ in recovery of $\mathscr{B}$ at z_o from data at the distinct points $z_1,\dots,z_n$ satisfies

$$\lambda_{min}^{\frac{1}{2}} \le E_{z_o}(\mathscr{B}) \le \lambda_{max}^{\frac{1}{2}}$$

where $\lambda_{min}, \lambda_{max}$ are the smallest and largest eigenvalues of the matrix $(k(z_i,z_j))_{i,j=0,1,\dots,n}$.

Proof. Let u_o be the vector $(1,0,\dots,0)$ in $\mathbb{C}^{n+1}$, e the vector $(\eta_o,\eta_1,\dots,\eta_n)$ defined by (4.8), K the linear operator $\mathbb{C}^{n+1} \to \mathbb{C}^{n+1}$ defined by the matrix $(k(z_i,z_j))$. K is hermitian positive-definite, let $K^{\frac{1}{2}}$ be its positive square root. By (4.8), $Ke = u_o$, hence $K^{\frac{1}{2}}e = K^{-\frac{1}{2}}u_o$. By (4.9)

$$(4.10)\qquad E_{z_o}(\mathscr{B}) = |K^{-\frac{1}{2}}u_o|^{-1}$$

where $|v|$ denotes the Euclidean norm of the vector $v \in \mathbb{C}^{n+1}$. The assertion follows from (4.10).

Remark 4.1. Up to this point we have assumed that the interpolation nodes $z_1,\dots,z_n$ are distinct. In more general recovery problems we admit equality of some of the z_i, say $z_1 = \dots = z_m$. With the usual convention, the data vector F then stands for

$(f(z_1), f'(z_1, \ldots, f^{(m-1)}(z_1), \ldots)$ ("Hermite data"). To simplify the exposition assume $m = n$. We assert that the $\mathcal{A}$-spline interpolating Hermite data has the form

$$\tilde{s} = \sum_{i=0}^{n-1} \gamma_i k^{(i)}(\cdot, z_1)$$

where $k^{(i)}(\cdot,z) = \partial^i k(\cdot,z)/\partial \bar{z}^i$. In fact if $g \in \mathcal{A}$ is such that $g^{(i)}(z_1) = 0$ for $i = 0,1,\ldots,n-1$ then since $(g, k^{(i)}(\cdot,z_1))_{\mathcal{A}} = g^{(i)}(z_1) = 0$

$$||\tilde{s} + g||^2 = ||\tilde{s}||^2 + \|g\|^2 .$$

Thus, $\tilde{s}$ has the same minimizing property with respect to Hermite data that s in (2.4a) has with respect to Lagrange data (2.4b). We write again $\tilde{S}F$ or $\tilde{S}f$ for $\tilde{s}$ and conclude that the spline operator $\tilde{S}$ is the optimal recovery scheme of $\mathcal{B}$ from Hermite interpolation data.

5. $\mathcal{A}(D_R)$-SPLINES FROM INTERPOLATION DATA

Assume $R > 1$ fixed and let D_R be the open disk $\{z \in \mathbb{C}: |z| < R\}$. We denote by $\mathcal{A}(D_R)$ the space of functions f analytic in D_R for which $\int |z|^2 |dz|$ is a bounded function of r in $0 \le r < R$. We express this latter condition as

$$\int_{\partial D_R} |f|^2 < \infty . \tag{5.1}$$

This is justified because under the above condition the limit of $f(re^{i\theta})$ as $r \to R - 0$ exists for almost all $\theta \in \mathbb{R}$ and is locally square-integrable. f in (5.1) may be thought of as this limit.

Using the integral (5.1) we define the norm

$$\|f\|_{\mathcal{A}(D_R)} = \left(\frac{1}{2\pi R} \int_{\partial D_R} |f|^2\right)^{\frac{1}{2}} \tag{5.2}$$

on $\mathcal{A}(D_R)$ and the corresponding inner product $(\cdot,\cdot)_{\mathcal{A}(D_R)}$. $\mathcal{A}(D_R)$ is known to be a Hilbert space.

Let $e_\nu (\nu = 0,1,\ldots)$ denote the function defined by $e_\nu(z) = z^\nu$. Clearly

$$(e_\mu, e_\nu)_{\mathcal{A}(D_R)} = R^{2\mu} \delta_{\mu\nu} \tag{5.3}$$

and the sequence $\{R^{-\nu} e_\nu\}_{\nu=0,1,\ldots}$ forms a complete orthonormal system in $\mathcal{A}(D_R)$. If $f = \Sigma\, \alpha_\nu e_\nu$ for some $f \in \mathcal{A}(D_R)$ then $f(z) = \Sigma\, \alpha_\nu z^\nu$ is the usual power series expansion of f.

Condition (5.1) is equivalent to

$$(5.4) \qquad \sum_{\nu=0}^{\infty} R^{2\nu}|\alpha_\nu|^2 < \infty$$

and if $g = \Sigma\, \beta_\nu e_\nu$ then

$$(5.5) \qquad (f,g)_{\mathscr{A}(D_R)} = \sum_{\nu=0}^{\infty} R^{2\nu}\alpha_\nu\bar{\beta}_\nu$$

$\mathscr{A}(D_R)$ has the reproducing kernel

$$(5.6) \qquad k_D(z,z_o) = \sum_{\nu=0}^{\infty} R^{-2\nu}z^\nu\bar{z}_o^\nu = \frac{1}{1 - z\bar{z}_o/R^2} .$$

Thus, the spline space $\mathcal{S}_D \subset \mathscr{A}(D_R)$, based on the distinct nodes $z_1,\ldots,z_n$ in D_R, is spanned by the functions

$$(5.7) \qquad s_{D,i}(z) = \frac{1}{1 - z\bar{z}_i/R^2} \qquad i = 1,\ldots,n$$

$s_{D,i}$ is, except for a constant factor, characterized as the rational function which vanishes at ∞ and has a simple pole at $R^2/\bar{z}_i$. This point, which is the inversion of the node z_i at the circle ∂D_R, may be called the <u>knot</u> of the spline s_i generated by the node z_i. The knots of the usual polynomial splines, which are discontinuities of high-order derivatives, are replaced here by poles of the analytic continuation of the spline to the exterior of D_R. We may state: <u>The spline space $\mathcal{S}_D \subset \mathscr{A}(D_R)$, based on the distinct nodes $z_1,\ldots,z_n$, is the span of rational functions that vanish at ∞ and have simple poles at the points $R^2/\bar{z}_1,\ldots,R^2/\bar{z}_n$.</u>

We derive an explicit formula for the $\mathscr{A}(D_R)$-spline S_Zf that interpolates f on $Z = (z_1,..,z_n)$. Since S_Zf has simple poles at $R^2/\bar{z}_i$ $(i = 1,\ldots,n)$ and vanishes at ∞,

$$(5.9) \qquad s_j(z) := S_Z\delta_j(z) = \frac{\prod_{i\neq j}(z - z_i)\prod_i (z_j - R^2/\bar{z}_i)}{\prod_i (z - R^2/\bar{z}_i)\prod_{i\neq j}(z_j - z_i)}$$

is the spline for the data vector $\delta_j = (\delta_{ij})_{(i=1,\ldots,n)}$. Therefore,

$$(5.10) \qquad S_Zf(z) = \sum_{j=1}^{n} f(z_j)\,\frac{\prod_{i\neq j}(z - z_i)\prod_i (z_j - R^2/\bar{z}_i)}{\prod_i (z - R^2/\bar{z}_i)\prod_{i\neq j}(z_j - z_i)}$$

is an explicit formula for S_Zf. It is seen that, as $R \to \infty$, S_Zf converges, uniformly on every bounded subset of $\mathbb{C}$, to the Lagrange interpolation polynomial

(5.11) $$L_Z f(z) = \sum_{j=1}^{n} f(z_j) \frac{\prod_{i\neq j} (z - z_i)}{\prod_{i\neq j} (z_j - z_i)} .$$

The minimal error $E_{z_o}(\mathscr{B})$ at $z_o \in D_R$ can also be given in explicit form. The spline c_{z_o} (see Lemma 2.2) is

(5.12) $$c_{z_o}(z) = \frac{(z_o - R^2/\bar{z}_o) \prod_i (z - z_i) \prod_i (z_o - R^2/\bar{z}_i)}{(z - R^2/\bar{z}_o) \prod_i (z - R^2/z_i) \prod_i (z_o - z_i)} ,$$

hence by Theorem 3.1

$$E_{z_o}(\mathscr{B}) = \frac{\prod_i |z_o - z_i|}{(1 - |z_o|^2/R^2) \prod_i |1 - z_o\bar{z}_i/R^2|}$$

$$\times \frac{1}{2\pi R} \left\{ \int_{\partial D_R} \frac{\prod_i |z - z_i|^2}{|1 - z\bar{z}_o/R^2|^2 \prod_i |1 - z\bar{z}_i/R^2|^2} \right\}^{-\frac{1}{2}} .$$

Using $z = Re^{i\theta}$ in the integral, then $\Pi_i |z - z_i|/\Pi|1 - z\bar{z}_i/R^2| = R^n$, and $|1 - z\bar{z}_o/R^2|^2 = R^{-2}(R^2 - 2\,Rr_o\cos\theta + r_o^2)$, hence

(5.13) $$E_{z_o}(\mathscr{B}) = \frac{R^{-n}}{(1 - |z_o|^2/R^2)^{3/4}} \prod_i \frac{|z_o - z_i|}{|1 - z_o\bar{z}_i/R^2|} .$$

This is a remarkable simple expression for the minimal error in the general case. One of its consequences is the following limit relation. We write $\mathscr{B}_R$ for $\mathscr{B}$ to indicate the dependence on R.

(5.14) $$\lim_{R\to\infty} R^n E_{z_o}(\mathscr{B}_R) = \prod_{i=1}^{n} |z_o - z_i| .$$

Thus

(5.15) $$E_{z_o}(\mathscr{B}_R) = O(R^{-n}) , \quad R \to \infty .$$

More detailed results are obtained for the special case of symmetrically distributed interpolation nodes. Suppose $0 < r_o < R$

(5.16) $$z_j = r_o\varepsilon_n^j , \quad \varepsilon_n = e^{2\pi i/n} \qquad j = 0,1,\ldots,n - 1$$

and write Z_* for this special set of nodes. A simple calculation shows that (5.10) becomes

(5.17a) $$S_{Z_*}f(z) = \frac{1 - (r_o/R)^{2n}}{1 - (r_o z/R^2)^n} \; L_{Z_*}f(z)$$

where $L_{Z_*}f$ is the Lagrange interpolation polynomial

(5.17b) $$L_{Z_*}f(z) = \frac{1}{n} \sum_{j=0}^{n-1} \frac{f(z_j)}{z_j^{n-1}} \; \frac{z^n - r_o^n}{z - z_j} \; ,$$

or also

(5.17c) $$L_{Z_*}f(z) = \sum_{j=0}^{n-1} \hat{f}_j (z/r_o)^j \; , \quad \hat{f}_j = \frac{1}{n} \sum_{k=0}^{n-1} f(r_o \varepsilon_n^k) \varepsilon_n^{-jk} \; .$$

These formulas allow for extremely simple computation of $S_{Z_*}f$ (see Appendix). The fractional factor in (5.17a) is, in most cases, so close to 1 that it can be ignored.

We also derive an explicit expression for the partial fraction expansion of the rational function $S_{Z_*}f$. It has simple poles at the points $\zeta_j = (R^2/r_o)\varepsilon^j$ $(j = 0, \ldots, n - 1)$ and

$$\operatorname*{res}_{\zeta_j} S_{Z_*}f = \frac{1 - (r_o/R)^{2n}}{-n\zeta_j^{-1}} \; L_{Z_*}f(\zeta_j) \; .$$

Therefore,

(5.18) $$S_{Z_*}f(z) = [1 - (r_o/R)^{2n}] \; \frac{1}{n} \; \frac{\zeta_j L_{Z_*}f(\zeta_j)}{\zeta_j - z} \; .$$

This formula is, except for the factor $[1 - (r_o/R)^{2n}]$, the discrete analogue (with the integral evaluated by the trapezoidal rule) of the Cauchy integral for $L_{Z_*}f$ along the circle $|\zeta| = R^2/r_o$. Like (5.17), it is easily programmed for computation. Moreover, it leads to particular simple expressions for the optimal recovery of the expansion coefficients of f at some $z_o \in D_R$ (and of other linear functionals, see Corollary 3.1). If we set $f(z) = \sum_{\nu=0}^{\infty} \alpha_\nu(z_o)(z - z_o)^\nu$ and $S_{Z_*}f(z) = \sum_{\nu=0}^{\infty} \alpha_{\nu_*}(z_o)(z - z_o)^\nu$ then

(5.19) $$\alpha_{\nu_*}(z_o) = (1 - (r_o/R)^{2n})\frac{1}{n} \sum_{j=0}^{n-1} \frac{\zeta_j L_{Z_*}f(\zeta_j)}{(\zeta_j - z_o)^{\nu+1}}, \quad \nu = 0,1,\ldots$$

For the integral $\int_0^{z_o} f(\zeta)d\zeta$ we find the optimal quadrature formula

(5.20) $$\int_0^{z_o} S_{Z_*}f = (1 - (r_o/R)^{2n})\frac{1}{n} \sum_{j=0}^{n-1} \zeta_j L_{Z_*}f(\zeta_j)\log(1 - z_o/\zeta_j).$$

For the Fourier coefficient $(f,e_\nu)_{\mathscr{A}(D_R)} = \alpha_\nu(0)R^{2\nu}$ we find the optimal value

(5.21) $$(S_{Z_*}f,e_\nu)_{\mathscr{A}(D_R)} = \alpha_{\nu_*}(0)R^{2\nu} = \hat{f}_\nu r_o^{\nu}(R/r_o)^{2\nu}$$

$$\nu = 0,1,\ldots;\ \nu' \equiv \nu(n),\ 0 \le \nu' \le n-1.$$

Let the minimal error at z_o be denoted as $E_{z_o}^{(n)}(\mathscr{B}_R)$ to indicate the dependence on n as well as on R. By (5.13)

(5.22) $$E_{z_o}^{(n)}(\mathscr{B}_R) = \frac{R^{-n}}{(1-|z_o|^2/R^2)^{3/4}} \cdot \frac{|z_o^n - r_o^n|}{|1 - z_o^n r_o^n/R^{2n}|} .$$

We note the two limit relations

(5.23) $$\lim_{R\to\infty} R^n E_{z_o}^{(n)}(\mathscr{B}_R) = |z_o^n - r_o^n| .$$

(5.24) $$\lim_{n\to\infty} \sqrt[n]{E_{z_o}^{(n)}(\mathscr{B}_R)} = \begin{cases} |z_o|/R & \text{if } |z_o| > r_o \\ r_o/R & \text{if } |z_o| < r_o \end{cases} .$$

If $|z_o| = r_o$, but $z_o^n \neq r_o^n$ for any n (i.e. z_o never coincides with an interpolation node) then

(5.25) $$\underline{\lim}_{n\to\infty} \sqrt[n]{E_{z_o}^{(n)}(\mathscr{B}_R)} = 0 ,\quad \overline{\lim}\ \sqrt[n]{E_{z_o}^{(n)}(\mathscr{B}_R)} = 2 .$$

We also note convenient bounds for the case $|z_o| > r_o$ and $(r_o/|z_o|)^n < \frac{1}{3}$, $(r_o|z_o|/R^2)^n < \frac{1}{3}$:

(5.26) $$\frac{1}{2}\frac{(|z_o|/R)^n}{(1-|z_o|^2/R^2)^{3/4}} < E_{z_o}^{(n)}(\mathscr{B}_R) < 2\,\frac{(|z_o|/R)^n}{(1-|z_o|^2/R^2)^{3/4}}.$$

6. $\mathscr{A}(D_R)$-SPLINES SATISFYING INITIAL CONDITIONS. HYPEROPTIMAL λ-RULES.

In the first part of this section we obtain an explicit formula for the $\mathscr{A}(D_R)$-spline operator for optimal recovery of the ball in $\mathscr{A}(D_R)$ from Hermite data

(6.1) $$f(z_1),\ f'(z_1),\ldots,\ f^{(n-1)}(z_1),\quad z_1 \in D_R .$$

Let the spline be called $\tilde{S}_{z_1}f$. By Remark 4.1

(6.2a) $$\tilde{S}_{z_1}f = \sum_{i=0}^{n-1} \gamma_i k^{(i)}(\cdot,z_1)$$

where k is the reproducing kernel of $\mathscr{A}(D_r)$ (see 5.6) and the γ_i must be determined from the conditions

$$(\tilde{S}_{z_o} f)^{(i)}(z_1) = f^{(i)}(z_1), \quad i = 0,1,\ldots,n-1 . \tag{6.2b}$$

$S_{z_1} f$ is a rational function with one pole of order n at $R^2/\bar{z}_1$ and vanishing at ∞. Hence we may set

$$\tilde{S}_{z_1} f(z) = \frac{\sum_{k=0}^{n-1} \beta_k (z - z_1)^k}{(1 - z\bar{z}_1/R^2)^n} . \tag{6.3}$$

To determine the β_k we observe that $\tilde{S}_n f$ is analytic at z_1, hence

$$\tilde{S}_{z_1} f(z) = \sum_{\nu=0}^{\infty} \tilde{\alpha}_{\nu,n}(z_1)(z - z_1)^{\nu} . \tag{6.4}$$

From (6.2b) we know the first n coefficients:

$$\tilde{\alpha}_{\nu,n}(z_1) = f^{(\nu)}(z_1)/\nu! \ , \quad \nu = 0,1,\ldots,n-1 . \tag{6.5}$$

Comparing coefficients in (6.3) and (6.4), we find the β_k. With some simplifications the result is

$$\begin{aligned} \tilde{S}_{z_1} f(z) &= \frac{R^2 - z_1\bar{z}_1}{R^2 - z\bar{z}_1} \sum_{k=0}^{n-1} \tilde{\beta}_k (z - z_1)^k , \\ \tilde{\beta}_k &= \sum_{j=0}^{k} \binom{n}{k-j} \left(\frac{-\bar{z}_1}{R^2 - z\bar{z}_1} \right)^{k-j} \frac{f^{(j)}(z_1)}{j!} \end{aligned} \tag{6.6}$$

These formulas are easily programmed for computation.

The spline c_{z_o} (see Lemma 2.2) is

$$c_{z_o}(z) = \left(\frac{z - z_1}{z_o - z_1}\right)^n \left(\frac{1 - z_o z_1/R^2}{1 - z\bar{z}_1/R^2} \right)^n \frac{1 - z_o z_o/R^2}{1 - z\bar{z}_o/R^2} . \tag{6.7}$$

From this the minimal error for recovery is calculated by Theorem 3.1:

$$E_{z_o}^{(n)}(\mathscr{B}_R) = \frac{R^{-n}}{(1 - |z_o|^2/R^2)^{3/4}} \left| \frac{z_o - z_1}{1 - z_o\bar{z}_1/R^2} \right|^n . \tag{6.8}$$

We note the limit relations

$$\lim_{R \to \infty} R^n E_{z_o}^{(n)}(\mathscr{B}_R) = |z_o - z_1|^n \tag{6.9}$$

(6.10) $$\lim_{n\to\infty} \sqrt[n]{E_{z_o}^{(n)}(\mathscr{B}_R)} = |z_o - z_1|/R|1 - z_o\bar{z}_1/R^2| \ .$$

In the second part of this section we return to the case where the data vector is $(f(z_1),\ldots,f(z_n))$, with $z_1,\ldots,z_n$ distinct points in D_R. We first establish a convenient formula for the minimal error E_λ in recovery of the unit ball $\mathscr{B}$ at the functional $\lambda \in \mathscr{A}(D_R)'$ and then minimize E_λ by choice of the data points z_i. This inquiry was stimulated by work of Wilf's, Eckhardt's and Engels' on optimal quadrature formulas [13-17]* .

If S is the spline projector of (5.10) then

(6.11) $$E_\lambda := \sup_{f\in\mathscr{B}} |\lambda f - \lambda Sf|$$

Let s_i $(i = 1,\ldots,n)$ be the spline (5.9), which interpolates the data $\delta_{i1},\ldots,\delta_{in}$ at $z_1,\ldots,z_n$, and set $\sigma_i = \lambda s_i$. Then

(6.12) $$\tilde{\lambda}f := \lambda Sf = \sum_{i=1}^{n} \sigma_i f(z_i)$$

is the "optimal λ-rule" for computing λf, given the data $f(z_i)$, $f \in \mathscr{B}$. Using the kernel k of (5.6) and the conjugate functional λ^*, defined by

$$\lambda^*(f) = \overline{\lambda(\bar{f})},$$

we have

(6.13) $$f(z) = (f,k(\cdot,z)), \qquad \lambda f = (f,\lambda_z^* k(\cdot,z)) \ .$$

(The subscript $\mathscr{A}(D_R)$ is omitted from the inner product symbol, and the subscript z in λ_z^* indicates that λ^* acts on the function $z \mapsto k(\cdot,z)$.) By (6.12) and (6.13)

(6.14) $$\tilde{\lambda}f = (f, \sum_{i=1}^{n} \bar{\sigma}_i k(\cdot,z_i)) \ .$$

A simple calculation now gives

(6.15) $$\begin{aligned} E_\lambda^2 &= \|\lambda_z^* k(\cdot,z) - \sum_{i=1}^{n} \bar{\sigma}_i k(\cdot,z_i)\|^2 \\ &= \lambda_u \lambda_v^* k(u,v) - \sum_{i,j=1}^{n} \sigma_i\bar{\sigma}_j k(z_i,z_j) \\ &= \lambda_u \lambda_v^* k(u,v) - \sum_{i=1}^{n} \sigma_i \lambda_z^* k(z_i,z) \ , \end{aligned}$$

*The author wishes to express his thanks to Dr. U. Eckhardt for calling his attention to these papers and for providing him with preprints of the unpublished ones.

where we have made use of the fact that $k(\cdot,z_j)$ is a spline, $Sk(\cdot,z_j) = k(\cdot,z_j)$, hence by (6.12)

$$\lambda_z k(z,z_j) = \sum_{i=1}^{n} \sigma_i k(z_i,z_j), \quad \lambda_z^* k(z_i,z) = \sum_{j=1}^{n} \bar{\sigma}_j k(z_i,z_j). \tag{6.16}$$

Using the function

$$K = \lambda_z^* k(\cdot,z) \ , \tag{6.17}$$

(6.15) can also be written in the compact form

$$E_\lambda^2 = \lambda K - \tilde{\lambda} K \ , \tag{6.18}$$

which says that E_λ^2 is equal to the error in computing λK by the approximation $\tilde{\lambda} K$.

We now assume that λ is a <u>real functional</u>, by which we mean

$$\lambda(f) \in \mathbb{R} \quad \text{if} \quad f(x) \in \mathbb{R} \quad \text{for} \quad x \in \mathbb{R} \cap D_R \ .$$

We also assume that there are distinct real nodes $z_i = x_i$ $(i = 1,\ldots,n)$ in D_R that make E_λ a minimum, so that

$$\partial E_\lambda^2 / \partial x_\ell = 0 \ , \qquad \ell = 1,\ldots,n \ . \tag{6.19}$$

Then the numbers $\sigma_i = \lambda s_i$ and $\lambda_z^* k(x_i,z) = \lambda_z k(x_i,z)$ $(i = 1,\ldots,n)$ are real. If condition (6.19) is satisfied, we say the optimal λ-rule

$$\tilde{\lambda} f = \lambda S f = \sum_{i=1}^{n} \sigma_i f(x_i) \tag{6.20}$$

is <u>hyperoptimal</u> provided none of the σ_i is 0. The proviso is natural since if some σ_i vanish the corresponding terms in (6.20) should be dropped and n reduced. In the remainder of this section we establish conditions equivalent to (6.19), from which the nodes x_i for a hyperoptimal λ-rule can be determined.

In the following k_1, k_2 denote the derivatives of the functions $z \mapsto k(z,\cdot)$, $\bar{z} \mapsto k(\cdot,z)$, and σ_{ij} $(i,j = 1,\ldots,n)$ denotes the derivative of the function $x_j \mapsto \sigma_i$ at x_j. Using (6.15), condition (6.19) becomes

$$\sum_{i,j=1}^{n} \sigma_i \sigma_{j\ell} k(x_j,x_i) + \sigma_\ell \lambda_z k_1(x_\ell,z) = 0, \quad \ell = 1,\ldots,n \ . \tag{6.21}$$

From the identity (6.16) one obtains

(6.22)
$$\sum_{j=1}^{n} \sigma_{j\ell}k(x_j,x_i) + \sigma_\ell k_1(x_\ell,x_i) = 0 , \quad i \neq \ell$$
$$\sum_{j=1}^{n} \sigma_{j\ell}k(x_j,x_\ell) + \sigma_\ell k_1(x_\ell,x_\ell) + \sum_{j=1}^{n}\sigma_j k_2(x_j,x_\ell) -$$
$$- \lambda_z k_2(z,x_\ell) = 0 , \quad i,\ell = 1,\ldots,n$$

From (6.21) and (6.22) it follows that

$$\sigma_\ell[\sum_{i=1}^{n}\sigma_i k_2(x_i,x_\ell) - \lambda_z k_2(z,x_\ell)] = 0$$

and since $\sigma_\ell \neq 0$, by (6.12):

$$\tilde{\lambda}k_2(\cdot,x_\ell) = \lambda k_2(\cdot,x_\ell), \quad \ell = 1,\ldots,n . \tag{6.23}$$

This equation says that if $\tilde{\lambda}$ is a hyperoptimal λ-rule then it evaluates the numbers $\lambda k_2(\cdot,x_\ell)$ (and, of course, also $\lambda k(\cdot,x_\ell)$) ($\ell = 1,\ldots,n$), accurately. The path from (6.19) to (6.23) can be reversed, hence $\tilde{\lambda}$ is a hyperoptimal λ-rule if and only if the x_ℓ satisfy equations (6.23). Let X denote the n-tuple $(x_1,\ldots,x_n)$ and Q_X the linear space of rational functions which vanish at ∞ and whose only singularities are poles of order ≤ 2 at the points R^2/x_i $(i = 1,\ldots,n)$. Q_X is spanned by the functions $k(\cdot,x_i)$, $k_2(\cdot,x_i)$ $(i = 1,\ldots,n)$, hence $\tilde{\lambda}$ is a hyperoptimal λ-rule if and only if

$$\tilde{\lambda}g = \lambda g, \quad g \in Q_X \tag{6.24}$$

The functions s_i, s_i^2 $(i = 1,\ldots,n)$ also form a basis of the space Q_X. If $\tilde{\lambda}$ is hyperoptimal then $\lambda s_i^2 = \tilde{\lambda}s_i^2 = \Sigma_j\sigma_j s_i^2(x_j) = \sigma_i$. Thus,

$$\lambda s_i = \lambda s_i^2 = \sigma_i , \qquad i = 1,\ldots,n \tag{6.25}$$

is still another necessary and sufficient condition that the λ-rule $\tilde{\lambda}$ with the weights σ_i be hyperoptimal. From (6.25) also follows that if λ is not only a real, but a <u>positive</u> functional, i.e.

$$\lambda f \geq 0 \quad \text{if} \quad f(x) \geq 0 \quad \text{for} \quad x \in \mathbb{R} \cap D_R ,$$

then the weights σ_i of a hyperoptimal rule are positive.

Still another characterization of hyperoptimality results from the following considerations. Let Hf denote the $\mathscr{A}(D_R)$-spline that interpolates f on the set $(z_1,z_1,z_2,z_2,\ldots,z_n,z_n)$. The data vector is then $(f(z_1),f'(z_1),f(z_2),f'(z_2),\ldots,f(z_n),f'(z_n))$ (see Remark 4.1). Hf is a linear combination of the splines $k(\cdot,z_j)$, $k_1(\cdot,z_j)$ $(j = 1,\ldots,n)$, thus Hf is a rational function that vanishes at ∞ and whose only singularities are poles of order ≤ 2 at the points $R^2/\bar{z}_i$ $(i = 1,\ldots,n)$. Let Q_Z denote the linear space of these rational functions. The functions $s_i^2(z)$, $(z - z_i)s_i^2(z) := t_i(z)$ $(i = 1,\ldots,n)$ also form a basis of Q_Z. Using this fact and the interpolation conditions

$$(6.26)\qquad (Hf)(z_i) = f(z_i), \quad (Hf)'(z_i) = f'(z_i), \quad i = 1,\ldots,n$$

it follows that

$$(6.27)\qquad (Hf)(z) = \sum_{i=1}^{n} [f(z_i) - 2s_i'(z_i)(z - z_i)f(z_i) + (z - z_i)f'(z_i)]s_i^2(z) .$$

and that

$$(6.28)\qquad \tilde{\tilde{\lambda}}f := \lambda Hf = \sum_{i=1}^{n} [\lambda(s_i^2)f(z_i) - 2s_i'(z_i)\lambda(t_i)f(z_i) + \lambda(t_i)f'(z_i)]$$

is the optimal λ-rule for computing λf, given the data $f(z_i)$, $f'(z_i)$, $f \in \mathscr{B}$.

Now assume $z_i = x_i$ $(i = 1,\ldots,n)$ are the nodes of a hyperoptimal λ-rule $\tilde{\lambda}$ with the weights $\sigma_i = \lambda(s_i) = \lambda(s_i^2)$. Then $\lambda(t_i) = \tilde{\lambda}(t_i) = 0$ by (6.12), hence

$$(6.29)\qquad \tilde{\tilde{\lambda}}f = \tilde{\lambda}f, \quad f \in \mathscr{A}(D_R) \quad .$$

If conversely the nodes $z_i = x_i$ are such that $\tilde{\tilde{\lambda}}f = \tilde{\lambda}f$ for every $f \in \mathscr{A}(D_R)$ (or only for every $f \in Q_X$) then (6.24) is satisfied, hence $\tilde{\lambda}$ is hyperoptimal.

We summarize some of these results in

Theorem 6.1. Suppose $\lambda \in \mathscr{A}(D_R)'$ is a real functional; $x_1,\ldots,x_n$ are real distinct points in D_R; $s_1,\ldots,s_n$ are the splines (5.4) with $z_i = x_i$; and $\sigma_i = \lambda s_i \neq 0$ $(i = 1,\ldots,n)$. The optimal λ-rule, given the data $f \in \mathscr{B}$, $f(x_i)$ is

$$\tilde{\lambda}f = \sum_{i=1}^{n} \sigma_i f(x_i)$$

and the minimal error is

$$E_\lambda = (\lambda K - \tilde{\lambda}K)^{\frac{1}{2}} ,$$

where $K = \lambda_z^* k(\cdot,z)$. The λ-rule $\tilde{\lambda}$ is hyperoptimal (i.e. $\partial E_\lambda/\partial x_i = 0$, $i = 1,\ldots,n$) if and only if $\tilde{\lambda}g = \lambda g$ for every rational function g that vanishes at ∞ and has no other singularities but poles of order ≤ 2 at the points R^2/x_i. This is the case if and only if $\lambda t_i = 0$, where $t_i(z) = (z - x_i)s_i^2(z)$.

Also, $\tilde{\lambda}$ is hyperoptimal if and only if $\tilde{\lambda}f = \tilde{\tilde{\lambda}}f$ for all $f \in \mathcal{A}(D_R)$, where $\tilde{\tilde{\lambda}}$ is the optimal λ-rule for the data $f \in \mathcal{B}$ $f(x_i)$, $f'(x_i)$. If λ is a positive functional then the weights σ_i of the hyperoptimal λ-rule are positive.

Remark. The special case $\lambda f = \int_{-1}^{1} f$ is treated in [13-17]. For this functional the hyperoptimal λ-rule is the Wilf quadrature formula for $R = 1$ (really $R \to 1$), and is the Gauss quadrature formula for $R \to \infty$.

7. $\mathcal{A}(A_R)$-SPLINES

1. Assume $R > 1$ fixed and let A_R be the open annulus $\{z \in \mathbb{C} : R^{-1} < |z| < R\}$. We denote by $\mathcal{A}(A_R)$ the space of functions f analytic in A_R for which $\int_{|z|=r} |f(z)|^2 |dz|$ is a bounded function of r in $R^{-1} < r < R$. As in Section 5, we express this condition as

$$(7.1) \qquad \int_{\partial A_R} |f|^2 < \infty .$$

Under this condition the functions

$$(7.2) \qquad \phi_o(\theta) = \lim_{r \to R-0} f(re^{2\pi i\theta}), \quad \phi_i(\theta) = \lim_{r \to R+0} f(re^{2\pi i\theta})$$

exist for almost all $\theta \in \mathbb{R}$ and are locally square-integrable.

Suppose $f \in \mathcal{A}(A_R)$ has the expansion $f(z) = \sum_{\nu=-\infty}^{\infty} \alpha_\nu z^\nu$, convergent for $z \in A_R$. Then

$$\int_0^1 |\phi_o|^2 = \sum_{\nu=-\infty}^{\infty} |\alpha_\nu|^2 R^{2\nu} , \quad \int_0^1 |\phi_i|^2 = \sum_{\nu=-\infty}^{\infty} |\alpha_\nu|^2 R^{-2\nu}$$

and (7.1) is equivalent to

$$(7.3) \qquad \sum_{\nu=-\infty}^{\infty} |\alpha_\nu|^2 R^{2|\nu|} < \infty .$$

We introduce the inner product norm

$$(7.4) \qquad \|f\|^2_{\mathcal{A}(A_R)} = \sum_{\nu=-\infty}^{\infty} |\alpha_\nu|^2 R^{2|\nu|}$$

and this makes $\mathcal{A}(A_R)$ a Hilbert space. This norm is not easily expressed in terms of $\int_0^1 |\phi_o|^2$ and $\int_0^1 |\phi_i|^2$.

Clearly the system $\{R^{-|\nu|}e_\nu\}_{\nu=0,\pm1,\ldots}$ forms a complete orthonormal system in $\mathcal{A}(A_R)$. It follows that this space has the reproducing kernel

$$(7.4)\qquad k_A(z,z_o) = \sum_{\nu=-\infty}^{\infty} R^{-2|\nu|} z^\nu \bar{z}_o^\nu = \frac{1}{1 - R^{-2}z\bar{z}_o} - \frac{1}{1 - R^2 z\bar{z}_o}$$

Thus, the spline space $\mathcal{S}_A \subset \mathcal{A}(A_R)$, based on the distinct nodes $z_1,\ldots,z_n$ in A_r, is spanned by the functions

$$(7.5)\qquad s_{A,i}(z) = \frac{z}{(z - R^2/\bar{z}_i)(z - R^{-2}/\bar{z}_i)}, \quad i = 1,\ldots,n .$$

$s_{A,i}$ is, except for a constant factor, characterized as the rational function which vanishes at 0 and ∞ and has simple poles at $R^2/\bar{z}_i$ and $R^{-2}/\bar{z}_i$, which are the inversions of the node z_i at ∂A_R. An equivalent condition to vanishing at 0 is that the quotient of the residues at $R^2/\bar{z}_i$, $R^{-2}/\bar{z}_i$ is $-R^4$. <u>The spline space $\mathcal{S}_A \subset \mathcal{A}(A_R)$, based on the nodes $z_1,\ldots,z_n$, is the space of rational functions that vanish at ∞, have simple poles at the points $R^2/\bar{z}_i$, $R^{-2}/\bar{z}_i$ $(i = 1,\ldots,n)$ with residues whose quotient for each pair is $-R^4$.</u>

The $\mathcal{A}(A_R)$-spline that interpolates f on $Z = (z_1,\ldots,z_n)$ is given by

$$(7.6a)\qquad S_Z f(z) = \sum_{i=1}^{n} \gamma_i \frac{z}{(z - R^2/\bar{z}_i)(z - R^{-2}/\bar{z}_i)}$$

where the γ_i are determined from the conditions

$$(7.6b)\qquad S_Z f(z_j) = f(z_j), \qquad j = 1,\ldots,n .$$

This requires matrix inversion and no practical explicit inverse has been found for the general case.

For the special case of symmetrically distributed interpolation nodes

$$(7.7)\qquad Z_* = (r_o\varepsilon^j)_{j=0,1,\ldots,n-1}, \quad \varepsilon = e^{2\pi i/n}, \quad R^{-1} < r_o < R$$

we obtain simple explicit formulas for the spline interpolant and for the error function. We set

$$\begin{aligned} S_{Z_*}f(z) &= \sum_{i=0}^{n-1} \gamma_i \left(\frac{1}{1 - R^{-2}r_o^{-1}z\varepsilon^i} - \frac{1}{1 - R^2 r_o^{-1} z\varepsilon^i}\right) \\ &= \sum_{i=0}^{n-1} \gamma_i \sum_{\nu=0}^{\infty} \left[(R^{-2}r_o^{-1}z\varepsilon^i)^\nu + (R^{-2}r_o z^{-1}\varepsilon^{-i})^{\nu+1}\right] \end{aligned}$$

$$= \sum_{\nu=0}^{n-1} \Big[\Big(\sum_{i=0}^{\infty} \gamma_i \varepsilon^{i\nu} \Big) (R^{-2} r_o^{-1} z)^{\nu} + \Big(\sum_{i=0}^{n-1} \gamma_i \varepsilon^{-i(\nu+1)} \Big) (R^{-2} r_o z^{-1})^{\nu+1} \Big] .$$

Since $S_{Z_*} f(r_o \varepsilon^k) = f(r_o \varepsilon^k)$ we obtain

$$\hat{f}_j = \frac{1}{n} \sum_{k=0}^{n-1} f(r_o \varepsilon^k) \varepsilon^{-kj} = \mu_j \sum_{i=0}^{n-1} \gamma_i \varepsilon^{ij}$$

$$\mu_j = \sum_{\nu \equiv j(n)} (r_o/R)^{2\nu} + \sum_{\nu \equiv -j(n)} (r_o R)^{-2\nu} , \quad j = 0,1,\ldots,n-1 \tag{7.9}$$

$$= \frac{(R^{2n-2\ell} - R^{2\ell-2n}) r_o^{2\ell} + (R^{2\ell} - R^{-2\ell}) r_o^{2(\ell-n)}}{(R^{2n} + R^{-2n}) - (r_o^{2n} + r_o^{-2n})} .$$

With $\sum_{i=0}^{n-1} \gamma_i \varepsilon^{ji} = \hat{f}_j / \mu_j$ substituted in (7.8), one finds

$$S_{Z_*} f(z) = \frac{(R^{2n} + R^{-2n}) - (r_o^{2n} + r_o^{-2n})}{(R^{2n} + R^{-2n}) - ((r_o z)^n + (r_o z)^{-n})} \sum_{j=0}^{n-1} \hat{f}_j \frac{a_j(z)}{a_j(r_o)}$$

$$a_j(z) = (R^{2n-2j} - R^{2j-2n})(r_o z)^j + (R^{2j} - R^{-2j})(r_o z)^{j-n} \tag{7.10}$$

$$\hat{f}_j = \frac{1}{n} \sum_{k=0}^{n-1} f(r_o \varepsilon^k) \varepsilon^{-kj} \qquad j = 0,1,\ldots,n-1 .$$

Formula (7.10) is easily programmed for computing. Except for the fractional factor (which is indistinguishable from 1 in most cases), $S_{Z_*} f(z)$ equals

$$L_{Z_*} f(z) = \sum_{j=0}^{n-1} \hat{f}_j \, a_j(z)/a_j(r_o) , \tag{7.11}$$

which is the unique linear combination of the simple functions $a_0, a_1, \ldots, a_{n-1}$ that interpolates f on Z_*.

We also give the explicit partial fractions expansion of $S_{Z_*} f$. There are simple poles at the points $\zeta_j = R^2/r_o \varepsilon^j$ and $\zeta_j^{*} = R^{-2}/r_o \varepsilon^j$ $(j = 0,1,\ldots,n-1)$ with residues

$$\operatorname*{res}_{\zeta_j} S_{Z_*} f = \frac{(R^{2n} + R^{-2n}) - (r_o^{2n} + r_o^{-2n})}{-n(R^{2n} - R^{-2n}) \zeta_j^{-1}} L_{Z_*} f(\zeta_j) . \tag{7.12}$$

Therefore

$$S_{Z_*}f(z) = \frac{(R^{2n} + R^{-2n}) - (r_o^{2n} + r_o^{-2n})}{R^{2n} - R^{-2n}} \tag{7.13}$$
$$\times \frac{1}{n} \sum_{j=0}^{n-1} \left(\frac{\zeta_j L_{Z_*} f(\zeta_j)}{\zeta_j - z} - \frac{\zeta_j' L_{Z_*} f(\zeta_j')}{\zeta_j' - z} \right) .$$

This formula is, except for the constant factor in front, the discrete analogue of the Cauchy integral for L_{Z_*} on the boundary of the annulus $R^{-2}/r_o < |z| < R^2/r_o$ containing the knots ζ_j and $\zeta_j' (j = 0,1,\ldots,n-1)$. From it we obtain a simple formula for the optimal recovery of the Taylor expansion coefficients of f at some $z_o \in A_R$. If $f(z) = \sum_{\nu=0}^{\infty} \alpha_\nu(z_o)(z - z_o)^\nu$ and $S_n f(z) = \sum_{\nu=0}^{\infty} \alpha_{\nu_*}(z_o)(z - z_o)^\nu$ then

$$\alpha_{\nu_*}(z_o) = \frac{R^{2n} + R^{-2n} - (r_o^{2n} + r_o^{-2n})}{R^{2n} - R^{-2n}} \tag{7.14}$$
$$\times \frac{1}{n} \sum_{j=0}^{n-1} \left(\frac{\zeta_j L_{Z_*} f(\zeta_j)}{(\zeta_j - z)^{\nu+1}} - \frac{\zeta_j' L_{Z_*} f(\zeta_j')}{(\zeta_j' - z)^{\nu+1}} \right) .$$

2. Let $\mathcal{B}_R$ be the unit ball in $\mathcal{A}(A_R)$. The minimal error in recovery of $\mathcal{B}_R$ at $z_o \in A_R$ is, by Theorem 3.1,

$$E_{z_o}(\mathcal{B}_R) = \|c_{z_o}\|^{-1}_{\mathcal{A}(A_R)} \tag{7.15}$$

where c_{z_o} is the $\mathcal{A}(A_R)$-spline which interpolates 0 on Z_* and 1 at z_o. By the characterization of the $\mathcal{A}(A_R)$-splines given above it follows that

$$c_{z_o}(z) = \gamma \frac{z(z^n - r_o^n)q(z)}{[(R^{2n}+R^{-2n})-((r_o z)^n+(r_o z)^{-n})](z-R^2/\bar{z}_o)(z-R^{-2}/\bar{z}_o)} \tag{7.16}$$

where γ is a constant and q is a polynomial in z^{-1} of degree n:

$$q(z) = \sum_{\ell=0}^{n} \beta_\ell (z\bar{z}_o)^{-\ell} \quad . \tag{7.17}$$

The β_ℓ are determined (except for a common factor) from the conditions

$$\operatorname*{res}_{R^2/r_o\varepsilon^k} c_{z_o} + R^4 \operatorname*{res}_{R^{-2}/r_o\varepsilon^k} c_{z_o} = 0 \ , \quad k = 0,1,\ldots,n-1 \tag{7.18}$$

and γ is then found from $c_{z_o}(z_o) = 1$.

The remainder of this formidable calculation is carried out only for the case $r_o = 1$, i.e., $Z_* = (\varepsilon^j)$. (7.18) gives

$$\sum_{\ell=0}^{n} \beta_\ell \bar{z}_o^{-\ell}[R^{2\ell-n}(R^2\varepsilon^k - 1/R^2\bar{z}_o) + R^{n-2\ell}(R^{-2}\varepsilon^k - R^2/\bar{z}_o)]\varepsilon^{-k\ell} = 0, \qquad (7.19)$$

$$k = 0,1,\ldots,n-1 .$$

Set

$$R^j + R^{-j} = r_j . \qquad (7.20)$$

Multiplying (7.19) by ε^{jk} and summing over k gives

$$(7.21) \qquad \begin{aligned} r_{n-2j-4}\beta_{j+1} - r_{n-2j+2}\beta_j &= 0 , \quad j = 1,\ldots,n-1 \\ r_{n-4}\beta_1 - r_{n+2}\beta_o - r_{n-2}\beta_n\bar{z}_o^{-n} &= 0 \end{aligned}$$

By choosing β_1 as $(r_n r_{n-2} r_{n-4})^{-1}$ one finds

$$(7.22) \qquad \begin{aligned} \beta_j &= 1/r_{n-2j+2}r_{n-2j}r_{n-2j-2} , \quad j = 1,\ldots,n-1 \\ \beta_o &= \frac{r_{n+2} - r_{n-2}\bar{z}_o^{-n}}{r_{n-2}r_n r_{n+2}^2} . \end{aligned}$$

Thus q in (7.16) is explicitly determined, and so is γ:

$$\gamma = (r_{2n} - z_o^n - z_o^{-n})(z_o - R^2/\bar{z}_o)/z_o(z_o^n - 1)q(z_o) . \qquad (7.23)$$

We use the result to calculate the asymptotic value of $E_{z_o}(\mathscr{B}_R)$ for $R \to \infty$. For this we use

<u>Lemma 7.1</u>. If $f_R \in \mathscr{A}(A_R)$ for each $R > 0$, $f_R = f_{R^{-1}}$ and

$$\lim_{R\to\infty} R^{-\kappa}|f_R(Re^{i\theta})| = K(\theta)$$

$$\lim_{R\to 0} R^{\lambda}|f_R(Re^{i\theta})| = L(\theta)$$

then

$$\text{(i)} \qquad \lim_{R\to\infty} R^{-2\kappa}\| f_R \|^2_{\mathscr{A}(A_r)} = \frac{1}{2\pi}\int_0^{2\pi} K^2 d\theta \qquad \text{if } \kappa > \lambda$$

$$\text{(ii)} \qquad \lim_{R\to\infty} R^{-2\lambda}\| f_R \|^2_{\mathscr{A}(A_R)} = \frac{1}{2\pi}\int_0^{2\pi} L^2 d\theta \qquad \text{if } \kappa < \lambda$$

$$\text{(iii)} \qquad \lim_{R\to\infty} R^{-2\kappa}(\|f_R\|^2_{\mathscr{A}(A_R)} + |\alpha_o(R)|^2) = \frac{1}{2\pi}\int_0^{2\pi}(K^2 + L^2)dr \qquad \text{if } \kappa = \lambda .$$

Here $\alpha_o(R) = (2\pi i)^{-1} \int_{|z|=1} z^{-1} f_R(z)dz.$

Proof. Suppose $f_R(z) = \sum_{\nu=-\infty}^{\infty} \alpha_\nu(R)z^\nu$ for $z \in A_R$. From the hypothesis it follows that

$$\lim_{R \to \infty} R^{-2\kappa} \frac{1}{2\pi} \int_0^{2\pi} |\Sigma\, \alpha_\nu(R)R^\nu e^{i\nu\theta}|^2 d\theta = \frac{1}{2\pi} \int_0^{2\pi} K^2 d\theta \ ,$$

hence

$$\lim_{R \to \infty} R^{-2\kappa}\, \Sigma|\alpha_\nu(R)|^2 R^{2\nu} = \frac{1}{2\pi} \int_0^{2\pi} K^2 d\theta \ . \tag{7.24}$$

Similarly,

$$\lim_{R \to 0} R^{2\lambda}\, \Sigma|\alpha_\nu(R)|^2 R^{2\nu} = \frac{1}{2\pi} \int_0^{2\pi} L^2 d\theta \ ,$$

hence since $\alpha_\nu(R) = \alpha_\nu(R^{-1})$,

$$\lim_{R \to \infty} R^{-2\lambda}\, \Sigma|\alpha_\nu(R)|^2 R^{-2\nu} = \frac{1}{2\pi} \int_0^{2\pi} L^2 d\theta \ . \tag{7.25}$$

If $\kappa > \lambda$ then (7.25) gives

$$\lim_{R \to \infty} R^{-2\kappa} \Sigma|\alpha_\nu(R)|^2 R^{-2\nu} = 0 \tag{7.26}$$

and adding this to (7.24) gives

$$\lim_{R \to \infty} R^{-2\kappa}\, \Sigma|\alpha_\nu(R)|^2 (R^{2\nu} + R^{-2\nu}) = \frac{1}{2\pi} \int_0^{2\pi} K^2 d\theta \ ,$$

hence also

$$\begin{aligned} \lim_{R \to \infty} R^{-2\kappa} \|f_R\|^2_{\mathcal{A}(A_R)} &= \lim_{R \to \infty} R^{-2\kappa}\, \Sigma|\alpha_\nu(R)|^2 R^{2|\nu|} \\ &= \frac{1}{2\pi} \int_0^{2\pi} K^2 d\theta \ . \end{aligned} \tag{7.27}$$

Here we have used $\lim_{R \to \infty} R^{-2\kappa}|\alpha_o(R)|^2 = 0$, which follows from (7.26). Thus (i) is proved and (ii) follows similarly. If $\kappa = \lambda$ then we arrive at

$$\lim_{R \to \infty} R^{-2\kappa}\, \Sigma\, |\alpha_\nu(R)|^2 (R^{2\nu} + R^{-2\nu}) = \frac{1}{2\pi} \int_0^{2\pi} (K^2 + L^2) d\theta \ ,$$

which implies

$$\begin{aligned} \lim_{R \to \infty} R^{-2\kappa} &\left(\Sigma\, |\alpha_\nu(R)|^2 R^{2|\nu|} + |\alpha_o(R)|^2\right) \\ &= \frac{1}{2\pi} \int_0^{2\pi} (K^2 + L^2) d\theta \ . \end{aligned} \tag{7.28}$$

This proves (iii) since $\alpha_o(R) = (2\pi i)^{-1} \int_{|z|=1} z^{-1}f(z)dz$.

Using this lemma we now prove

Theorem 7.1. If $E_{z_o}(\mathscr{B}_r)$ denotes the minimal error for recovery of the ball $\mathscr{B}_R \subset \mathscr{A}(A_R)$ at $z_o \neq 0$ from data on the set $(e^{2\pi ik/n})_{k=0,1,\ldots,n-1}$ then

$$\lim_{R\to\infty} R^{n/2}E_{z_o}(\mathscr{B}_R) = 2^{-\frac{1}{2}}|z_o|^{-n/2}|z_o^n - 1| .$$

Proof. For simplicity we assume n even, $n = 2m$. Let q_R be the function defined in (7.14), with the β_ℓ given in (7.22). One sees readily that

$$\begin{aligned}
&\lim_{R\to\infty} R^{m+4}|q_R(Re^{i\theta})| = \lim_{R\to\infty} R^{m+4}|\beta_m R^{-m}z_o{}^{-m}| = \frac{1}{2}|z_o|^{-m} \\
(7.29)\quad &\lim_{R\to 0} R^{m-4}|q_R(Re^{i\theta})| = \lim_{R\to 0} R^{m-4}|\beta_m R^{-m}z_o{}^{-m}| = \frac{1}{2}|z_o|^{-m} \\
&\lim_{R\to\infty} R^4|q_R(z_o)| = \lim_{R\to\infty} R^4|\beta_m z_o{}^{-2m}| = \frac{1}{2}|z_o|^{-n} .
\end{aligned}$$

If γ_R denotes the constant of (7.23) then, using (7.29) one finds

$$(7.30)\qquad \lim_{R\to\infty} R^{2n+6}|\gamma_R^{-1}| = \frac{1}{2}|z_o|\;|z_o^n - 1|\;|z_o|^{-n} .$$

Finally, define

$$f_R(z) = \frac{z(z^n - 1)q_R(z)}{[R^{2n} - (z^n + z^{-n})](z - R^2/\bar{z}_o)(z - R^{-2}/\bar{z}_o)}$$

Then using (7.29), one obtains

$$\begin{aligned}
&\lim_{R\to\infty} R^{3m+6}|f_R(Re^{i\theta})| = \frac{1}{2}|z_o|^{-m+1} \\
(7.31)\quad & \\
&\lim_{R\to 0} R^{-3m-6}|f_R(Re^{i\theta})| = \frac{1}{2}|z_o|^{-m+1}
\end{aligned}$$

f_R satisfies the hypotheses of Lemma 7.1. By (7.31), Case (iii) applies and since $\alpha_o(R) = 0$ we conclude

$$(7.32)\qquad \lim_{R\to\infty} R^{-3m-6}\|f_R\|_{\mathscr{A}(A_R)} = 2^{-\frac{1}{2}}|z_o|^{-m+1} .$$

By (7.15), (7.16), (7.30) and (7.32)

$$\begin{aligned}
\lim_{R\to\infty} R^m E_{z_o}(\mathscr{B}_R) &= \lim_{R\to\infty} R^m|\gamma_R^{-1}|\;\|f_R\|^{-1}_{\mathscr{A}(A_R)} \\
&= 2^{-\frac{1}{2}}|z_o|^{-m}|z_o^n - 1| ,
\end{aligned}$$

which proves the theorem.

3. We now investigate the dependence of the minimal error on n, the number of data points. To indicate this dependence I denote the error as $E_{z_o}^{(n)}(\mathscr{B}_R)$. Although we do have an explicit expression for $E_o^{(n)}(\mathscr{B}_R)$ I did not find it useful for the study of the asymptotic behavior of the error as $n \to \infty$. Indeed I use the result of Theorem 3.2

$$(7.33) \qquad E_{z_o}^{(n)}(\mathscr{B}_R) = [k_{z_o}(z_o) - S_n k_{z_o}(z_o)]^{\frac{1}{2}} ,$$

where the reproducing kernel is that of (7.4) and S_n is the spline projector denoted as S_{Z^*} above, for the set of nodes $Z_* = (\varepsilon^j)$. We use the series expansions

$$(7.34) \qquad \begin{aligned} k_{z_o}(z) &= \sum_{\nu=-\infty}^{\infty} (z\bar{z}_o)^{\nu} R^{-2|\nu|} \\ S_n k_{z_o}(z) &= \sum_{\nu=-\infty}^{\infty} b_{\nu}(z)\bar{z}_o^{\nu} R^{-2|\nu|} , \end{aligned}$$

where

$$(7.35) \qquad \begin{aligned} b_{\nu}(z) &= S_n e_{\nu} = \lambda_{\nu}^{-1} \sum_{\mu \equiv \nu(n)} z_o^{\mu} R^{-2|\mu|} \\ \lambda_{\nu} &= \sum_{\mu \equiv \nu(n)} R^{-2|\mu|} \\ &= \frac{R^{-2\nu} + R^{2\nu-2n}}{1 - R^{-2n}} \qquad \text{if } \nu = 0,1,\ldots,n-1 . \end{aligned}$$

Clearly,

$$(7.36) \qquad b_{\nu} = b_{\nu+n} , \quad \lambda_{\nu} = \lambda_{\nu+n} .$$

These formulas follow easily from (7.10) by using e_{ν} for f.

For the error in approximating e_{ν} by $S_n e_{\nu}$ we have

$$(7.37) \qquad \begin{aligned} z_o^{\nu} - b_{\nu}(z_o) &= \lambda_{\nu}^{-1} \sum_{\mu \equiv \nu} (z_o^{\nu} - z_o^{\mu}) R^{-2|\mu|} \\ &= \lambda_{\nu}^{-1}[z_o^{\nu}(1 - z_o^{n})R^{-2|\nu+n|} + z_o^{\nu}(1 - z_o^{-n})R^{-2|\nu-n|}] + \Delta_{\nu} , \end{aligned}$$

where we have introduced the remainder term Δ_{ν}. For simplicity we assume n is even, $n = 2m$, and temporarily, $|z_o| \geq 1$. To bound $E_{z_o}^{(n)}$ from above we consider first the terms with $|\nu| \leq m$.

I. If $|\nu| \leq m$ then by (7.35), $\lambda_{\nu} \geq R^{-2\nu}$ and one obtains for contribution of the Δ_{ν} terms to the difference $k_{z_o}(z_o) - S_n k_{z_o}(z_o)$, with some work:

(7.38) $$\sum_{\nu=-m+1}^{m} |z_o|^{\nu} R^{-2|\nu|} \Delta_\nu < 2|z_o|^n R^{-n} \frac{R^2 - R^{-2}}{R^2 + R^{-2} - |z_o|^2 - |z_o|^{-2}} \delta_n$$

where

$$\delta_n = \max\left[\frac{R^{-2n}}{1 - (|z_o|/R^2)^n}, \frac{(|z_o|/R)^{2n}}{1 - R^{-2n}}\right], \quad \text{hence}$$

(7.39)
$$\lim_{n \to \infty} \delta_n = 0 .$$

For the principal terms in (7.34) one finds similarly

$$\left|\sum_{\nu=-m+1}^{m} \lambda_\nu^{-1} \bar{z}_o^\nu R^{-2|\nu|}\left[z_o^\nu(1 - z_o^n) R^{-2|\nu+n|} + z_o^\nu(1 - z_o^{-n}) R^{-2|\nu-n|}\right]\right|$$

(7.40)
$$< 2|z_o|^n R^{-n} \frac{R^2 - R^{-2}}{R^2 + R^{-2} - |z_o|^2 - |z_o|^{-2}} .$$

Thus,

$$\left|\sum_{\nu=-m+1}^{m} \bar{z}_o^\nu R^{-2|\nu|}(z_o^\nu - b_\nu(z_o))\right|$$

(7.41)
$$< 2(|z_o|/R)^n \frac{R^2 - R^{-2}}{R^2 + R^{-2} - |z_o|^2 - |z_o|^{-2}} (1 + \delta_n).$$

II. For terms with $|\nu| > m$ the error is treated as due to truncation. One finds

$$\left|\sum_{|\nu| \geq m} \bar{z}_o^\nu R^{-2|\nu|}(z_o^\nu - b_\nu(z_o))\right| \leq \left|\sum_{|\nu| \geq m} |z_o|^{2\nu} R^{-2|\nu|}\right|$$

(7.42)
$$+ \sum_{|\nu| \geq m} |z_o|^2 |b_\nu(z_o)| R^{-2|\nu|} < 6\left(\frac{|z_o|}{R}\right)^n \frac{R^2}{R^2 - |z_o|^\nu}$$

From (7.41) and (7.42) we obtain

(7.43) $$E_z^{(n)}(\mathscr{B}_R)^2 < (|z|/R)^n\, 2(1 + \delta_n) \frac{R^2 - R^{-2}}{R^2 + R^{-2} - |z|^2 - |z|^{-2}} +$$

$$+ 6 \frac{R^2}{R^2 - |z|^2} .$$

Putting $|z| = r$ and observing that

$$\frac{R^2 - R^{-2}}{R^2 + R^{-2} - |z|^2 - |z|^{-2}} < \frac{R^2}{R^2 - |z|^{-2}} + \frac{R^{-2}}{|z|^{-2} - R^{-2}} = \frac{1 + (r/R)^2}{1 - (r/R)^2},$$

we have the slightly less accurate bound

$$\left[E_z^{(n)}(\mathscr{B}_R)\right]^2 < 8(1+\delta_n)(r/R)^n \; \frac{1+(r/R)^2}{1-(r/R)^2}, \qquad (7.44)$$

$$R > |z| = r \geq 1.$$

Next, we determine a lower bound for the error. We choose the point $z_1 = re^{-\pi i/n}$ (or $re^{(2k+1)\pi i/n}$, $k \in \mathbb{Z}$). By Theorem 3.2

$$\left[E_{z_1}^{(n)}(\mathscr{B}_R)\right]^2 \leq \| k_{z_1} - S_n k_{z_1} \|^2_{\mathscr{H}(A_R)} . \qquad (7.45)$$

The coefficient of z^m $(m = n/2)$ in the Laurent expansion of $k_{z_1} - S_n k_{z_1}$ is, by (7.34), $R^{-2m}(\bar{z}_1^m - b_m(\bar{z}_1))$. Because of the orthogonality of the terms in this expansion it follows that

$$\begin{aligned} E_{z_1}^{(n)}(\mathscr{B}_r) &> R^{-m}|\bar{z}_1^m - b_m(\bar{z}_1)| \\ &= R^{-m}\lambda_m^{-1}\left| (\bar{z}_1^m - \bar{z}_1^{-m})R^{-2m} + \sum_{\substack{\mu \equiv m(n) \\ |\mu| \geq 3m}} (\bar{z}_1^m - \bar{z}_1^{\mu})R^{-2|\mu|} \right| . \end{aligned} \qquad (7.46)$$

The infinite sum in (7.46) is treated again as a truncation error. One finds since $\lambda_m^{-1} > R^{-2m}$:

$$\begin{aligned} &E_{z_1}^{(n)}(\mathscr{B}_R) > 2(r/R)^m(1-\eta_n), \quad \text{where} \\ &\eta_n < \frac{2(r/R^2)^n}{1-(r/R^2)^n}, \quad \text{hence} \quad \lim_{n\to\infty} \eta_n = 0 . \end{aligned} \qquad (7.47)$$

Putting (7.44) and (7.47) together we have

$$\begin{aligned} (r/R)^{n/2} 2(1-\eta_n) &< \sup_{|z| = r} E_z^{(n)}(\mathscr{B}_R) \\ &< (r/R)^{n/2}\left[8(1+\delta_n)\frac{1+(r/R)^2}{1-(r/R)^2}\right], \\ &1 \leq r < R . \end{aligned} \qquad (7.48)$$

A symmetry argument shows that these bounds are valid with r replaced by r^{-1} if $R^{-1} < r \leq 1$. In particular, the error bounds are the same for $|z| = r$ and $|z| = r^{-1}$. By the maximum principle it follows that these bounds also hold for

$$E_r^{(n)}(\mathscr{B}_R) := \sup_{r^{-1} \leq |z| \leq r} E_z^{(n)}(\mathscr{B}_R), \quad 1 \leq r < R . \qquad (7.49)$$

In particular, we have

formula for the optimal recovery of the Taylor expansion coefficients of f at some $z_o \in E_R$. If $f(z) = \sum_{\nu=0}^{\infty} \alpha_\nu(z_o)(z - z_o)^\nu$ and $S_n f(z) = \sum_{\nu=0}^{\infty} \alpha_{\nu *}(z_o)(z - z_o)^\nu$ then

(8.23) $$\alpha_{\nu *}(z_o) = \frac{1 - R^{-2n}}{1 + R^{-2n}} \cdot \frac{1}{n} \frac{\sqrt{\zeta_j^2 - 1}}{(\zeta_j - z_o)^{\nu+1}} L_n f(\zeta_j).$$

We also have the strange optimal quadrature formula:

(8.24) $$\int_0^{z_o} S_n f = \frac{1 - R^{-2n}}{1 + R^{-2n}} \cdot \frac{1}{n} \sqrt{\zeta_j^2 - 1}\ L_n f(\zeta_j) \log(1 - z_o/\zeta_j)$$

for any $z_o \in E_R$.

The asymptotic values of the minimal error of Theorem 7.1 and 7.2 are readily applied to this case. For the error of recovery of $\mathscr{B}_R$ at any $z_o \in \mathbb{C}$ we have

(8.25) $$\lim_{R \to \infty} R^{\frac{n}{2}} E_{z_o}^{(n)}(\mathscr{B}_R) = 2^{-\frac{1}{2}} |z_o + \sqrt{z_o^2 - 1}|^{-\frac{n}{2}} |(z_o + \sqrt{z_o^2 - 1})^n - 1|.$$

For the maximum of the minimal error in the ellipse E_ρ with $\rho < R$ the result is

(8.26) $$\lim_{n \to \infty} \sqrt[n]{E_\rho^{(n)}(\mathscr{B}_R)} = (\rho/R)^{\frac{1}{2}}\ .$$

9. COMPARISON WITH n-WIDTHS

The concept of n-width of a set and its usefulness in approximation theory are by now well known (for survey articles see [8] and [9].) If $\mathscr{C}$ is a compact set in a normed vector space $\mathscr{X}$ then the <u>n-width</u> $(n = 0,1,2,\dots)$ <u>of $\mathscr{C}$ with respect to $\mathscr{X}$</u> is defined as follows:

(9.1) $$d_{\mathscr{X}}^{(n)}(\mathscr{C}) = \inf_{M \in \mathscr{M}_n} \sup_{f \in \mathscr{C}} \inf_{g \in M} \|f - g\|.$$

Here $\mathscr{M}_n$ is the set of n-dimensional subspaces of $\mathscr{X}$, $\inf_{g \in M} \|f - g\|$ is, of course, the distance of f from the subspace M, and by taking the supremum of these distances for all $f \in \mathscr{C}$ we get the "deviation" of $\mathscr{C}$ from M. $d_{\mathscr{X}}^{(n)}(\mathscr{C})$ measures the minimal deviation that $\mathscr{C}$ can have from an n-dimensional subspace of $\mathscr{X}$. This is clearly a lower bound for the error in approximating f by linear combinations of any n elements in $\mathscr{X}$, where the choice of them does not depend on f, but only on the class $\mathscr{C}$.

If the n-width of $\mathscr{C}$ is attained for some $M_* \in \mathscr{M}_n$ then M_* is said to be an <u>optimal manifold</u> for n-dimensional approximation of $\mathscr{C}$ in $\mathscr{X}$.

We will show that the spline spaces considered in Sections 5-8 come close to being optimal manifolds for certain function classes. Also, the error bounds found for the recovery schemes supply useful upper bounds for some n-widths that are difficult to calculate.

1. We consider first the space $\mathscr{A}(D_R)$ of Section 5 with the unit ball $\mathscr{B}_R$. Given r, $0 < r < R$, we define two new topologies on $\mathscr{A}(D_R)$, which are coarser than the initial one. One is that of $\mathscr{A}(D_r)$, with r replacing R; the other one is an $\mathscr{L}_\infty$-norm:

$$(9.2) \qquad \|f\|_{\mathscr{L}_\infty(D_r)} = \sup_{z \in D_r} |f(z)| \,.$$

In either topology the ball $\mathscr{B}_R$ is compact, hence the n-widths $d^{(n)}_{\mathscr{A}(D_r)}(\mathscr{B}_R)$ and $d^{(n)}_{\mathscr{L}_\infty(D_r)}(\mathscr{B}_R)$ are finite. Since $\mathscr{B}_R$ can be characterized by the quadratic inequality

$$(9.3) \qquad \mathscr{B}_R = \{f = \sum_{\nu=0}^{\infty} \alpha_\nu e_\nu : \Sigma |\alpha_\nu|^2 (R/r)^{2\nu} r^{2\nu} \le 1\} \,,$$

where r^ν is the $\mathscr{A}(D_r)$-norm of e_ν, it follows from Kolmogorov's theory of n-widths of "ellipsoids" (see [10], [9, Chapter 9.4]) that

$$(9.4) \qquad d^{(n)}_{\mathscr{A}(D_r)}(\mathscr{B}_R) = (r/R)^n$$

and that $M_*^{(n)} = \mathrm{sp}(e_0, e_1, \ldots, e_{n-1})$ is an optimal manifold for $\mathscr{B}_R$ in $\mathscr{A}(D_r)$. Since $\|f\|_{\mathscr{L}_\infty(D_r)} \ge \|f\|_{\mathscr{A}(D_r)}$ for any $f \in \mathscr{A}(D_R)$, (9.4) also gives a lower bound for the other n-width:

$$(9.5) \qquad d^{(n)}_{\mathscr{L}_\infty(D_r)}(\mathscr{B}_R) \ge (r/R)^n \,.$$

In (5.22) we found the error $E_z^{(n)}(\mathscr{B}_R)$ for recovery of $\mathscr{B}_R$ at z from n data values on the circle $\{|z| = r_0\}$. The maximum of $E_z^{(n)}(\mathscr{B}_R)$ for $\{|z| = r\}$ must be an upper bound for (9.5), and this is true for any r_0. Using the limit as $r_0 \to 0$ we obtain

$$(9.6) \qquad (r/R)^n \le d^{(n)}_{\mathscr{L}_\infty(D_r)}(\mathscr{B}_r) \le \frac{(r/R)^n}{[1 - (r/R)^2]^{3/4}} \,.$$

Clearly, the bounds of (9.6) are rather sharp for the $\mathscr{L}_\infty(D_r)$ n-width of $\mathscr{B}_R$. As far as I know the true value has not been found yet. If we denote the maximum of the recovery error $E_z^{(n)}(\mathscr{B}_R)$ for $\{|z| = r\}$ from data on $\{|z| = r_0\}$ by $E_r^{(n)}(\mathscr{B}_R)$ then we have by (5.24)

(9.7) $$\lim_{n \to \infty} \sqrt[n]{E_r^{(n)}(\mathcal{B}_R)} = r/R \quad \text{if} \quad r > r_o \ .$$

Thus, by (9.4)-(9.7), if $r > r_o$:

(9.8) $$\lim_{n \to \infty} \sqrt[n]{E_r^{(n)}(\mathcal{B}_R)} = \lim_{n \to \infty} \sqrt[n]{d^{(n)}_{\mathcal{L}_\infty(D_r)}(\mathcal{B}_R)} = \sqrt[n]{d^{(n)}_{\mathcal{A}(D_r)}(\mathcal{B}_R)}$$
$$= r/R \ .$$

This means that optimal recovery of the unit ball in $\mathcal{A}(D_R)$ from data on the set $\{r_o e^{2\pi ik/n}\}_{k=0,1,\ldots,n-1}$ is also an asymptotically optimal linear approximation if $r > r_o$. This is not so if $r < r_o$. Equation (6.10) shows that (9.8) also holds if $E_r^{(n)}(\mathcal{B}_R)$ refers to the error of recovery from Hermite data at the point 0.

2. We next consider the space $\mathcal{A}(A_R)$ of Section 7. Again we introduce two coarser topologies: that of $\mathcal{A}(A_r)$ for $1 \le r < R$ and that of $\mathcal{L}_\infty(A_r)$ with norm:

(9.9) $$\|f\|_{\mathcal{L}_\infty(A_r)} = \sup_{|z| \in A_r} |f(z)| .$$

As before, the ball $\mathcal{B}_R$ is compact in either topology. Proceeding as above one finds easily the $\mathcal{A}(A_r)$ - width:

(9.10) $$d^{(n)}_{\mathcal{A}(A_r)}(\mathcal{B}_R) = (r/R)^m \quad \text{for both} \quad n = 2m \quad \text{and} \quad n = 2m+1.$$

An optimal manifold for the case $n = 2m + 1$ is

(9.11) $$M_*^{2m+1} = sp(e_{-m},\ldots,e_o,\ldots,e_m)$$

$(e_k(z) = z^k)$. If $f = \sum \alpha_\nu e_\nu$ then

$$\|f\|^2_{\mathcal{A}(A_r)} = \sum_{\nu=-\infty}^{\infty} |\alpha_\nu|^2 r^{2|\nu|} < \sum_{\nu=-\infty}^{\infty} |\alpha_\nu|^2 (r^{2\nu} + r^{-2\nu})$$

$$= \frac{1}{2\pi r} \int_{|z|=r} |f(z)|^2 |dz| + \frac{1}{2\pi r} \int_{|z|=r^{-1}} |f(z)|^2 |dz|$$

$$\le \sup_{|z|=r} |f(z)|^2 + \sup_{|z|=r^{-1}} |f(z)|^2$$

$$\le 2 \sup_{r^{-1} \le |z| \le r} |f(z)|^2.$$

Therefore, $\|f\|_{\mathcal{L}_\infty(A_r)} \ge 2^{-\frac{1}{2}} \|f\|_{\mathcal{A}(A_r)}$ and (9.10) supplies a lower bound for the $\mathcal{L}_\infty(A_r)$ n-width:

$$d^{(n)}_{\mathscr{L}_\infty(A_r)}(\mathscr{B}_R) > 2^{-\frac{1}{2}}(r/R)^{[n/2]} \quad .$$

An upper bound is given by the upper bound (7.44) for the spline errors. Thus

$$(9.12) \quad 2^{-\frac{1}{2}}(r/R)^{[n/2]} < d^{(n)}_{\mathscr{L}_\infty(A_r)}(\mathscr{B}_R) < \left[8(1+\delta_n)\frac{1+(r/R)^2}{1-(r/R)^2}\right]^{\frac{1}{2}}(r/R)^{n/2}$$

For the asymptotic values we have, by (9.10)-(9.12) and (7.50):

$$(9.13) \quad \lim_{n\to\infty} \sqrt[n]{E_r^{(n)}(\mathscr{B}_R)} = \lim_{n\to\infty} \sqrt[n]{d^{(n)}_{\mathscr{L}_\infty(A_r)}(\mathscr{B}_R)}$$
$$= \sqrt[2[n/2]]{d^{(n)}_{\mathscr{A}(A_r)}(\mathscr{B}_R)} = (r/R)^{\frac{1}{2}} \quad .$$

Thus: <u>Optimal recovery of the unit ball in $\mathscr{A}(A_R)$ from data on the set $\{e^{2\pi ik/n}\}_{k=0,1,\ldots,n-1}$ is also an asymptotically optimal linear approximation</u>.

3. Little needs to be changed if $\mathscr{A}(A_R)$ is replaced by the space $\mathscr{A}(\tilde{A}_R)$ of Section 8. Using the above terminology we find in place of (9.10)

$$(9.14) \quad d^{(n)}_{\mathscr{A}(\tilde{A}_r)}(\mathscr{B}_R) = (r/R)^{n-1} \quad .$$

An optimal manifold is in this case

$$(9.15) \quad M_*^{(n)} = sp\{\tilde{e}_o, \tilde{e}_1, \ldots, \tilde{e}_{n-1}\}$$

$(\tilde{e}_k(z) = \frac{1}{2}(z^k + z^{-k}))$. (9.17) is replaced by

$$\lim_{n\to\infty} \sqrt[n]{E_r^{(2n-1)}(\mathscr{B}_R)} = \lim_{n\to\infty} \sqrt[n]{d^{(n)}_{\mathscr{L}_\infty(\tilde{A}_r)}(\mathscr{B}_R)}$$
$$= \sqrt[n-1]{d^{(n)}_{\mathscr{A}(\tilde{A}_r)}(\mathscr{B}_R)} = r/R \ , \ 1 \le r < R \ .$$

Consistent with previous notation $E_r^{(2n-1)}(\mathscr{B}_R)$ refers to the maximum on A_r of the error of recovery of $\mathscr{B}_R$ from data on the set $\{e^{2\pi ik/2n-1}\}_{k=0,1,\ldots,2n-2}$, of which there are only n independent ones.

For the space $\mathscr{A}(E_R)$ of functions analytic in the ellipse E_R the corresponding results are:

$$(9.17) \quad d^{(n)}_{\mathscr{A}(E_r)}(\mathscr{B}_R) = (r/R)^{n-1} \quad .$$

Here r and R refer to the sums of semiaxes in the confocal ellipses E_r and E_R, respectively. An optimal manifold is $M_*^{(n)} = sp(e_o, e_1, \ldots, e_{n-1})$. Also

$$(9.19) \qquad \lim_{n \to \infty} \sqrt[n]{E_r^{(2n-1)}(\mathscr{B}_R)} = \lim \sqrt[n]{d^{(n)}_{\mathscr{L}_\infty(E_r)}(\mathscr{B}_R)}$$

$$= \sqrt[n-1]{d^{(n)}_{\mathscr{A}(E_r)}(\mathscr{B}_R)} = r/R.$$

It should be observed that, whereas the optimal manifold of dimension n for $\mathscr{B}_R$ in $\mathscr{A}(E_r)$ is the space of polynomials of degree $\leq n - 1$, the n-dimensional spline space for optimal recovery is a space of rational functions of degree $(2n - 2, 2n - 1)$ (see (8.20)), with fixed denominators, and numerators that are spanned by 1 and fixed linear combinations of the Chebyshev polynomials T_j and T_{2n-1-j} $(j = 1, \ldots, n - 1)$.

10. STABILIZATION OF THE RECOVERY SCHEMES

It has been pointed out by numerical analysts (see, for example [11], [12]) that analytic continuation of functions only given by their values in some infinite compact set is not a well-posed problem. Small errors in the data can produce arbitrarily large errors of the continuation. If in the problem of Section 5, where $f \in \mathscr{B}_R$ is to be recovered in a disk D_r, the data $f(r_o e^{2\pi ik/n})$ are replaced by $f_\varepsilon(r_o e^{2\pi ik/n})$, where $f_\varepsilon(z) = f(z) + \varepsilon(z/r_o)^N$, with $\varepsilon > 0$ sufficiently small so that $f_\varepsilon \in \mathscr{B}_R$, then the change in the data is of absolute value ε, but the change in the continuation of the function at z, $|z| = r_1 > r_o$, is of absolute value $\varepsilon(r_1/r_o)^N$, which becomes arbitrarily large with N. Moreover, this is so no matter how large the number n of data is.

The sequence of spline values $\{S_n f(z)\}_{n=1,2,\ldots}$ represents a valid numerical method of analytic continuation from the data set to $z \in D_r$ provided it is true that

$$(10.1) \qquad \lim_{n \to \infty} \sup_{z \in D_r} |f(z) - S_n f(z)| = 0.$$

Indeed this convergence follows immediately from the error bound (5.24). However the proof requires that $f \in \mathscr{B}_R$ (more generally that $f \in M\mathscr{B}_R$ for some $M < \infty$) and consequently $\|S_n f\| \leq M$ for all n. In general, given discrete data do not allow this conclusion. Numerical data are necessarily of limited accuracy, if for no other reason but that they are given as finite decimals. One cannot be sure that the sequence of data vectors $\{F_n\}$ interpolate a function $f \in \mathscr{B}_R$. It is more reasonable to expect that they interpolate a bounded measurable function g that is uniformly close to the function f which is to be continued. But then the data must include some information that counteracts the instability of the continuation process. An explicit a priori bound for the function f (that is, $\|f\|_{\mathscr{A}} \le \mu$, μ a known number) is sufficient for this purpose and is often available. We show how, under these conditions, the splines can be modified so as to achieve stable analytic continuation (for a similar, but not identical procedure see [12]). We carry out the details only for the problem treated in Section 5. The function class to be recovered is the unit ball $\mathscr{B}_R$ in $\mathscr{A}(D_R)$, the data are given in the set $\{r_o\varepsilon_n^o, \ldots, r_o\varepsilon_n^{n-1}\}$ where $\varepsilon_n = e^{2\pi i/n}$, and the approximation is to be uniform over the disk D_r with $r_o < r < R$.

Specifically, we assume: Given are positive numbers ε, μ and a function $g \in \mathscr{L}_\infty(|z| = r_o)$ such that

$$\|f\|_{\mathscr{A}(D_r)} \le \frac{\mu}{2}\ . \tag{10.2}$$

$$\sup_{|z| = r_o} |f(z) - g(z)| \le \frac{\varepsilon}{2}\ . \tag{10.3}$$

The function f is to be recovered in the disk D_r, $r_o < r < R$, from the data vectors $G_n = \{g(r_o\varepsilon_n^o), \ldots, g(r_o\varepsilon_n^{n-1})\}$ $(n = 1,2,\ldots)$.

We define

$$\begin{aligned} s_{n,\varepsilon}(z) &= \frac{1 - (r_o/R)^{2n}}{1 - (r_o z/R^2)^n} \sum_{j=0}^{n-1} \hat{g}_j (z/r_o)^j \\ \hat{g}_j &= \frac{(1/n)\sum_{k=0}^{n-1} g(r_o\varepsilon_n^k)\varepsilon_n^{-jk}}{1 + (\varepsilon/2\mu)(R/r_o)^j}\ . \end{aligned} \tag{10.4}$$

It is seen that the denominator in $\hat{g}_j$ distinguishes $s_{n,\varepsilon}$ from $S_n g$ (see 5.17). We prove

<u>Theorem 10.1</u>.

$$\lim_{\varepsilon\to 0} \sup_{z\in D_r} |s_{n,\varepsilon}(z) - S_n f(z)| = 0$$

uniformly in n.

The proof is based on a few lemmas. We write for the $\mathscr{A}(D_\rho)$ norm more simply

$$\|f\|_\rho = \{\frac{1}{2\pi\rho} \int_{|z| = \rho} |f(z)|^2 |dz|\}^{\frac{1}{2}} \tag{10.5}$$

and write $\tilde{\varepsilon}$ for ε/μ.

Lemma 10.1. If $(r/R)^{2n} \leq 1 - (1 + \tilde{\varepsilon})^{-2}$ and (10.2), (10.3) hold then

(10.6) $$\| s_{n,\varepsilon} - S_n f \|_{r_o} < \varepsilon$$

(10.7) $$\| s_{n,\varepsilon} - S_n f \|_R < M = \mu + \varepsilon/2.$$

Proof. By the definition of $s_{n,\varepsilon}$ and $S_n f$, and writing z_k for $r_o \varepsilon_n^k$:

$$\| s_{n,\varepsilon} - S_n f \|_{r_o} \leq \frac{1 - (r_o/R)^{2n}}{1 - (r_o/R)^{2n}} \| \sum_{j=0}^{n-1} (\hat{g}_j - \hat{f}_j) e_j / r_o^j \|_{r_o}$$

(10.8)
$$= \{ \sum_{j=0}^{n-1} |\hat{g}_j - \hat{f}_j|^2 \}^{\frac{1}{2}}$$

$$= (1/n) \{ \sum_{j=0}^{n-1} (1 + \tilde{\varepsilon}(R/r_o)^j)^{-2}$$

$$\cdot | \sum_{k=0}^{n-1} [g(z_k) - f(z_k)(1 + \tilde{\varepsilon}(R/r_o)^j] \varepsilon_n^{-jk} |^2 \}^{\frac{1}{2}}$$

$$\leq (1/n) \sum_{j=0}^{n-1} (1 + \tilde{\varepsilon}(R/r_o)^j)^{-2} | \sum_{k=0}^{n-1} [g(z_k) - f(z_k)] \varepsilon_n^{-jk} |^2 \}^{\frac{1}{2}}$$

$$+ (1/n) \sum_{j=0}^{n-1} (1 + \tilde{\varepsilon}(R/r_o)^j)^{-2} \tilde{\varepsilon}(R/r_o)^{2j} | \sum_{k=0}^{n-1} f(z_k) \varepsilon_n^{-jk} |^2 \}^{\frac{1}{2}}$$

$$= I + II.$$

By (10.3):

(10.9)
$$I < (1/n) \{ \sum_{j=0}^{n-1} | \sum_{k=0}^{n-1} [g(z_k) - f(z_k)] \varepsilon_n^{-jk} |^2 \}^{\frac{1}{2}}$$

$$= (1/n) \{ \sum_{k=0}^{n-1} |g(z_k) - f(z_k)|^2 \}^{\frac{1}{2}} \leq \varepsilon/2 .$$

If $f(z) = \sum_{\nu=0}^{\infty} \alpha_\nu z^\nu$ and $|z| = r_o$ then $(1/n) \sum_{k=0}^{n-1} f(z_k) \varepsilon_n^{-jk} = \Sigma_{\nu \equiv j(n)} \alpha_\nu r_o^\nu$, hence

$$|(1/n) \Sigma f(z_k) \varepsilon_n^{-jk}|^2 \leq \sum_{\nu \equiv j(n)} |\alpha_\nu R^\nu|^2 \sum_{\nu \equiv j(n)} (r_o/R)^{2\nu}$$

$$= (r_o/R)^{2j} [1 - (r_o/R)^{2n}]^{-1} \sum_{\nu \equiv j(n)} |\alpha_\nu R^\nu|^2 .$$

Thus, if the condition of the Lemma is satisfied

$$\text{(10.10)} \qquad II < \frac{\tilde{\varepsilon}}{1+\tilde{\varepsilon}} [1 - (r_o/R)^{2n}]^{-\frac{1}{2}} [\sum_{\nu=0}^{\infty} |\alpha_\nu R^\nu|^2]^{\frac{1}{2}}$$

$$\leq \tilde{\varepsilon} \| f \|_{\mathcal{A}(D_R)} \leq \frac{\varepsilon}{2} .$$

By (10.8) - (10.10), (10.6) is proved.

To prove (10.7) we proceed as above and obtain

$$\text{(10.11)} \qquad \| s_{n,\varepsilon} - S_n f \| \leq III + IV$$

where III and IV are the terms corresponding to I and II above. Using the above inequalities and the conditions of the Lemma, we have

$$\text{(10.12)} \qquad III = (1/n)\{\sum_{j=0}^{n-1} (R/r_o)^{2j}(1 + \tilde{\varepsilon}(R/r_o)^j)^{-2}$$

$$\sum_{k=0}^{n-1} [g(z_k) - f(z_k)]\varepsilon_n^{-jk}|^2\}^{\frac{1}{2}}$$

$$< \varepsilon/2\tilde{\varepsilon} = \mu/2.$$

$$IV \leq \{\sum_{j=0}^{n-1}(1 + \tilde{\varepsilon}(R/r_o)^j)^{-2}\tilde{\varepsilon}^2(R/r_o)^{2j} \sum_{\nu \equiv j(n)} |\alpha_\nu R^\nu|^2\}^{\frac{1}{2}}$$

$$\times [1 - (r_o/R)^{2n}]^{-\frac{1}{2}}$$

$$< (1 + \tilde{\varepsilon})\{\sum_{\nu=0}^{\infty} |\alpha_\nu R^\nu|^2\}^{\frac{1}{2}} = (1 + \tilde{\varepsilon})\| f \|_{\mathcal{A}(D_R)} \leq (\mu + \varepsilon)/2.$$

By (10.11) - (10.13) we obtain (10.7), so the Lemma is proved.

From (10.6) and (10.7) we derive now that $\| s_n - S_n f \|_r$ is small with ε. For this we use an L_2-version of Hadamard's 3-Circle Theorem.

<u>Lemma 10.2</u>. Suppose $0 < r_1 < r_2 < r_3$. If f is analytic in the closed annulus $\{z: r_1 \leq |z| \leq r_3\}$ then

$$\text{(10.14)} \qquad \| f \|_{r_2}^{\log(r_3/r_1)} \leq \| f \|_{r_1}^{\log(r_3/r_2)} \| f \|_{r_3}^{\log(r_2/r_1)} .$$

<u>Proof</u>. If $f(z) = \sum_{\nu=-\infty}^{\infty} \alpha_\nu z^\nu$ then $\| f \|_r^2 = \Sigma\, |\alpha_\nu|^2 r^{2\nu}$. The function ϕ, defined by $\phi(z) = \sum_{\nu=-\infty}^{\infty} |\alpha_\nu|^2 z^{2\nu}$ is also analytic in the above annulus, hence by Hadamard's Theorem

$$M_2^{2\log(r_3/r_1)} \leq M_1^{2\log(r_3/r_2)} M_3^{2\log(r_2/r_1)}$$

where $M_k^2 = \sup_{|z| \le r_k} |\phi(z)|$. But $M_k = \|f\|_{r_k}$ $(k = 1,2,3)$, hence the Lemma is proved.

The above proof shows that the Lemma remains true if f is analytic in the open annulus and the limits of $\int_{|z|=r} |f|^2 |dz|$ as $r \to r_1 + 0$, $r \to r_3 - 0$ exist.

If Lemma (10.2) is applied to (10.6) and (10.7), one obtains

$$(10.15) \qquad \|s_{n,\varepsilon} - S_n f\|_r < \varepsilon^{\frac{\log(R/r)}{\log(R/r_o)}} M^{\frac{\log(r/r_o)}{\log(R/r_o)}} \quad \text{if } r_o \le r < R$$

or, if we set

$$r/r_o = (R/r_o)^\alpha ,$$

$$(10.16) \qquad \|s_{n,\varepsilon} - S_n f\|_r < \varepsilon^{1-\alpha} M^\alpha \quad .$$

This inequality shows that the $\mathscr{A}(D_r)$-norm of $s_{n,\varepsilon} - S_n f$ becomes small with ε. We now show that this is also true for the sup norm over D_r. For this purpose we prove

Lemma 10.3. Suppose $0 < r_o < r < r_1$, and f is analytic in the (open) disk D_{r_1} with $\|f\|_{r_1} < \infty$. Then for every τ, $0 < \tau < 1$

$$(10.17) \qquad \sup_{z \in D_r} |f(z)| \le \left(\frac{r_1}{r_1 - r} + \frac{\tau}{1-\tau} \frac{r}{r_1 - r} \right) \|f\|_{r_o}^{\frac{\log(r_1/r_\tau)}{\log(r_1/r_o)}} \|f\|_{r_1}^{\frac{\log(r_\tau/r_o)}{\log(r_1/r_o)}}$$

where $r_\tau = \tau r + (1 - \tau) r_1$.

Proof. For $|z| = r$ we have

$$|f(z)| = \left| \frac{1}{2\pi i} \int_{|\zeta|=r_\tau} \frac{f(\zeta)}{\zeta - z} \right|$$

$$\le \left\{ \frac{1}{2\pi r} \int_{|\zeta|=r_\tau} |f(\zeta)|^2 |d\zeta| \right\}^{1/2} \left\{ \frac{r_\tau}{2\pi} \int_{|\zeta|=r_\tau} |\zeta - z|^{-2} |d\zeta| \right\}^{1/2}$$

$$\le \|f\|_{r_\tau} \frac{r_\tau}{r_\tau - r} \quad .$$

Using Lemma 10.2 with $r_1 = r_o$, $r_2 = r_\tau$, $r_3 = r_1$, we get (10.17).

We return to the proof of Theorem 10.1. If Lemma 10.3 is applied to the function $s_{n,\varepsilon} - S_n f$, with $r_1 = R$, and (10.6) and (10.7) are used, one finds

$$\sup_{z \in D_r} |s_{n,\varepsilon}(z) - S_n f(z)| \tag{10.18}$$
$$\leq \left(\frac{R}{R-r} + \frac{\tau}{1-\tau}\frac{r}{R-r}\right)\varepsilon^{\frac{\log(R/r_\tau)}{\log(R/r_o)}} M^{\frac{\log(r_\tau/r_o)}{\log(R/R_o)}},$$

$$r_\tau = \tau r + (1-\tau)R.$$

If we set

$$\frac{\tau r + (1-\tau)R}{r_o} = \left(\frac{R}{r_o}\right)^{\beta_\tau},$$

then (10.18) takes the simpler form

$$\sup_{z \in D_r} |s_{n,\varepsilon}(z) - S_n f(z)| \leq \left(\frac{R}{R-r} + \frac{\tau}{1-\tau}\frac{r}{R-r}\right)^{1-\beta_\tau} M^{\beta_\tau}. \tag{10.19}$$

Each τ between 0 and 1 provides an upper bound for the difference on the left.

By taking the limit of (10.19) for $\varepsilon \to 0$, we seem to obtain the conclusion of Theorem 10.1. However, this argument ignores the condition (see Lemma 10.1)

$$\rho_n := [1 - (r_o/R)^{2n}]^{-\frac{1}{2}} \leq 1 + \tilde{\varepsilon} \tag{10.20}$$

By taking n sufficiently large ρ_n can be made smaller than $1 + \varepsilon/\mu$ for a given $\varepsilon > 0$, but not for all $\varepsilon > 0$. Yet this difficulty is easily overcome. Condition (10.20) was used only in (10.10) and (10.12). Without it (10.10) and (10.12) give

$$II < \frac{\varepsilon}{2}\frac{\rho_n}{1+\tilde{\varepsilon}}, \quad IV < \frac{\mu}{2}\rho_n$$

and Lemma 10.1 remains true with ε, M replaced by

$$\varepsilon_1 = \frac{\varepsilon}{2}\left(1 + \frac{\rho_n}{1+\tilde{\varepsilon}}\right), \quad M = \frac{\mu}{2}(1 + \rho_n). \tag{10.21}$$

Clearly $\lim_{n\to\infty} \rho_n = 1$. If N is so chosen that, for example, $\rho_n < 1.1$ for $n \geq N$ then $\varepsilon_1 < 1.05\varepsilon$ and $M_1 < 1.05M$ and the inequality (10.20) holds for all $n \geq N$ if ε, M are replaced by the slightly larger ε_1, M_1. This completes the proof of Theorem 10.1.

REFERENCES

[1] Golomb, M. and H. F. Weinberger. Optimal Approximation and Error Bounds. Symposium on Numerical Approximation, R. E. Langer ed., Madison 1959, pp. 117-190.

[2] deBoor, C. R., and R. E. Lynch. On splines and their minimum properties, J. Math Mech., 15(1966), 953-969.

[3] Miccheli, C. A., T. J. Rivlin and S. Winograd. The optimal recovery of smooth functions, Numer. Math. 26(1976), 19- .

[4] Micchelli, C. A. and Allan Pinkus. On a best estimator for the class $\mathcal{M}^r$ using only function values. MRC Technical Summary Report, February 1976.

[5] Knauff, W. and R. Kress. Optimale Approximation linearer Funktionale auf periodischen Funktionen. Numer. Math. 22(1974).

[6] Knauff, W. and R. Kress, Optimale Approximation mit Nebenbedingungen an lineare Funktionale auf periodischen Funktionen. Numer. Math.

[7] Golomb, M., Approximation by periodic spline interpolants on uniform meshes. J. Approximation Theory 1(1968), 26-65.

[8] Tikhomirov, V. M. Widths of sets in function spaces and the theory of best approximation (Russian). Uspekhi Math. Nauk, t.XV, 3(93) (1960), 81-120.

[9] Lorentz, G. G., Approximation of Functions. New York, Chicago, San Francisco, Toronto, London, 1966.

[10] Kolmogorov, A., Ueber die beste Annaeherung von Funktionen einer gegebenen Funktionenklasse. Annals of Math. 37 (1936), 107-111.

[11] Douglas, J., A numerical method for analytic continuation. Boundary Problems in Differential Equations, Madison, 1960, pp. 179-186.

[12] Miller, Keith. Least squares method for ill-posed problems with a prescribed bound. SIAM J. Math. Anal. 1(1970), 52-74.

[13] Wilf, H. S., Exactness conditions in numerical quadrature. Numer. Math. 6 (1964), 315-319.

[14] Eckhardt, U., Einige Eigenschaften Wilfscher Quadraturformeln. Numer. Math. 12(1968), 1-7.

[15] Engels, H., Über allgemeine Gauss'sche Quadraturformeln. Computing 10(1972), 83-95.

[16] __________, Eine Familie interpolatorischer Quadraturformeln mit ableitungsfreien Fehlerschranken. Author's preprint.

[17] Engels, H. and Eckhardt, U., The determination of nodes and weights in Wilf quadrature formulas. To appear in Abhandlungen Math. Sem. Un. Hamburg.

OPTIMAL DEGREE OF APPROXIMATION BY SPLINES

Karl Scherer

University of Bonn, Institute of Applied Mathematics

Wegelerstrasse 6, 5300 Bonn

1. THE PROBLEM

A description of spline - or piecewise polynomial functions - involves three characteristics: the nature of their pieces, their location and the smoothness at the connections. Here the following classes of spline functions are considered: Given a partition $\Pi = \{x_i\}_{i=o}^N$ of knots x_i of the interval $[a,b]$

(1) $a = x_o < x_1 < \dots < x_N = b$,

and integers n,k with $-1 \le k \le n-2$ we set

(2) $Sp(\Pi,n,k) = \{s(x): \; s(x)|_{(x_i,x_{i+1})} \in P_n, \; s^{(k)}(x) \text{ continuous}\}$,

where P_n denotes the space of all polynomials of degree $\le n-1$ (In case $k = -1$ a spline function $s(x)$ may be discontinuous at the knots its value there being defined as the limit from the right.).

Correspondingly we define the best approximation of $L_p(a,b)$ integrable functions f

(3) $E_p(f;\Pi,n,k) = \inf\{||f-s||_p : s \in Sp(\Pi,n,k)\}, \; 1 \le p \le \infty$

In case $p = \infty$ we consider the class of continuous functions on $[a,b]$. The norm $||\cdot||_p$ denotes the usual L_p-metric.

Now, given a sequence $\{\Pi_l\}_{l=o}^{\infty}$ of partitions, we say the degree of (best) approximation of $f \in L_p(a,b)$ is $\alpha(>o)$ if

(4) $E_p(f;\Pi_l,n,k) \leq C\,\overline{\Pi}_l^{\alpha}$

for some constant $C>o$ where for a partition Π we have set

(5) $\overline{\Pi} = \max_i (x_{i+1}-x_i)$, $\underline{\Pi} = \min_i (x_{i+1}-x_i)$.

There are well known estimates in spline theory of the form

(6) $E_p(f;\Pi,n,n-2) \leq C\,\overline{\Pi}^j\,||f^{(j)}||_p$,

$C = C(n,p)$ being independent of Π and f, if f belongs to the Sobolev space $W_p^j(a,b)$ of functions with absolutely continuous $(j-1)$-th derivative and j-th derivative in $L_p(a,b)$, $j=1,\ldots,n$. These estimates show that the degree of (best) approximation for the classes $W_p^j(a,b)$ is j. They may indeed be realized by a bounded linear projection from $L_p(a,b)$ to $Sp(\Pi,n,n-2)$ (cf. [2] for $p = \infty$ and [7] for $1 \leq p < \infty$).

One may now ask after the sharpness or optimality of such estimates. More precisely we want to discuss the following questions

I. For a given $\alpha>o$ what is the optimal class K of functions which have this degree of best approximation?

II. Is there an optimal degree α of best approximation for all classes of functions?

The best way to solve these questions is to try to prove some sort of inverse estimates to (4), (6), i.e. estimates from above by the best approximation (3). This will be shown in the next section where the known inverse theorems are presented. We shall see that the answer depends on the nature of the sequence $\{\Pi_l\}_{l=o}^{\infty}$ of partitions as well as on the parameters n and k. But before doing so we must refine the estimate (6) somewhat. To this end we introduce the n-th modules of continuity in the L_p-metric $(o<t<\infty)$

$$w_n(f;t)_p = \begin{cases} \sup\limits_{o<|h|\leq t} \{ \int\limits_{\substack{x,x+nh \\ \epsilon(a,b)}} |\Delta_h^n f(x)|^p dx\}^{1/p} & , 1\leq p\leq\infty \\ \sup\limits_{o<|h|\leq t} \sup\limits_{\substack{x,x+nh \\ \epsilon(a,b)}} |\Delta_h^r f(x)| & , p = \infty \end{cases}$$

where $\Delta_h^n f(x)$ is the n-th forward difference with increment h.

Then one can prove (cf. [7])

THEOREM 1: There is a constant $C>o$ independent of f and Π such that

$$(7) \quad E_p(f;\Pi,n,n-2) \leq Cw_n(f;\overline{\Pi})_p \qquad (1\leq p\leq\infty).$$

This result generalizes and refines estimate (6) as one can see by the introduction of the following Lipschitz-classes of functions

$$(8) \quad Lip(\alpha,n;p) = \{f\epsilon L_p(a,b): \; w_n(f;t)_p = O(t^\alpha), t\to o+\}.$$

Note that in view of the well known fact

$$(9) \quad t^{-n} w_n(f;t)\to o, t\to o+ \Longrightarrow f\epsilon P_n$$

only the Lipschitz-classes $Lip(\alpha,n;p)$ with $o<\alpha\leq n$ are meaningful. For these we know by Theorem 1

$$(10) \quad f\epsilon Lip(\alpha,n;p) \Longrightarrow E_p(f;\Pi_1,n,k) \leq C\overline{\Pi}_1^\alpha$$

where $o<\alpha\leq n$, $-1\leq k\leq n-2$.

But in view of

$$(11) \quad w_n(f;t)_p \leq 2^{n-j}\, t^j ||f^{(j)}||_p \quad , f\epsilon W_p^j(a,b)$$

for $j = 1,\dots,n$ the class $Lip(j,n;p)$ is larger than $W_p^j(a,b)$, hence (10) is stronger than (6).

It will turn out below that within certain ranges of α assertion (10) is invertible so that the classes $Lip(\alpha,n;p)$ give the answer to problem I.

2. INVERSE THEOREMS

Basic for the following inverse theorems is the following observation: Suppose x and $x+nh$ lie in one segment (x_i, x_{i+1}) of a partition Π of $[a,b]$, then for any $s \in Sp(\Pi,n,-1)$

$$(11)\qquad |\Delta_h^n f(x)| = |\Delta_h^n [f(x) - s(x)]| \le 2^n ||f-s||_\infty$$

$$\le 2^n E_\infty(f;\Pi,n,-1)$$

since then s takes the values of <u>one</u> polynomial $\in P_n$ for $x+rh$, $r=o,\dots,n$. For $nh/\underline{\Pi}$ small enough this almost would give the inverse of (7) (for $p=\infty$) if the argument would not fail in a neighbourhood of the knots. This motivates the following (cf. [8])

DEFINITION: A sequence $\{\Pi_l\}_{l=1}^\infty$ of partitions is called <u>mixed</u> if there exists a constant $d>o$ such that

$$(12)\qquad \sup_{l\ge k} \operatorname{dist}(x,\Pi_l) \ge d\overline{\Pi}_k$$

uniformly in $x \in [a,b]$ and $k, k\to\infty$.

Thus "mixed" guarantees that for each $k, k\to\infty$, and each $x \in [a,b]$ one can find a partition Π_l, $l=l(x)\ge k$ (low l would be of no use in (11)) such that x, $x+nh$ lie in <u>one</u> segment of Π_l, provided h is small enough, namely $nh \le d\overline{\Pi}_k$. Hence (11) immediately leads to

$$w_n(f;\frac{d\overline{\Pi}_k}{n})_\infty \le 2^n \sup_{l\ge k} E_\infty(f;\Pi_l,n-1),$$

and after some elementary manipulation to the well known result (e.g. [8])

THEOREM 2: Assume the sequence $\{\Pi_l\}_{l=1}^\infty$ is mixed and $\overline{\Pi}_{l+1} \le \overline{\Pi}_l$. Then there holds

$$(13)\qquad w_n(f;t)_\infty \le C \sup_{\overline{\Pi}_l \le t} E_\infty(f;\Pi_l,n,-1)$$

for some constant C independent of f (= continuous in $[a,b]$) and t.

It is not hard to show that the sequence of equidistant partitions satisfies the conditions of this theorem which allows now an answer to the initial questions I and II:

COROLLARY 1: Under the assumption of Theorem 2 one has for $-1\leq j\leq n-2$, $o<\alpha\leq n$

a) $E_\infty(f;\Pi_k,n,j) \leq C\overline{\Pi}_k^{\alpha} \iff f\in Lip(\alpha,n;\infty)$

b) $E_\infty(f;\Pi_k,n,j)\overline{\Pi}_k^{-n} \to o, k\to\infty \implies f\in P_n$

Here we have only used definition (8) of the Lipschitz-spaces and property (9).

In the case $1\leq p<\infty$ the proof of a theorem similar to Theorem 2 is not so simple. I take the opportunity here to present a slight generalization of the result in [13]

THEOREM 3: Let the sequence $\{\Pi_l\}_{l=1}^{\infty}$ of partitions, $\Pi_l = \{x_i^{(l)}\}_{i=o}^{N_l}$, have uniformly bounded mesh ratios $\overline{\Pi}_l/\underline{\Pi}_l$ and suppose it satisfies the following Σ-mixing condition:

$$(14)\qquad \sum_{l\geq k}\alpha_l \ \mathrm{dist}(x_i^{(k)},\Pi_l)^{np+1} \geq C\overline{\Pi}_k^{\,np+1}\ ,\quad \sum_{l\geq k}\alpha_l <\infty$$

uniformly in i and k, $k\to\infty$. Then for $f\in L_p(a,b)$, $1\leq p<\infty$, there holds

$$w_n(f;\overline{\Pi}_k)_p \leq C \sup_{l\geq k} E_p(f;\Pi_l,n,-1)$$

with C being independent of f and k.

For the proof one needs the following (concerning a proof see [9])

LEMMA: For any $s\in Sp(\Pi,n,-1)$ and $\varepsilon>o$ there holds

$$(15)\qquad w_n(s;\varepsilon)_p \leq C(1+\varepsilon/\underline{\Pi})^{1-1/p}\varepsilon^{1/p} \max_{o\leq j\leq n-1} \{\sum_i|\varepsilon^j[s]_i^j|^p\}^{1/p}$$

for some C only depending on p and n, where $[s]_i^j = s^{(j)}(x_i+) - s^{(j)}(x_i-)$ denotes the jump of the j-th derivative of s at the knot x_i of the partition Π.

Proof of Theorem 3: By familiar properties of the moduli of continuity one has

$$(16)\quad w_n(f;\overline{\Pi}_k)_p \le w_n(f-s_k;\overline{\Pi}_k)_p + w_n(s_k;\overline{\Pi}_k)_p$$

$$\le 2^n\, E_p(f;\Pi_k,n-1) + (1+\overline{\Pi}_k/\underline{\Pi}_k)^n\, w_n(s_k;\overline{\Pi}_k)_p.$$

By (15) it follows

$$w_n(s_k;\overline{\Pi}_k)_p \le C \max_{o\le k\le n-1} \{\sum_i \overline{\Pi}_k^{jp+1}\, |[s_k-s_l]_i^j|^p\}^{1/p}$$

for any $s_l \epsilon Sp(\Pi_l,n,-1)$ and $dist(x_i^{(k)},\Pi_l) = \phi_i(k,l) > o$. Applying Markov's inequality on the intervals $(x_i^{(k)} \pm \phi_i(k,l),\ x_i^{(k)})$ to $|s_k^{(j)}(x_i^{(k)}\pm) - s_l^{(j)}(x_i^{(k)})|$ and denoting by $||\ ||_{p,i}$ the Lp-norm with respect to the interval $(x_i^{(k)} - \phi_i(k,l), x_i^{(k)} + \phi_i(k,l))$ gives

$$w_n(s_k,\overline{\Pi}_k)_p \le C \max_{o\le j\le n-1} \{\sum_i \overline{\Pi}_k^{jp+1}\, \phi_i(k,l)^{-jp-1}\, ||s_k-s_l||_{p,i}^p\}^{1/p}.$$

and hence by (14)

$$w_n(s_k,\overline{\Pi}_k)_p \le C\, \{\sum_i \sum_{l\ge k} \alpha_l\, ||s_k-s_l||_{p,i}^p\}^{1/p}$$

$$\le C\{\sum_{l\ge k} \alpha_l\, ||s_k-s_l||_p^p\}^{1/p} \le C \sup_{l\ge k} ||s_l-f||_p$$

which, together with (16), establishes Theorem 3 (for simplicity each constant is denoted by C).

It follows that for sequences of partitions with uniformly bounded mesh ratio and the mixing property (14) we can answer questions I and II for the L_p-metric, $1\le p<\infty$, as in Corollary 1. Again the sequence of equidistant partitions is an example for which this applies. In fact, by a lemma in [6] it can be shown that (14) holds with

$$\alpha_l = \begin{cases} 1, & l=k \\ 1/l, & k+1\le l\le 2k \\ 0, & \text{otherwise} \end{cases}$$

However, there are obviously sequences of partitions, which cannot be mixed in any sense at all, namely when $\Pi_{l+1} \subset \Pi_l$, for $l = 1,2,\ldots$. In this case one can prove

THEOREM 4: Let $\{\Pi_l\}_{l=1}^{\infty}$ be a sequence of nested partitions, i.e. $\Pi_{l+1} \subset \Pi_l$ for $l = 1,2,\ldots$ and $\Pi_o = \{a,b\}$. For each $f \in L_p(a,b), 1 \le p \le \infty$, and $-1 \le j \le n-2, n>o$, we have

$$(17) \quad w_n(f;\underline{\Pi}_k)_p \le C\Pi_k^{j+1+1/p} \sum_{l=1}^{k} \underline{\Pi}_l^{-j-1-1/p}\{E_p(f;\Pi_l,n,j)+E_p(f;\Pi_{l-1},n,j)\}$$

where C is a constant only depending on n and p.

Sketch of the proof: In view of $Sp(\Pi_l,n,k) \subset Sp(\Pi_{l+1},nk)$ we apply the classical Bernstein-argument for inverse approximation theorems. We choose $s_l \in Sp(\Pi_l,n,k)$ such that $||f-s_l||_p = E_p(f;\Pi_l,n,j)$, write $f = f-s_k + \sum_{l=1}^{k}(s_l-s_{l-1}) + s_o$ and obtain

$$(18) \quad w_n(f;\underline{\Pi}_k)_p \le C\{E_p(f;\Pi_k,n,j) + \sum_{l=1}^{k} \overline{\Pi}_k^{j+1+1/p}||s_l^{(j+1)}||_\infty\}$$

Then we apply a Markov-type inequality to $s_l^{(j+1)} - s_{l-1}^{(j+1)}$ (on a segment of Π_l) in order to pass over to the difference s_l-s_{l-1} and by the triangle inequality to (17).

As a consequence we obtain an answer for question I.

COROLLARY 2: Under the assumptions of Theorem 4 let $\{\Pi_l\}$ have a uniformly, bounded mesh ratio $\overline{\Pi}_l/\underline{\Pi}_l$. Then one has for $f \in Lp(a,b), 1 \le p \le \infty$, and $o<\alpha < j+1+1/p$:

$$E_p(f;\Pi_k,n,j) \le C\Pi_k^{-\alpha} \iff f \in Lip(\alpha,n;p)$$

Thus the answer is the same as in Corollary 1 but with the restriction $o<\alpha<j+1+1/p$. Without this restriction the corollary becomes false as is demonstrated by examples in [9]. Concerning question II the answer is negative since, according to the theorem of Bernstein, in view of $Sp(\Pi_l,n,k) \subset Sp(\Pi_{l+1},n,k)$ there exist non-trivial functions with an arbitrarily high degree of approximation.

What can we say about arbitrary sequences of partitions? In [9] it is shown that the answer is the same as in the case of sequences with nested partitions. This follows from (cf. [9]).

THEOREM 5: Let $\{\Pi_l\}_{l=1}^{\infty}$ be a sequence of partitions with $\underline{\Pi}_{l+1} \le \underline{\Pi}_l$ and $\Pi_o = \{a,b\}$.
Then the statement of Theorem 4 remains true.

The proof involves a combination of the arguments in Theorem 3 and 4. Instead of (18) one uses

$$w_n(f;\underline{\Pi}_k)_p \le C\{E_p(f;\Pi_k,n,j) + w_n(s_k;\underline{\Pi}_k)_p\},$$

the lemma from above and successive application for $l = k,\dots,1$ of

$$(19) \qquad \max_{j+1\le r\le n-1} \{\sum_i |\underline{\Pi}_l^{r-1-j}|[s_l]_i^r|^p\}^{1/p} \le$$

$$\le C\underline{\Pi}_l^{-j-1-1/p}\{E_p(f;\Pi_l,n,j) + E_p(f;\Pi_{l-1},n,j)\} +$$

$$+ \max_{j+1\le r\le n-1} \{\sum_i |\underline{\Pi}_l^{r-1-j}[s_{l-1}]_i^j|^p\}^{1/p}$$

Besides other technical details the key step of the proof in the simplest case $j = n-2$ of (19) is the observation that

$$(20) \qquad [s_l]_i^{n-1} = T_l^{(n-1)}(b_i+) - T_l^{(n-1)}(a_i-) + [s_{l-1}]_{j_i}^{n-1},$$

where $T_l = s_l - s_{l-1}$ and the a_i, b_i are right end or left end points, respectively, of open intervals of length $\ge\underline{\Pi}_l/4$ which are <u>free</u> of knots of Π_l and Π_{l-1}. This allows then an appropriate application of a Markov type inequality to the terms of (20). Concerning further details the reader is referred to [9].

Theorem 5 does not give an answer to question II after a possible optimal (or saturation) order of approximation. The above discussion showed that this may depend from whether the sequence of partitions is mixed or nested. Nevertheless there is the following result which shows that at least for smooth functions there is a limit in the degree of approximation.

THEOREM 6: Let the $(n-1)$-th derivation of f be absolutely continuous on $[a,b]$ and its n-th derivation be in $L_p(a,b), 1\le p<\infty$

or continuous if $p = \infty$. Then $E_p(f;\Pi_l,n,-1)\underline{\Pi}_l^{-n} \to o, l \to \infty$ for a sequence $\{\Pi_l\}_{l=1}^{\infty}$ of partitions with $\lim_{l\to\infty} \overline{\Pi}_l = o$ implies that f is a polynomial of degree $n-1$.

Proof: I take the occasion to give a corrected version of the proof in [13] restricted to the case of polynomial splines. Inequality (15) and the same argumentation as in (16) give for any $o<\varepsilon<\underline{\Pi}_k$

$$(21) \quad w_n(f;\underline{\Pi}_k)_p \leq 2^n E_p(f;\Pi_k,n,-1) +$$

$$+ C(1+\underline{\Pi}_k/\varepsilon)^n \varepsilon^{1/p} \max_{o\leq j\leq n-1} \{\sum_i |\varepsilon^j [s_k]_i^j|^p\}^{1/p}$$

Now let $u_i(x)$ be the Taylor-polynomial of f at $x_i^{(k)} - \varepsilon$ of degree $n-1$ or degree n for $1\leq p<\infty$ or $p = \infty$, respectively. Then denoting by $||\cdot||_{p,i}$ the L_p-norm on $(x_i-\varepsilon, x_i+\varepsilon)$ we have

$$||f - u_i||_{p,i} \leq \begin{cases} \varepsilon^n \,||f^{(n)}||_{p,i} & ,1\leq p<\infty \\ \varepsilon^n \Phi(\varepsilon) & ,p=\infty \end{cases}$$

for some function $\Phi(x)$ with $\lim_{x\to o+} \Phi(x) = 0$.

Again a Markov-type inequality gives

$$(22) \quad \max_{o\leq j\leq n-1} \{\sum_i \varepsilon^{jp+1} |[s_k]_i^j|^p\}^{1/p} \leq C \{\sum_i ||s_k-u_i||_{p,i}^p\}^{1/p}$$

$$\leq C[E_p(f;\Pi_k,n,-1) + \varepsilon^n\{\sum_i ||f^{(n)}||_{p,i}^p\}^{1/p}]$$

for $1\leq p<\infty$, and for $p = \infty$

$$(23) \quad \max_{o\leq j\leq n-1} \max_i \varepsilon^j |[s_k]_i^j| \leq C \max_i ||s_k-u_i||_{\infty,i}$$

$$\leq C\,[E_\infty(f;\Pi_k,n,-1) + \varepsilon^n \Phi(\varepsilon)].$$

Inserting (22), (23) into (21) yields

$$(24)\quad w_n(f;\underline{\Pi}_k)_p \leq C \begin{cases} (1+\underline{\Pi}_k/\varepsilon)^n E_p(f;\Pi_k,n-1)+\underline{\Pi}_k^n\{\sum_i \|f^{(n)}\|_{p,i}^p\}^{1/p}, & p<\infty \\ (1+\underline{\Pi}_k/\varepsilon)^n E_\infty(f;\Pi_k,n,-1)+\underline{\Pi}_k^n\Phi(\varepsilon) & ,p=\infty \end{cases}$$

Now by assumption $E_p(f;\Pi_k,n,-1) = \underline{\Pi}_k\Psi(\underline{\Pi}_k)$ where $\Psi(x)$ is some positive function with $\lim_{x\to o+}\Psi(x) = 0$. Setting $\varepsilon = \underline{\Pi}_k\Psi(\underline{\Pi}_k)^{1/2n}$ we have for $k\to\infty$

$$(1+\underline{\Pi}_k/\varepsilon)^n E_p(f;\Pi_k,n,-1) \leq C\Psi(\underline{\Pi}_k)^{1/2}\underline{\Pi}_k .$$

Since furthermore $\operatorname{meas}\{\bigcup_i (x_i^{(k)}-\varepsilon,x_i^{(k)}+\varepsilon)\} \leq 2[b-a]\Psi(\underline{\Pi}_k)^{1/2n}\to o$, $k\to\infty$ the result follows from (24).

We conclude with the remark that most of the above results have been carried over to the more general case of spline functions which are pieced together by null-solutions of a differential operator $\Lambda = \sum_{i=o}^{n} a_i(x)\frac{d^i}{dx^i}$ with $a_n(x)\geq a>o$ (cf. [10], [13]). Inverse theorems have also been obtained in case of best approximation by splines when the knots are not fixed with respect to the functions to be approximated but only their number (see [12], [5], [3], [1]). Also the multidimensional case has been treated (cf. [11], [4]).

REFERENCES

[1] J.BERGH - J.PEETRE: On the spaces $Vp(o<p\leq\infty)$. Boll.Un.Mat. Ital. IV Ser 10 (1974), 632-648

[2] C.de BOOR: On uniform approximation by splines. J.Approx. Theory 1 (1968), 219-262

[3] Ju.A.BRUDNYI: Spline Approximation and functions of bounded variation. Soviet Math.Dokl. Vol.15 (1974), No.2

[4] Ju.A.BRUDNYI: Piecewise polynomial approximation, embedding theorem and rational approximation. Proc.Colloq. Approximation Theory, Bonn 1976, Springer Lecture Notes in Math. No.554

[5] H.G.BURCHARD: On the degree of convergence of piecewise polynomial approximation on optimal meshes. Preprint 1974, to appear

[6] G.BUTLER - F.RICHARDS: An L_p-saturation theorem for splines. Canad.J.Math. 24 (1972), 957-966

[7] R.DEVORE: Degree of approximation. To appear in Proc.Symp. Approximation Theory, University of Texas, Austin, 1976

[8] R.DEVORE - F.RICHARDS: Saturation and inverse theorems for spline approximation. Spline Functions Approx.Theory, Proc.Symp.Univ.Alberta, Edmonton 1972, ISNM 21 (1973), 73-82

[9] R.DEVORE - K.SCHERER: A constructive theory for approximation by splines with an arbitrary sequence of knot sets. Proc.Colloq.Approximation Theory, Bonn 1976, Springer Lecture Notes in Math. No. 554

[10] H.JOHNEN - K.SCHERER: Direct and inverse theorems for best approximation by Λ-splines. Proc.Symp. Spline Functions, Karlsruhe 1975, Springer Lecture Notes in Math. 501, pp. 116-131

[11] M.J.MUNTEANU - L.L.SCHUMAKER: Direct and inverse theorems for multi-dimensional spline approximation. Indiana Univ.Math.J. 23 (1973), 461-470

[12] V.A.POPOV: Compt. rend. Acad. bulg. Sci. 26 (1973), 10,1297

[13] K.SCHERER: Some inverse theorems for best approximation by Λ-splines. To appear in Proc.Symp.Approximation Theory, University of Texas, Austin 1976

MINIMAL PROJECTIONS

Carlo Franchetti

University of Florence

Istituto di Matematica U. Dini - V. le Morgagni 67/A

1 - Let X be a normed space, Y a closed subspace of X and denote by A the set of all projections $P : X \to Y$. A projection Q is minimal if $||Q|| \leq ||P||$, $P \in A$. If $||Q|| = 1$, Q is minimal since for any P, $||P|| \geq 1$. A may be empty, if not Y is called complemented, this is the case when the dimension or the codimension of Y is finite. When dim $Y < \infty$ there is always a minimal element in A, but codim $Y < \infty$ does not imply this conclusion (see [3]). We will discuss the problem of characterizing minimal element in A or in some proper subset. This problem is trivial when X is an Hilbert space or Y is 1-dimensional.

2 - This paragraph is an attempt to motivate our problem in two ways: one from approximation theory and the other from the theory of Banach spaces.

Suppose we want to approximate elements of X with elements of a subspace Y. Projections give a natural procedure for constructing approximations: if we regard Px as an approximation in Y of $x \in X$ then the error is $||x-Px||$. We have

$$||x - Px|| = ||(x-y) - P(x-y)|| = ||(I-P)(x-y)|| \leq$$

$$\leq ||I - P|| \; ||x - y|| \text{ hence}$$

$$||x-Px|| \leq ||I-P|| \inf_{y \in Y} ||x-y|| = ||I-P|| d(x,Y) \leq (1+||P||) d(x.Y)$$

So in some sense the "best" projection $P \in A$ would be one for which $||I-P||$ is a minimum, one for which $||P||$ is a minimum is only a "good" one. However it may happen that $||I-P||$ is a minimum iff $||P||$ is a minimum which is in particular true when $||I-P|| = 1 + ||P||$.

The last equation holds when $X=C(Q)$, with Q Hausdorff compact with no isolated points, and $\dim Y<\infty$; for a proof (see [5]).

To explain the second motivation we first give a short outline of the theory of projection constants.

A Banach space X is a G_λ-space, $\lambda\geq 1$, if

a) for every superspace Z of X there exists a projection $P:Z\to X$ with $||P|| \leq \lambda$.

The (relative) projection constant of X in Z is defined by

$$\lambda(X,Z) = \inf \{||P|| : P \text{ projects } Z \text{ onto } X\} ,$$

if X is not complemented in Z we put $\lambda(X,Z) = \infty$. Q is minimal in the set A of all projections from Z onto X iff $||Q|| = \lambda(X,Z)$.

The (absolute) projection constant of X is defined by

$$\lambda(X) = \sup \{\lambda(X,Z) , Z \text{ is a superspace of } X\} .$$

It is clear that $1\leq\lambda(X)\leq\infty$. To understand the relevance of this parameter $\lambda(X)$ we recall the property a) is equivalent to each of the following two properties:

b) For every superspace Z of X, for every Banach space Y and for every linear operator $T:X\to Y$ there exists an extension $\tilde{T}:Z\to Y$ such that $||\tilde{T}||\leq\lambda||T||$.

c) For every paid of Banach spaces Z, Y, $Z\supset Y$ and for every $T:Y\to X$ there exists an extension $\tilde{T}:Z\to X$ such that $||\tilde{T}||\leq\lambda||T||$.

A Banach space X is a G_λ-space for every $\lambda>\lambda(X)$. For a discussion of G_λ-spaces see [7], [14].

If $\dim X=\infty$, then X reflexive or separable $\Rightarrow\lambda(X)=\infty$ (Grothendieck).

If $\dim X = n$, then $\lambda(X)\leq \sqrt{n}$ ([13], [11]); this asymptotic estimate for $\lambda(X)$ is the best possible, for ex. $\lambda(l_1(n))\sim \sqrt{n}$ ([12]).

Given a Banach space X, of special interest are those spaces Z, $Z\supset X$ such that $\lambda(X,Z) = \lambda(X)$. For ex. this equation holds when Z is a G_1-space or when $\dim X<\infty$ and $Z = C(Q)$ with Q Hausdorff compact (see [13]). So any X with $\dim X<\infty$ is a G_λ-space also for $\lambda = \lambda(X)$ since minimal projections exist.

If $Z=C[a,b]$ and $\dim X<\infty$, for a minimal projection Q we have:

$$||Q|| = \lambda(X,Z) = \lambda(X) , \quad (X \text{ is a } G_{||Q||}\text{-space}).$$

$$||I - Q|| = 1+||Q|| , \quad (||I - Q|| \text{ is a minimum}).$$

Finally we note that for many Z the subspaces X for which $\lambda(X,Z)=1$ have been characterized. For ex. if $Z=L_p(T,\Sigma,\mu)$ $1\leq p<\infty$ then $\lambda(X,Z)=1$ iff X is isometric to a space $L_p(T',\Sigma',\mu')$, ([1]).

Investigating minimal projections from Z onto a subspace X is very different in the two cases $\lambda(X,Z)=1$, $\lambda(X,Z)>1$; we will consider the second one.

3 - We review here the few examples of (non trivial) minimal projections already known. Consider the Banach space C of continuous 2π-periodic functions with the usual sup-norm. The operator P_n which maps any $x \in C$ into the partial sum x_n of its Fourier series is a projection from C onto the subspace V_n of trigonometrical polynomials whose degree does not exceed n. Minimality of P was established by Lozinski in 1948, unicity was proved in 1969 ([4]). We have $\lambda(V_n,C)=\lambda(V_n)=||P_n|| = \lambda_n$, where λ_n are the classical Lebesgue constants. If everything is interpreted with the L,-norm we have $||P_n||= 1$, since P_n is orthogonal.

If Z is one of the sequence spaces c_o or l_1 and if X is an hyperplane in Z, then $\lambda(X,Z)$ and minimal projections (when exist) can be explicitily computed ([3]): the same is done if $Z=l_1(n)$ or $Z=l_\infty(n)$ ([9]).

4 - We discuss now some recently discovered minimal projections.

i) The space $X=L_1(T,\Sigma,\mu)$.

Assume that (T,Σ,μ) is a finite nonatomic measure space, that Y is a finite dimensional and smooth subspace of X(Y is smooth iff $y \in Y$, $y \neq 0 \Rightarrow y(t) \neq 0$ μ-almost everywhere on T). Minimal elements of A are characterized in [8]. If Y is also rotund we have unicity, if Y is the subspace of linear polynomials there is also a unique minimal projection which is explicitily computed. The main results depend upon the use of a suitable "Lebesgue function". Recall that if L is any operator mapping C(Q) into itself the classical Lebesgue function Λ_L may be defined by $\Lambda_L(t)=||\hat{t}\circ L||$, where $\hat{t}$ is the evaluation functional at $t \in Q$ ($\hat{t}(x) = x(t)$, $x \in C(Q)$); we have $\Lambda_L \in C(Q)$ and $||\Lambda_L||=||L||$. A "Lebesgue function" for an operator M on an arbitrary Banach space should be a function $\Lambda_M(\cdot)$ such that $||M|| = ||\Lambda_M||_\infty$. In our L_1-case the Lebesgue function for a projection $P:X \to Y$ is defined by

$$\Lambda_P = \sup_{v \in V} |P^* v| ,$$

where P^* is the adjoint operator of P and V is a countable set in L_∞ of norm-one elements such that $\sup_{v \in V}(y,v)=||y||$ for any $y \in Y$. It is easily seen that $\Lambda_P \in L_\infty$ and that $||\Lambda_P||_\infty=||P||$.

The following result is of interest:

THEOREM - If P is a minimal projection onto Y (under the same hypotheses), then the Lebesgue function Λ_P is constant μ-a.e. on T.

ii) The sequence spaces c_o and l_1.

Assume that $X=c_o$ or $X=l_1$, Y is a subspace of X with dim $Y<\infty$ and P is a projection onto Y. The Lebesgue functions are defined by:

$\wedge_p(t)=||\hat{t}oP||$, if $X = c_o$; $\wedge_p(t) = ||\hat{t}oP^*||$ if $X=l_1$.

Minimal projections are characterized ([9]) for any Y if $X=c_o$; if $X=l_1$ then Y satisfies a Haar type condition. For n-dimensional Haar subspaces of l_1 the Lebesgue function of a minimal projection is constant with the exception of at most n-1 points.

iii) The Rademacher projection.

Consider the topological space T consisting of 2^n disjoint intervals. If I_j is the j-th interval, then we define functions $y_1,\ldots,y_n$ in C(T) by setting $y_i(t) = (-1)^*$ on I_j with k = $[(j-1)2^{i-n}]$. Here $1\le i\le n$, $1\le j\le 2^n$ and $[\cdot]$ denotes the greatest integer. These functions are the classical Rademacher functions suitably modified for our purpose. The y_i are an orthonormal system in the L_2-sense, that is

$$< y_i , y_j> = \int_T y_i(t)y_j(t)\, dt = \delta_{ij} .$$

Call Y the linear span of the y_i and define the "Rademacher projection" $P:C(T)\to Y$ setting:

$$Px = \sum_{i=1}^{n} <x,y_i> y_i , \quad x\in C(T) .$$

P is minimal in A([10]). For the proof we use two results. Grünbaum ([12]) proved in 1960 that the projection constant of $l_1(n)$ is:

$$(1) \quad \lambda(l_1(n)) = \frac{2m+1}{2^{2m}} \binom{2m}{m} , \quad m = \left[\frac{n-1}{2}\right] .$$

Rademacher proved in 1922 ([16]) that the Lebesgue function of this projection is constantly equal to the number in (1), so $||P||= ||\wedge_p|| = \lambda(l_1(n))$. Now remark that Y is isometric to $l_1(n)$, hence $\lambda(Y,C(T)) = \lambda(T) = \lambda(l_1(n)) = ||P||$ and therefore P is minimal.

5 - We have seen that the Fourier and the Rademacher projection are orthogonal in the L_2-sense, also both possess a constant Lebesgue function. We consider here the role of orthogonality in the problem of minimal projections. Assume that X=C(Q), with Q Hausdorff compact; we say that a projection P from X onto a closed subspace Y is orthogonal ([2]) if there exists a nonnegative nonzero functional $f\in X^*$ such that

$$(2) \quad f[(x-Px)] = 0 \quad y\in Y , \quad x\in X .$$

If we put $<a,b>_f = f(ab)$, then $<\cdot,\cdot>_f$ is a pseudo inner product on X, it becomes a true inner product if $x \neq 0$ implies $f(x^2)>0$.

The set N of all orthogonal projections from X onto Y has a minimal element ([2]). For the general problem of existence of minimal projections in restricted classes see [2] and also [15]. Besides the classical orthogonal projections defined using polynomials orthogonal with respect a positive weight w $(f(x) = \int_Q x(t)w(t)dt)$, N includes also Lagrange interpolatory projections and many other projections.

Given a subspace Y of X with dim $Y<\infty$ let us call Y^2 the linear span of all products y_1y_2 with $y_i \in Y$. For a positive functional $f \in X^*$ assume that $<\cdot,\cdot>_f$ defines a true inner product on Y ($y \in Y$, $y \neq 0 \Rightarrow f(y^2)>0$). Let us take in Y an orthonormal basis $\{y_1,\ldots,y_n\}$ (with respect to $<\cdot,\cdot>_f$); the projection P_f defined by

$$(3)\quad P_f x = \sum_{i=1}^{n} <x;y_i>_f y_i = \sum_{i=1}^{n} f(xy_i)y_i$$

is orthogonal in the sense that (2) is satisfied. Observe that to any $f \in X^*$ such that $f \geq 0$; $y \in Y$, $y \neq 0 \Rightarrow f(y^2)>0$ it corresponds one and only one projection which satisfies (3), this (orthogonal) projection is given by (3). One can see that the Lebesgue function of P_f ($\Lambda_{P_f}(t) = ||\hat{t}oP_f||$) is given by $\Lambda_{P_f}(t) = f(|\Sigma_i y_i(t)y_i|)$ hence $||P_f|| = \sup_t f(|\Sigma_i y_i(t)y_i|)$.

Suppose now that a positive functional $g \in X^*$ is such that $g/Y^2 = f/Y^2$, then $<\cdot,\cdot>_f = <\cdot,\cdot>_g$ on Y; in other words f and g induce on Y the same Hilbert space structure, P_g can be represented in the form $P_g x = \Sigma g(xy_i)y_i$, where $\{y_1,\ldots,y_n\}$ is a $<\cdot,\cdot>_f$-orthonormal basis in Y. All this suggests the

DEFINITION ([10]). Let f be a positive functional with the property that $y \in Y$, $y \neq 0 \Rightarrow f(y^2)>0$. A_f denotes the set

$$A_f = \{P_g : g \in X^*,\ g \geq 0,\ g/Y^2 = f/Y^2\}.$$

A_f is a subset of N which has a minimal element. Generally is a rather "large" set, but in some special cases it reduces to a singleton; the structure of A_f is investigated in [10] where, in particular, a simple characterization theorem for minimal elements is given. Statements on minimal elements of A_f depend upon the critical set K_f of the Lebesgue function ($K_f = \{t \in Q : \Lambda_{P_f}(t) = ||P_f||\}$); when f is supported on all Q (when $x \neq 0 \Rightarrow f(x^2)>0$) we have special results. As an example of the theory we refer here a necessary and a sufficient condition for minimality in A_f.

THEOREM 1 Assume that Q is infinite, $1 \in Y$, Y^2 is an Haar subspace of C(Q), supp f=Q, $||P_f||>1$ and P_f minimal in A_f then K_f is infinite.

THEOREM 2 Assume that $1 \in Y^2$ and that Λ_{P_f} is constant, then P_f is minimal in $\mathcal{A}_f$.

Theorem 2 shows that the Fourier and the Rademacher projections are minimal in their classes $\mathcal{A}_f$ (actually both are minimal in the total class). Let us now consider for $n \geq 2$ the Fourier-Chebyshev projection S_n (S_n is obtained taking the Chebyshev orthogonal polynomials). In [6] it is proved that S_n is not minimal in the total class since Λ_{S_n} has only two critical points; theorem 1 implies the nonminimality of S_n in $\mathcal{A}_f$.

We hope that the use of orthogonal projections can make easier to understand the nature of the (absolute) minimal projections, for ex. from $C[a,b]$ onto the polynomial subspaces π_n (even for $n=2$ minimal projections are unknown). A progress would be to prove or disprove that among minimal projections there is one which is finitely supported (representable in the form $Px = \Sigma x(t_i)y_i$, where $y_i \in \pi_n$ and $t_i \in [a,b]$).

REFERENCES

[1] S.J. Bernau - H.E. Lacey. The range of a contractive projection on an L_p-space. Pacif.J.Math. 53(1974), 21-41.

[2] J. Blatter - E.W. Cheney. On the existence of extremal projections. J. of Approximation Theory 6 (1972), 72-79.

[3] J. Blatter - E.W. Cheney. Minimal projections on hyperplanes in sequence spaces. Ann. Mat. Pura e Appl. (IV) Vol. 101 (1974), 215-227.

[4] E.W. Cheney, C.R. Hobby, P.D. Morris, F. Schurer, D.E. Wulbert. On the minimal property of the Fourier projection. Trans.Am. Math. Soc. 143 (1969), 249-258.

[5] E.W. Cheney - K.H. Price. Minimal projections, in Approximation Theory, A. Talbot ed. Academic Press, New York (1970),261-289.

[6] E.W. Cheney - T.J. Rivlin. Some polynomials approximation operators. Math. Z. 145 (1975), 33-42.

[7] M. Day. Normed linear spaces. Academic Press, New York (1962).

[8] C. Franchetti - E.W. Cheney. Minimal projections in $\mathcal{L}_1$-spaces. Duke Math. Journ. 43 (1976).

[9] C. Franchetti - E.W. Cheney. Minimal projections of finite rank in sequence spaces. To appear.

[10] C. Franchetti - E.W. Cheney. Orthogonal projections in spaces of continuous functions. To appear.

[11] D.J.H. Garling - Y. Gordon. Relations between some constants associated with finite dimensional Banach spaces. Israel J. Math. 9 (1971), 346-361.

[12] B. Grünbaum. Projection constants. Trans. Am. Math. Soc. 95 (1960) 451-465.

[13] M.I. Kadets - M.G. Snobar. Some functionals over a compact Minkowskii space. Math. Notes 19 (1971), 694-696.

[14] J. Lindenstrauss. Extension of compact operators. Memoirs of the A.M.S. 48 (1964).

[15] P.D. Morris - E.W. Cheney. On the existence and characterization of minimal projections. J. reine angew. Math. 270 (1974), 61-76.

[16] H. Rademacher. Einige Sätze über Reihen von allgemeinen Orthogonalfunktionen. Math. Annalen 87 (1922), 112-138.

ESTIMATION PROBLEMS IN CRYSTALLOGRAPHY

Robert Schaback

Lehrstühle für Numerische und Angewandte Mathematik

3400 Göttingen, Lotzestr. 16-18

One of the most important problems in crystallography is the determination of the unit cell of a crystalline substance and the relative positions of atoms within that cell. This contribution is mainly concerned with the estimation of unit cell dimensions from data obtained by x-ray diffraction of a polycrystalline substance. Compared to the number of parameters to be estimated and to the desired accuracy, the given information turns out to be rather limited and relatively noisy. Therefore some deeper insight into the underlying problem is necessary. This reveals a nonlinear mixed-integer approximation problem and gives some hints for the numerical solution of the problem. By combination of common approximation methods and combinatorial strategies of branch- and bound-type a numerical estimation method for a high-speed computer can be developed. In practice this method provides the user with a set of estimates which fit into the noise limits of the data, as is shown by a series of test examples. A reduction of the number of possible estimates can be made by introducing additional restrictions involving more crystallographic information. The purpose of this contribution is not to give a purely crystallographical or a purely mathematical description of the topic but to try to arouse some interest for crystallographical questions among mathematicians working with high-speed computers.

1. INTRODUCTION AND PROBLEM FORMULATION

As in every branch of the natural sciences, there are of course lots of estimation problems arising in crystallography. In this contribution we want to concentrate on the most important one, i.e. the estimation of the parameters describing the geometrical structure of crystals and their complicated inner lattice of atoms.

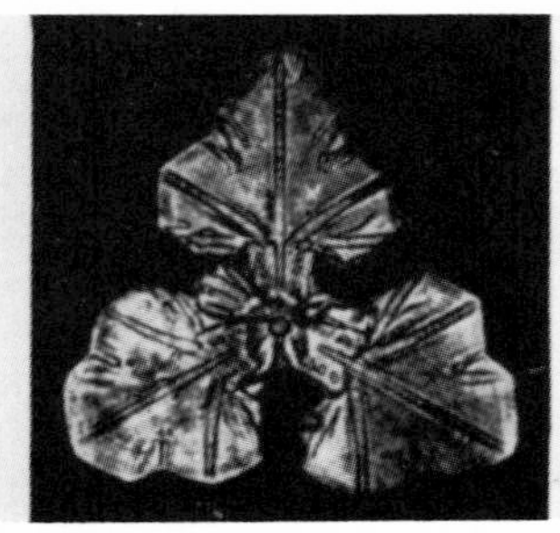

Fig.1: Crystal of snow [42]

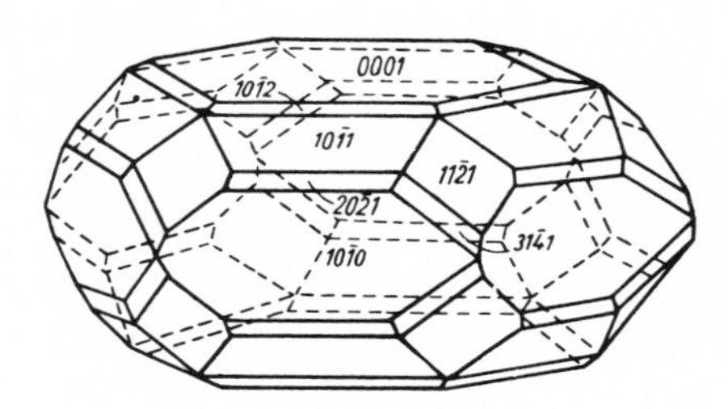

Fig.2: Crystal of apatite, $Ca_5[(OH,Cl,F)|(PO_4)_3]$,[42]

For sake of preciseness we should first note that a crystal may be defined as a solid composed of atoms, ions or molecules arranged in a pattern periodic in three dimensions [17,42]. Since in practice the periods are of length $\sim 10^{-8}$ cm, one can consider the pattern as being infinite in any direction. The periodicity implies that there are three linear independent vectors a_i, i = 1,2,3, such that the transformations

$$x \to x + a_i \ , \ i = 1,2,3 \ , \ x \in \mathbb{R}^3$$

map the pattern onto itself, and such that the lengths of the a_i can not be shortened. Then the parallelepiped spanned by the a_i is called a <u>unit cell</u> of the crystal. By periodic repetition of the unit cell we get the <u>point lattice</u> of the crystal.

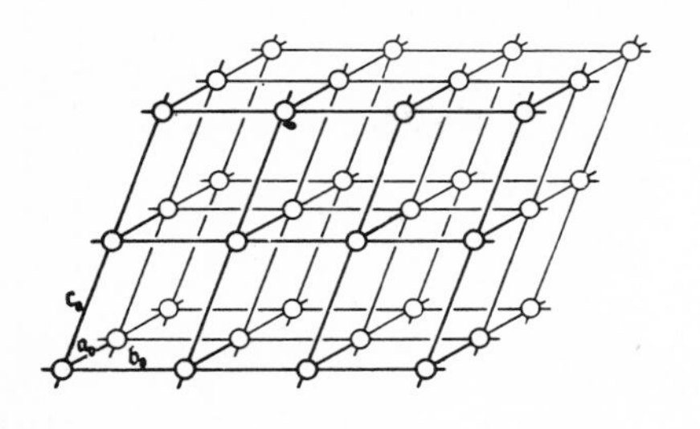

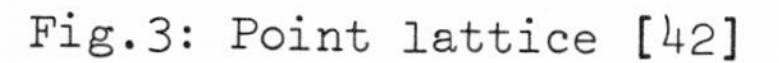
Fig.3: Point lattice [42]

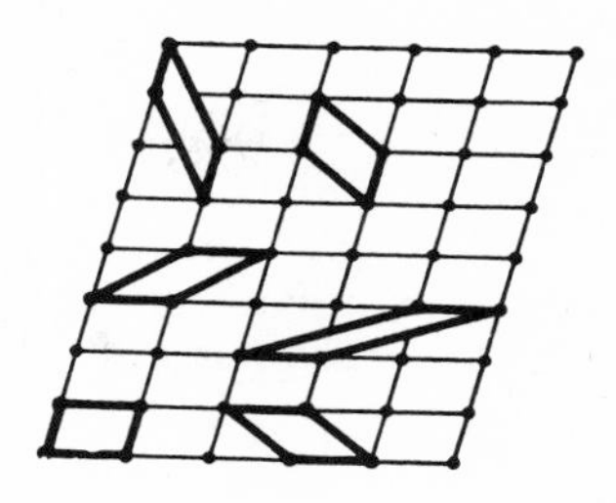

Fig.4: Unit cells [42]

Fig.4 shows that the unit cell is a purely mathematical tool for describing the domain of periodicity of the arrangement of atoms in the crystal. We now can formulate

Estimation Problem 1. Determine the shape of a unit cell of a given crystalline substance.

Before we turn to methods for solving this problem (which will be the main part of this paper) we should look upon what follows a solution of the problem. The obvious next step is

Estimation Problem 2. Determine the sort and the number of atoms in the unit cell.

Whereas the first part of this is a chemical task which usually is completed before the crystallographer starts his work, the second part consists of a simple calculation:

$$S = \frac{\rho \cdot V}{1.66020} \tag{1}$$

gives the sum S of the atomic weights of the atoms in the unit cell as a function of the density ρ of the substance (in g/cm^3) and the volume V of the unit cell (in Å^3). From the chemical formula of the substance one gets the sum s of atom weights of atoms in the chemical unit. Then S/s must be approximately equal to an integer number giving the number of chemical units in the unit cell (e.g. 4 for Na Cl, 8 for diamond). We note for further use that (1) provides a check for correctness of unit cell volumes.

One now knows how many atom positions (say n) within the unit cell have to be estimated. Since n may be fairly large (see Fig.6) and there are 3n unknown real parameters, one first tries to reduce the number of variables by

<u>Estimation Problem 3.</u> Determine the symmetries of the contents of the unit cell.

This task consists of finding the correct group of symmetry transformations out of 230 possibilities [6,42]. The choice depends on the shape of the unit cell (e.g. a cubic cell may admit many more symmetry transformations than a general triclinic cell), on the number and the sorts of atoms in the unit cell (e.g. symmetry transformations must map atoms of the same sort onto each other), and on certain peculiarities arising in the data from x-ray diffraction of specimens of the substance (i.e. systematic extinctions of diffracted beams due to inner symmetries of the unit cell [42]). Even if all the above information is available, there usually are several possibilities left. The problem of finding the correct symmetry group (which is, like problem 2, more a decision problem than an estimation problem) thus comes out to be one of ruling out impossible alternatives and testing the remaining possibilities with respect to the outcome of

<u>Estimation Problem 4.</u> Determine the relative positions of atoms (or ions or molecules) within the unit cell.

This problem is usually tackled by assuming a certain structure, calculating its x-ray diffraction properties and modifying the structure until its calculated x-ray diffraction properties fit the observed x-ray diffraction data. We shall discuss this method in the third section of this paper.

The total problem of determining the structure of a crystal usually is considered as being solved if the above problems have led to a structure which is in good agreement with all available measurements [10]. Since problems 1,3 and 4 need not have unique solutions (in the sense of mathematics) some doubts to this argument seem to be appropriate.

2. DETERMINATION OF UNIT CELL DIMENSIONS

The most important means for measuring crystallographic data is x-ray diffraction [3,10,17,39,42,69,74]. These methods either use a single-crystal specimen (von Laue, rotating-crystal, Weissenberg, divergent-beam and retigraph methods) or a specimen in powder form (Debye-Scherrer method, improved by Seeman & Bohlin and Guinier). In case a single crystal is available, one of the above methods will give complete information on the unit cell in a comparatively easy way. If not, one can only try to analyze the data supplied by a Debye-Scherrer-type method, which will be done the sequel.

A powder specimen diffracts a monochromatic beam of x-rays into a(practically finite) number of cones of rays (see e.g. [3,10,17,39, 42,69]). The relation of the (measured) angles $2 \cdot \theta_i$ of the cones to the unit cell (spanned by vectors a,b,c) of the crystal is the following:

There are integer numbers h_i, k_i, l_i such that

$$q_i = (4 \cdot \sin^2 \theta_i)/\lambda^2 = \|h_i a^* + k_i b^* + l_i c^*\|^2$$

where the Euclidean norm is taken and a*,b*,c* are "reciprocal" to a,b,c:

$$a^* = (b \times c)/V, \; b^* = (c \times a)/V, \; c^* = (a \times b)/V,$$

where V = (a x b,c) denotes the volume of the unit cell. It is easy to see that the determination of the unit cell of the given crystal is equivalent to the determination of the "reciprocal" unit cell spanned by a*,b*,c*. If, in addition, we introduce A as the Gram matrix corresponding to a*,b*,c*, we finally arrive at the following

Estimation Problem 1'. Given a vector $q \in \mathbb{R}^N$ with positive components, determine a real 3 x 3 - matrix A and an integer 3 x N - matrix M such that

(2) the diagonal of $M^T A M$ equals q,

$$\text{i.e. } q_k = \sum_{i,j=1}^{3} a_{ij} \cdot m_{ik} \cdot m_{jk} \text{ if } A = (a_{ij}),\; M = (m_{ik}),\; q = (q_k),$$

(3) A is positive definite.

Because of periodicity, A contains all information on the (reciprocal) unit cell in the form of 6 parameters (3 vector lenghts and 3 angles). As was already pointed out in [61], the above problem will have many symmetrical solutions, if any, and because one has to allow for noise in q_k (about 0.1%, but unevenly distributed) it has always solutions fitting into arbitrarily small noise limits. Thus an additional restriction is needed:

(4) $|m_{jk}| \leq c$ (c chosen between 4 and 18 in practice),

which eliminates practically impossible cases of diffraction of arbitrarily high order or by extremely improbable sets of parallel lattice planes.

To account for random noise we reformulate the task to get

Estimation Problem 1". Given a positive vector $q \in \mathbb{R}^N$, find a real 3 x 3 - matrix A and an integer 3 x N - matrix M such that

(5) $f(A,M) = \| q - \operatorname{diag}(M^T A M) \|^2_{L_2}$

is minimized subject to (3) and (4).

Since this problem consists of the simultaneous estimation of six real parameters and $3 \cdot N$ integer parameters occurring nonlinearly in a given object function, one is led to try to separate the two sets of variables and to estimate them iteratively:

Algorithm 1. Start with an estimate A^o for A.

Iteration. Step 1: Given A^i, minimize $f(A^i,M)$ with respect to M under the restricion (4).

Step 2: Given a solution M^{i+1} of Step 1, minimize $f(A,M^{i+1})$ with respect to A under the restriction (3). Re-iterate with the result A^{i+1}.

The algorithm is discussed in detail in [61], where we proved the following

Theorem. If all M^i have rank 3, then

a) Step 2 always has a unique solution.
b) Step 1 always has a solution
c) The algorithm is finite if the result of Step 1 is calculated deterministically from A^i (i.e. equal A's yield equal results).

A practical way of handling Step 1 (Step 2 is a standard linear least-squares problem) is described in detail in [61]. Since the publication of [61] E.Larisch has carried out a large number of test runs with Algorithm 1. We shall comment on two observations.

First, if one starts the algorithm with "true" cell parameters from the standard reference file issued by the American Society for Testing Materials (ASTM) one can frequently observe that the iteration stops with a different set of cell parameters and a reasonable improvement of the least-squares sum f(A,M). We found out later (see below) that this effect occurred if the noise in the data set was larger than the critical noise level.

We therefore conclude that serious doubts are appropriate concerning the use of the ASTM file as a standard reference (see also [17], p.386: "... the user must decide which pattern in the file is the most reliable").

Second, the algorithm may give a different "solution" if a perturbation of 0.5% in the initial estimate is applied ([45]). This implies that the algorithm needs fairly precise starting parameters and performs only a refinement of an estimate.

Therefore a good routine for getting started is needed. Since the testing of many candidates must be expected to get the "correct" estimates, the algorithm should make as sure as possible that the "correct" estimate is not overlooked. This is done by carefully placing the intersection between a statistical and a deterministic part of the computation in order to allow as little randomness as possible and to keep deterministic time-consuming calculations within reasonable bounds. After a series of tests of different approaches E.Larisch [45] finally developed the following

Algorithm 2. Step 1: Determine all 2-dimensional sublattices (planes) which can be formed out of the first n_1 data values and calculate their "fitting quality" with respect to the first n_2 data values. This "zone-finding" step is very similar to the corresponding method in [38,66,70,71].

Step 2: Scan the output of Step 1 for higher symmetries of the crystal, i.e. test whether special assumptions on cell parameters lead to reasonably good fittings. Calculate such estimates and their fitting quality with respect to all data values.

Step 3: Take the m best planes from Step 1. For each pair of planes, estimate the two remaining parameters by checking the distribution of the parameters when some of the unfitted data values are assumed to stem from systematically varied grid points. Calculate the fitting quality of the estimates with respect to all data values.

Step 4: Perform some additional checks on the output of steps 1 and 2 (e.g. test for evidence of parallel planes to the already constructed planes).

The output of this algorithm may consist of up to 5000 estimates for the cell parameters, all of which are refined by a subsequent run of Algorithm 1. The CPU time of Algorithm 2 on the IBM 370/165 of the Regionales Hochschulrechenzentrum Bonn varied between 50 seconds and 30 minutes with an average of 6.4 minutes whereas the refinement of all possibilities by Algorithm 1 varied between 6 seconds and 18 minutes with an average of 4.4 minutes. Since the length of the calculation can by no means be estimated beforehand, one has to incorporate a restart facility into the program.

The test runs in [45] provided a series of examples throwing some light on the dependence of the result on the noise in the data. The first and unexpected observation is that in any case, even for arbitrarily manipulated data, the algorithm is able to give (usually several) estimates with an error not exceeding $0.5 \cdot 10^{-7}$ but only rarely lower than $0.35 \cdot 10^{-8}$ (the error is measured by a weighted sum of squares, the weights of which are related to the experimental situation and the quality of the observations; since the weights are normalized to give a sum of 1, the error can be used for comparison between different test examples). This implies that the noise level in the data (measured in the same way as the error with respect to the true solution) must be lower than $0.5 \cdot 10^{-7}$ in order to make a better estimate from the data by any method whatsoever. On the other hand, only estimates with error less than $0.35 \cdot 10^{-8}$ can be considered as dependent on the data.

The outcome of the test runs can be simply grouped by the magnitude of the error E_o of the "known solution" taken as a measure for the noise in the data.

Group 1. $E_o \leq 0.2 \cdot 10^{-7}$ (i.e. somewhat better than maximum noise level). In all nine examples tested the correct cell was found (among others), except for equivalence transformations of the unit cell. In seven cases the optimal solution had an error below $0.35 \cdot 10^{-8}$. Alternative estimates with equally low errors differed considerably with respect to the unit cell volume.

Group 2. $0.2 \cdot 10^{7} < E_o < 0.5 \cdot 10^{-7}$ (at noise level). The two examples in this group were not solved. The first example had an error reduction from $0.34 \cdot 10^{-7}$ to $0.79 \cdot 10^{-8}$, but the resulting estimates were false; the second example has a "known solution" which fits seven data values extremely badly.

Group 3. $E_o \geq 0.5 \cdot 10^{-7}$ (worse than the critical noise level). In 4 of the 10 examples in this group the correct cell was found, but this must be contributed to luck.

At first glance this result seems rather positive; nevertheless it must be borne in mind that even for the examples in Group 1 the correct solution may occur between estimates with equally low errors. At this stage it can only be recommended to concentrate on cases with noise level $E_o \leq 0.35 \cdot 10^{-8}$ and to ignore estimates with larger errors.

On the other hand, one should try to achieve some reduction of the noise by major modifications of the problem. This can be done by taking more data into account or by introducing more restrictions on the estimates.

A first possibility is to use (1) for the elimination of estimates giving impossible cell volumes. A second possibility arises from the observation that often the correct cell belongs to a special class of cells (monoclinic, orthorhombic etc.) but occurs among general cell estimates with equally good or lower errors or it has an error within the above noise level, which was obtained for general cells. Therefore the error of special estimates should be compared with the (yet unknown) noise level of estimates belonging to the corresponding class of cells. This would lead to higher levels of admittable noise and the significance of special estimates would be increased. There is some experimental evidence [46] that for data with noise below these levels the check on the cell volume using (1) will finally cut down the number of reasonable estimates to very few, including the correct solution.

A third possibility arises in conjunction with the second possibility: for some special classes of cells there may be subclasses having systematic absences of certain (m_{1k}, m_{2k}, m_{3k}) triplets depending on the interior symmetries of the unit cell (see e.g. [42]). For instance, if a,b,c are mutually perpendicular (this implies that the cell belongs to the rhombic, tetragonal, or cubic class) and the transformation $x \to x + (a+b+c)/2$ maps the crystal pattern onto itself, the pattern is called "body-centered" and only triplets with $m_{1k} + m_{2k} + m_{3k}$ even are possible. It should be tried to evaluate noise levels for these subclasses as well as for the classes themselves in order to gain a further increase in admittable noise.

A fourth possibility is to use the remaining estimates as a starting point for further investigations of the inner structure of the crystal. This means that estimates should be discarded if they turn out to be incapable of fitting additional diffraction data, mainly the observed intensities of diffracted beams when trying to solve Estimation Problem 4.

For comparative testing of other methods [4,5,37,38,49,56,64,65,70,71], and especially the computer-oriented methods of Visser ([66], based on [56,38,70,71]) and Haendler & Cooney ([8], based on [38]) we unfortunately had not enough common examples. The strategy described above uses elements of both methods in Step 1 of

Algorithm 2, but enlarges their scope by taking more possibilities into account. Therefore it can be assumed that Algorithms 1 and 2 have a smaller probability of missing the correct solutions (of course at the cost of considerably increased computing time).

3. ESTIMATION OF THE RELATIVE POSITIONS OF ATOMS WITHIN THE UNIT CELL

The theory of 3-dimensional point lattices (see e.g.[42]) implies that to each triplet $m = (m_1 m_2 m_3)$ of integers there corresponds a set of parallel lattice planes in the lattice. The diffraction of an x-ray beam by that set of planes occurs at an angle governed by Bragg's law (which gives the data used in section 2 of this paper) and with the (complex) amplitudes

$$(6) \qquad F_m = \int_C \rho(x)\, e^{2\pi i(x,m)} dx ,$$

(called the structure factor corresponding to the triplet m), where $\rho(x)$ is the (periodical) electron density within the crystal and the domain of integration is the unit cell C(e.g.[10],[39],[42]). This means that F_m is the Fourier transform of ρ and we have

$$(7) \qquad \rho(x) = \frac{1}{V} \sum_{m \in Z^3} F_m\, e^{-2\pi i(x,m)} ,$$

where V is the volume of the unit cell C. If the complex numbers F_m could be estimated, one would get ρ and the atom positions as extrema of ρ immediately from (7), (see Fig.8).

So we have

<u>Estimation Problem 4'.</u> Determine F_m for $m \in Z^3$.

The data for Estimation Problem 1 consists of measurements of diffraction angles θ_k for a series of diffracted beams; by solving Problem 1 via (2) it is known which beam belongs to which triplet $m_k = (m_{1k}, m_{2k}, m_{3k})$ and its intensity must therefore be the square of the absolute value $|F_{m_k}|$ of the complex amplitude F_{m_k}. Thus (after certain corrections due to experimental conditions) the measurements of intensities yield estimates for $|F_m|$ which can be considered as data for Problem 4'. The need of information on the phases ϕ_m of $F_m = |F_m| e^{i\phi_m}$ is called the <u>phase problem</u> in the determination of crystal structures [10,11]. We arrive at

Estimation Problem 4". Determine ϕ_m for $m \in Z^3$.

Unfortunately one can not expect in general to have more purely mathematical information for solving Estimation Problem 4 than the information about the unit cell C and the absolute values $|F_m|$, which leaves no direct information about the phases ϕ_m. Thus one needs additional assumptions from crystallography to arrive at mathematically reasonable estimation problems.

The simplest way of making additional assumptions is to consider $\rho(x)$ as being a superposition of weighted Dirac measures located at the positions x_j, $j = 1,\ldots,n$ of the n atoms in the unit cell. This implies

$$(8) \qquad |F_m| e^{i\phi_m} = \sum_{j=1}^{n} f_j \, e^{2\pi i (x_j, m)}$$

where the atomic scattering factors f_j can be estimated and some of the coordinates of x_j may already be known from symmetry considerations. In general, for N observed diffracted beams, (8) provides a set of N nonlinear equations for maximally $3 \cdot n$ unknowns. Thus it seems reasonable to try either to solve (8) by a standard method for equation solving or to approximate directly the vector of $|F_m|$ values by the absolute values of the right-hand sides of (8) in the L_2 sense, varying the x_j. The well-known nonuniqueness and stability problems arising in such approaches [7-9] make it necessary to start the process with a reasonably good estimate for the x_j, obtained by one of the methods described below.

Another iterative method for getting the atom positions x_j is to start with an estimate for the x_j's, determine an estimate for the ϕ_m's from (8) and calculate new estimates for the x_j's by moving them towards the extrema of $\rho(x)$, where $\rho(x)$ is calculated from (7) with coefficients $|F_m| e^{i\phi_m}$ using the measured value $|F_m|$ rather than the value given by (8). Within these procedures one can start with a smaller number of heavier atoms and add the lighter atoms later. Because the phases frequently are dominated by the contributions of the heavier atoms, this yields only smaller changes of the right-hand side of (8). Estimates for the positions of smaller atoms may be derived from new extrema showing up in plots of ρ calculated as in the second method. Another possible method is the global minimization (in the L_2 sense) of the function

$$\delta(x) = \sum_{m \in Z^3} (|F_m| - |\tilde{F}_m|) \, e^{2\pi i (m,x) + i\phi_m}$$

over the unit cell C, where $|F_m|$ is taken from the observations and $|\tilde{F}_m| e^{i\phi_m}$ is calculated by evaluating the right-hand side of (8).

This implies that δ depends only on the relative atom positions x_j. The shape of δ may also be used as a final check on the calculated estimates, because it resembles the errors in the $|F_m|$ <u>and</u> the ϕ_m.

The L_2 methods described above (and many other versions) have been used frequently for refinement of estimates in crystallographic research [1,2,10,12,16 22,23,31,44,50,55,59,60]. Since no fail-safe standard method has yet emerged, a thorough mathematical investigation by means of approximation theory seems worthwhile. In addition, numerical difficulties are great because of the large number of free parameters (sometimes more than 2000). Computing times may be enormous (50 minutes per iteration cycle with 1603 phase variables on an IBM 360/91[60]).

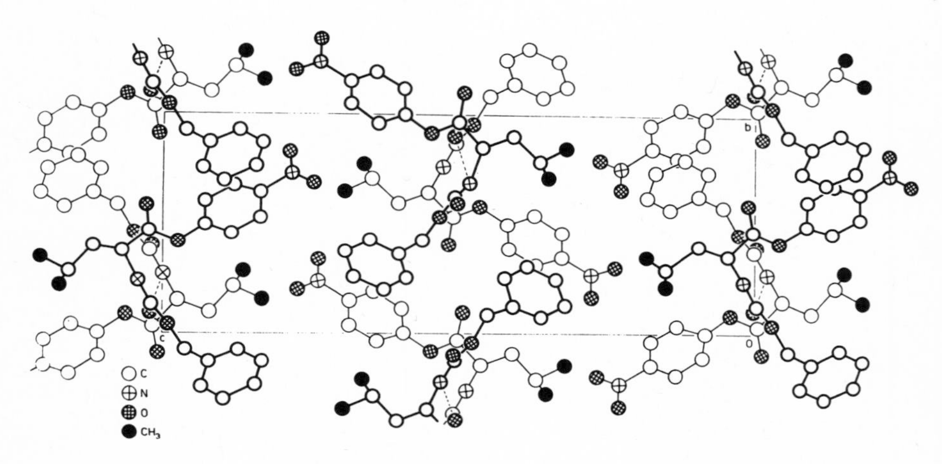

Fig.5: Structure of Carbobenzoxy-L-lencyl-p-nitrophenyl Ester [15]

Like in section 2 of this paper the iterative processes available for solving Estimation Problem 4 tend to calculate only local optima. So we finally must comment on corresponding global methods, which are of independent interest because of their close relationship to Fourier series theory.

The Patterson function P(y) is defined by the convolution

$$P(y) = \int_V \rho(x) \cdot \rho(x+y)\, dx \tag{9}$$

and has the series representation

$$(10) \qquad P(y) = \frac{1}{V^2} \sum_{m \in Z^3} |F_m|^2 \cos(2\pi(m,y))$$

which can be evaluated directly from the observed data. It is easily seen from (9) that P(y) is large if there are peaks of ρ having the distance vector y. This implies that distance vectors of atoms may be read off a plot of P(y), as shown in Fig.7.

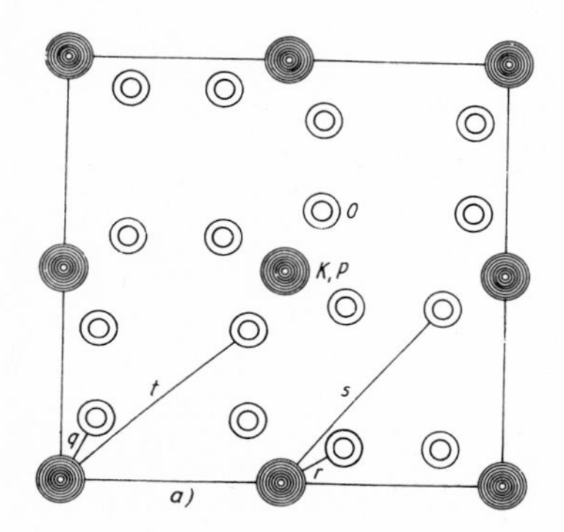

Fig.6: Structure of KH_2PO_4 [42] (2-dimensional projection)

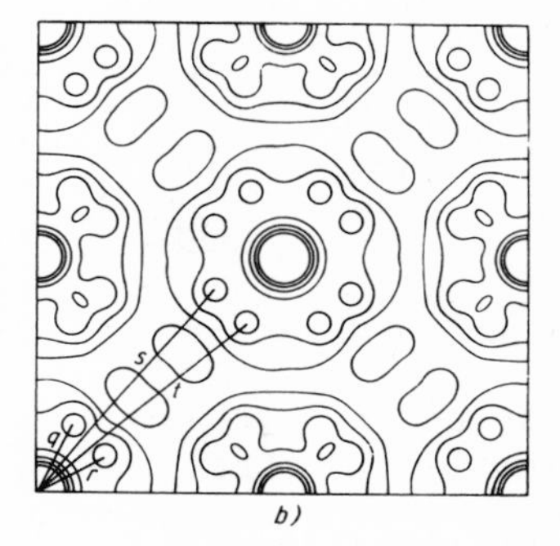

Fig.7: Correspondent plot of the Patterson function [42]

Refinements of this approach have been given by various authors [1,2,11,24,47,54,55,58,72], including computer programs and interactive graphical methods, but no generally satisfactory automatic method has yet been designed to calculate reliable global estimates for atomic distances from the intensities $|F_m|^2$ via the Patterson function alone. The second and perhaps most important class of methods of estimating phases is based on inherent relations between several values of F_m for varying m. One of the simplest [57] is given by the observation that the shape of ρ^2 should be very similar to that of ρ. This implies the convolution relation

$$(11) \qquad F_m \approx \frac{1}{V} \sum_{n \in Z^3} F_n F_{m-n}$$

and the influence of the absolute values on the phases is apparent. For instance, if $|F_{n_o} \cdot F_{m-n_o}|$ dominates the whole sum, then

$$\phi_m \approx \phi_{n_o} + \phi_{m-n_o} ,$$

or equivalently, if $|F_n \cdot F_m|$ is large

$$\phi_{n+m} \approx \phi_n + \phi_m .$$

Also, if the larger values of $|F_n|$ have correct phase estimates, (11) will estimate the phases of F_m with smaller absolute values. By separating (11) into real and imaginary parts one gets the equation

$$(12) \qquad \tan \phi_m \approx \frac{\sum_{n\in Z^3} |F_n||F_{n+m}| \sin(\phi_n+\phi_{n+m})}{\sum_{n\in Z^3} |F_n||F_{n-m}| \cos(\phi_n+\phi_{n+m})}$$

known among crystallographers as "tangent formula".

In addition, there are formulae describing the probability distribution for estimating phases by (11) and (12), provided the probabilities of already assigned phases on the right-hand side are known (see e.g.[1,13,14,29,30,58]).

Of course there are problems arising when initial phase guesses have to be made. But as the choice of origin of the atom coordinate system influences the ϕ_m values, one can fix the origin by assigning arbitrary phases to the three largest values of $|F_m|$ having linear independent vectors m. Because of symmetry there may be additional possibilities of fixing certain phases, but a major tool for reducing the number of free parameters is the symbolic addition procedure [41,73]: One assigns formal names to certain phases and uses origin-independent linear relations ("structure invariants") like

$$(13) \qquad \phi_{m_1} + \phi_{m_2} + \phi_{m_3} = \text{const for } m_1 + m_2 + m_3 = o$$

to generate formal phases for a number of other $|F_m|$ values. The more symmetry a crystal has, the more relations like (13) are valid. Now several possibilities for the values of the formal names are tested; in each case (11) or (12) is taken to try to assign more phases as long as certain probability measures are satisfactory. Thus a number of sets of phase estimates is generated; for any set (before L_2-refinement is made) a Fourier synthesis (7) is made by plotting ρ and practically impossible estimates are sorted out by visual inspection.

There is a vast literature on the above strategy (see e.g. [1,2,13,14,18,19,21,25-27,29,30,33-36,40,41,67,68,72,73] and many other publications) and there are mainly two important program systems based on it (X-RAY-subroutines, [1,62,63], and MULTAN [20,43, 48,51,52]), which already solved a large number of moderately complex structure problems.

Of course, the scope of the method is limited by the applicability of relations based on (11), which has led to new efforts in developing phase estimates and to combining phase estimation with L_2-methods [e.g. 1,22,23,33,34,48]. Since the chemical components occuring in today's microbiological research still mainly are beyond the scope of this method, some further investigation of the above estimation problems seems worthwhile, both from the mathematical and the crystallographical point of view.

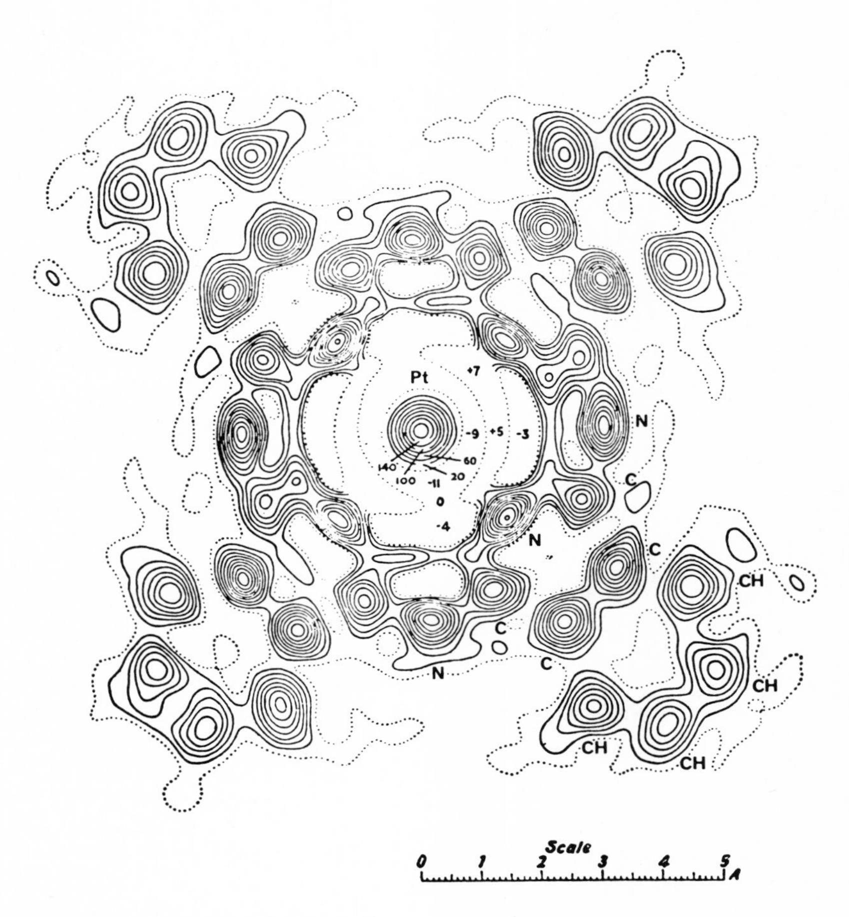

Fig.8: Electron density $\rho(x)$ of phthalocyanine [39]

REFERENCES

1. Ahmed,F.R.(ed.): Crystallographic Computing, Copenhagen: Munksgaard 1970

2. Ahmed,F.R.(ed.): Crystallographic Computing II, Copenhagen: Munksgaard (to appear)

3. Azaroff, L.V. and Buerger,M.J.: The Powder Method in X-ray Crystallography, New York: Mc Graw-Hill 1958

4. Babenko,N.F. and Brusentsev,F.A.: Indexing X-Ray Powder Patterns for a Crystal of an Arbitrary System, Kristallografiya 19(1974), 506-510

5. Barabash,I.A. and Davydov,G.V.: On the Indexing of Powder Patterns for Polycrystalline Materials of the Orthorhombic System, Acta Cryst.A 24(1968),608

6. Bradley,C.J. and Cracknell,A.P.: The mathematical theory of symmetry in solids. Representation theory for point groups and space groups. Oxford: Clarendon Press 1972

7. Braess,D.: Die Konstruktion der Tschebyscheff-Approximierenden bei der Anpassung mit Exponentialsummen, J. of Approx.Th.3(1970), 261-273

8. Braess,D.: Nichlineare Anpassung von Meßdaten nach der Methode der kleinsten Quadrate, Schriftenreihe des Rechenzentrums der Universität Münster 6(1974)

9. Braess,D.: Nonlinear Approximation in Lorentz,G.G.(ed.): Approximation Theory, London - New York: Academic Press (to appear)

10. Buerger,M.J.: Crystal-Structure Analysis, New York - London: Wiley & Sons 1960

11. Buerger,M.J.: Vector Space and its Application in Crystal-Structure Investigation, New York: Wiley & Sons 1959

12. Busing,W.R., Martin,K.O. and Levy,H.A.: ORFLS/ORFFE Oak Ridge National Laboratory Reports ORNL-TM-305/306, 1962/1964

13. Cochran,W.: Relations between the phases of structure factors, Acta Cryst.8(1955),473-478

14. Cochran,W. and Woolfson,M.M.: The theory of sign relations between structure factors, Acta Cryst. 8(1955),1-12

15. Coiro,V.M. and Mazza,F.: The Crystal Structure of the Conformational Analysis of Carbobenzoxy - L - lencyl - p - nitrophenyl Ester, Acta Cryst.B 30(1974),2607

16. Cruickshank,D.W.J., Pilling,D.E., Bujosa,A., Lovell,F.M. and Truter,M.R.: Computing Methods and the Phase Problem in X-ray Crystal Analysis, Oxford: Pergamon Press 1961

17. Cullity,B.D.: Elements of X-ray Diffraction, Reading, Mass.: Addison - Wesley 1956

18. Debaerdemaeker,T. and Woolfson,M.M.: On the Application of Phase Relationships to Complex Structures IV: The Coincidence Method Applied to General Phases, Acta Cryst.A 28(1972),477

19. Debaerdemaeker,T. and Woolfson,M.M.: The determination of structure invariants II, Acta Cryst.A 31(1975),401

20. Declerq,J.P., Germain,G., Main,P. and Woolfson,M.M.: On the application of Phase Relationships to Complex Structures V: Finding the solution, Acta Cryst.A 29(1973),231

21. Declerq,J.P. and Woolfson,M.M.: On the Application of Phase Relationships to Complex Structures VIII: An extension to the magic-integer approach, Acta Cryst.A 31(1975),367

22. Gassmann,J.: Least-squares Refinement of Phases in Direct and Reciprocal Spaces, Acta Cryst.A 32(1976),274

23. Gassmann,J.and Zechmeister,K.: Limits of Phase Expansion in Direct Methods, Acta Cryst.A 28(1972),270

24. Germain,G. and Woolfson,M.M.: Some Ideas on the Deconvolution of the Patterson Function, Acta Cryst.A 21(1966),845

25. Germain,G. and Woolfson,M.M.: On the Application of Phase Relationships to Complex Structures, Acta Cryst.B 24(1968),91

26. Germain,G., Main,P. and Woolfson,M.M.: On the Application of Phase Relationships to Complex Structures II: Getting a Good Start, Acta Cryst.B 26(1970),274

27. Germain,G., Main,P. and Woolfson,M.M.: On the Application of Phase Realtionships to Complex Structures III: The Optimum Use of Phase Relationships, Acta Cryst.A 27(1971),368

28. Haendler,H. and Cooney,W.: Computer Determination of Unit Cell from Powder-Diffraction Data, Acta Cryst.16(1963),1243

29. Hauptman ,H. and Karle,J.: The solution of the phase problem I: The centrosymmetrical crystal, Amer.Cryst.Assoc.Monograph No.3, Ann Arbor: Edwards Brothers 1953

30. Hauptman ,H. and Karle,J.: Phase determination from new joint probability distributions: Space group $P\bar{1}$, Acta Cryst.11(1958), 149-157

31. Hauptman ,H., Fisher,J. and Weeks,C.M.: Phase Determination by Least-Squares Analysis of Structure Invariants: Discussion of This Method as Applied on Two Androstane Derivations, Acta Cryst.B 27(1971),1550

32. Hauptman ,H.: Crystal structure determination, New York: Plenum Press 1972

33. Hauptman ,H.: A joint probability distribution of seven structure factors, Acta Cryst.A 31(1975),671

34. Hauptman ,H.: A new method in the probabilistic theory of the structure invariants, Acta Cryst.A 31(1975),680

35. Hoppe,W.: Phasenbestimmung durch Quadrierung der Elektronendichte im Bereich von 2 Å - bis 1,5 Å - Auflösung, Acta Cryst. A 15(1962),13

36. Hoppe,W. and Gassmann,J.: Phase Correction, a New Method to Solve Partially Known Structures, Acta Cryst.B 24(1968),97

37. Ishida,T. and Watanabe,Y.: Analysis of Powder Diffraction Patterns of Monoclinic and Triclinic Crystals, J.Appl.Cryst.4 (1971),311

38. Ito,T.: X-ray Studies on Polymorphism, Tokyo: Maruzen 1950

39. Jeffery,J.W.: Methods in X-ray Crystallography, London - New York: Academic Press 1971

40. Karle,J. and Hauptmann,H.: Phase determination from new joint probability distributions: Space group P1, Acta Cryst.11(1958), 264-269

41. Karle,J. and Karle,I.L.: The Symbolic Addition Procedure for Phase Determination for Centrosymmetric and Noncentrosymmetric crystals, Acta Cryst.21(1966),849

42. Kleber,W.: Einführung in die Kristallographie, 10.Auflage, Berlin: VEB Verlag Technik 1969

43. Koch,M.H.J.: On the Application of Phase Relationships to Complex Structures VI: Automatic Interpretation of Electron-Density Maps for Organic Structures, Acta Cryst.B 30(1974),67

44. Konnert,J.H.: A Restrained-Parameter Structure-Factor Least-Squares Refinement Procedure for Large Asymmetric Units, Acta Cryst.A 32(1976),614

45. Larisch,E.: Algorithmus zur Indizierung von Röntgen-Pulveraufnahmen, Diplomarbeit Münster 1975

46. Larisch,E.: Private communication, 1976

47. Lenstra,A.T.H. and Schoone,J.C.: An Automatic Deconvolution of the Patterson Synthesis by Means of a Modified Vector-Verification Method, Acta Cryst.A 29(1973),419

48. Lessinger,L.: On the Application of Phase Relationships to Complex Structures IX, MULTAN Failures, Acta Cryst.A 32(1976), 538

49. Louër,D. and Louër,M.: Méthode d' Essais et Erreurs pour l' Indexation Automatique des Diagrammes de Poudre, J.Appl.Cryst. 5(1972),271

50. Lynch,M.F., Harrison,J.M., Town,W.G. and Ash,J.E.: Computer Handling of Chemical Structure Information, London: Mac Donald, New York: American Elsevier 1971

51. Main,P.,Woolfson,M.M. and Germain,G.: MULTAN: A Computer Program for the Automatic Solution of Crystal Structures, Univ. of York 1971

52. Main,P., Woolfson,M.M., Lessinger,L., Germain,G. and Declerq, J.P.: MULTAN 74: A system of Computer Programmes for the Automatic Solution of Crystal Structures from X-Ray Diffraction Data 1974

53. Pawley,G.S.: Advances in Structure Research by Diffraction Methods, New York: Pergamon Press

54. Rae,A.D.: The Phase Problem and its Implications in the Least-Squares Refinement of Crystal Structures, Acta Cryst.A 30(1974), 761

55. Rollett,J.S.(ed.): Computing Methods in Crystallography, Oxford: Pergamon Press 1965

56. Runge,C.: Die Bestimmung eines Kristallsystems durch Röntgenstrahlen, Physik.Z.18(1917),509-515

57. Sayre,D.: The Squaring Method: A New Method for Phase Determination, Acta Cryst.5(1952),60

58. Sayre,D.: The double Patterson function, Acta Cryst.6(1953), 430

59. Sayre,D.: On Least-Squares Refinement of the Phases of Crystallographic Structure Factors, Acta Cryst.A 28(1972),210

60. Sayre,D.: Least-Squares-Refinement II: High-Resolution Phasing of a Small Protein,Acta Cryst.A 30(1974),180

61. Schaback,R.: Ein Optimierungsproblem aus der Kristallographie, in Collatz,L. und Wetterling,W.(ed.): Numerische Methoden bei Optimierungsaufgaben, ISNM 23, Basel - Stuttgart: Birkhäuser 1974,113-123

62. Stewart,J.M., Kundell,F.A. and Baldwin,J.C.: The X-RAY 70 system, Computer Science Center, Univ.of Maryland , College Park, Maryland 1970

63. Stewart,J.M., Kruger,G.J., Ammon,H.L., Dickinson,C. and Hall, S.R.: X-RAY system, Report TR-192, Computer Science Center, Univ.of Maryland, College Park, Maryland 1972

64. Vand,V. and Johnson,G.G.: Indexing of X-Ray Powder Patterns, Part I. The Theory of the Triclinic Case, Acta Cryst.A 24(1968), 543

65. Viswanathan,K.: A Systematic Approach to Indexing Powder Patterns of Lower Symmetry Using De Wolff's Principles, The American Mineralogist 53(1968),2047

66. Visser, J.W.: A Fully Automatic Program for Finding the Unit Cell from Powder Data, J.Appl.Cryst.2(1969),89-95

67. Viterbo,D. and Woolfson,M.M.: The Determination of Structure Invariants I. Quadrupoles and Their Uses, Acta Cryst.A 29(1973), 205

68. White,P.S. and Woolfson,M.M.: The Application of Phase Relationships to Complex Structures VII: Magic integers, Acta Cryst.A 31(1975),53

69. Wilson,A.J.C.: Mathematical Theory of X-Ray Powder Diffractometry, Eindhoven: Philips Technical Library 1963

70. de Wolff,P.M.: On the Determination of Unit-cell Dimensions from Powder Diffraction Patterns, Acta Cryst.10(1957),590-595

71. de Wolff,P.M.: Indexing of Powder Diffraction Patterns, Advances in X-ray Analysis 6,1-17

72. Woolfson,M.M.: Direct methods in Crystallography, Oxford: Pergamon Press 1961

73. Zachariasen,W.H.: A new analytical method for solving complex crystal structures, Acta Cryst.5(1952),68-73

74. Zachariasen,W.H.: Interpretation of Monoclinic X-ray Diffraction Patterns, Acta Cryst.16(1963),784

This work was supported by the Deutsche Forschungsgemeinschaft within the projects "Indizierung von Röntgen - Pulveraufnahmen" at the University of Münster and "Approximation auf DV - Anlagen" in the Sonderforschungsbereich 72 "Approximation und Optimierung" at the University of Bonn. Calculations were done on the IBM 360/50 of the Rechenzentrum of the University of Münster and the IBM 360/165 and IBM 360/168 of the Regionales Hochschulrechenzentrum Bonn.

ESTIMATION PROBLEMS IN DATA-TRANSMISSION SYSTEMS

G. Ungerboeck

IBM Zurich Research Laboratory

Säumerstrasse 4, 8803 Rüschlikon, Switzerland

ABSTRACT

The transfer of information over noisy and dispersive media has traditionally been, and still represents, an important subject of applied estimation and approximation theory. In this paper the specific problems encountered in synchronous data-transmission systems are reviewed. A first set of problems arises in timing- and carrier-phase tracking and in adaptive equalization, where continuous-valued parameters are to be estimated which may change slowly over time. A second set of problems deals with recovering the transmitted information from received noisy signals. It is shown that in the presence of severe signal distortion and/or redundant sequence coding the optimum receiver has to solve a dynamic programming problem.

I. INTRODUCTION

The field of information transfer over noisy and dispersive media has traditionally been closely related to estimation and approximation theory. It is primarily the receiver functions that follow in quite a direct manner from these areas. This is well illustrated by standard texts on communication theory, e.g., Van Trees,[1] and by many articles in the literature. In this paper the specific estimation and approximation problems encountered in the receiver section of synchronous data-transmission systems are reviewed. Today, this form of transmission is rapidly growing in importance, owing to increasing needs for efficient data communication over various media and to the expected massive introduction of PCM principles in telephony.

In synchronous data transmission, information is encoded in discrete signal values a_n which are transmitted at constant rate 1/T. There is little doubt that this is most efficiently accomplished by linear pulse-modulation schemes which, depending on the channel available, are used in baseband or linear carrier-modulation form. For this class of modulation schemes the signal at the receiver input can be written as

$$x_c(t) = \mathrm{Re}\left\{ \sum_n a_n\, h(t - \tau_s - nT)\, e^{j\omega_c t + j\varphi_c} \right\} + w_c(t), \qquad (1)$$

where

a_n ... discrete signal values (pulse amplitudes),
$h(t)$ ... signal element (elementary pulse waveform)
T ... signal spacing (1/T .. signaling rate),
τ_s ... timing phase,
ω_c ... carrier frequency in radians ($\omega_c = 2\pi f_c$),
φ_c ... carrier phase,
$w_c(t)$... additive channel noise,

and the term linear pulse modulation is due to $x_c(t)$ being linear in the a_n.

Each of the above quantities may a priori be unknown in the receiver, which ultimately must determine from $x_c(t)$ the values a_n that have been transmitted. Usually, however, at least 1/T and ω_c can be assumed to be known with sufficient precision so that uncertainities with respect to them can be absorbed in slow variations of τ_s and φ_c.

In baseband (BB) modulation ($\omega_c = 0$, $\varphi_c = 0$) a_n and $h(t) = h_r(t)$ are real-valued. In carrier modulation ($\omega_c \neq 0$) one must distinguish between vestigial-sideband (VSB) modulation, where a_n is real-valued and $h(t) = h_r(t) \pm j\, h_i(t)$ [$h_i(t) \simeq$ Hilbert transform of $h_r(t)$]* and double-sideband (DSB) modulation, where a_n may be real or complex-valued depending on whether one is dealing with amplitude modulation (AM), phase modulation (PM), or a combination thereof (AM/PM), and $h(t) \simeq h_r(t)$. These aspects are further summarized in Fig. 1 (see also Refs. 2-4).

---

*Half of the spectrum of $h_r(t)$ is cancelled by adding $(\pm)\, j\, h_i(t)$, since $H_i(f) \simeq -j\,\mathrm{sign}(f)\, H_r(f)$ leads to $H(f) \simeq 2\, H_r(f)$ for $f\,(\gtrless)\,0$ and to $H(f) \simeq 0$ for $f\,(\lessgtr)\,0$, with an appropriate spectral roll-off across $f = 0$ which on one side of $f = 0$ leaves a "vestigial sideband".

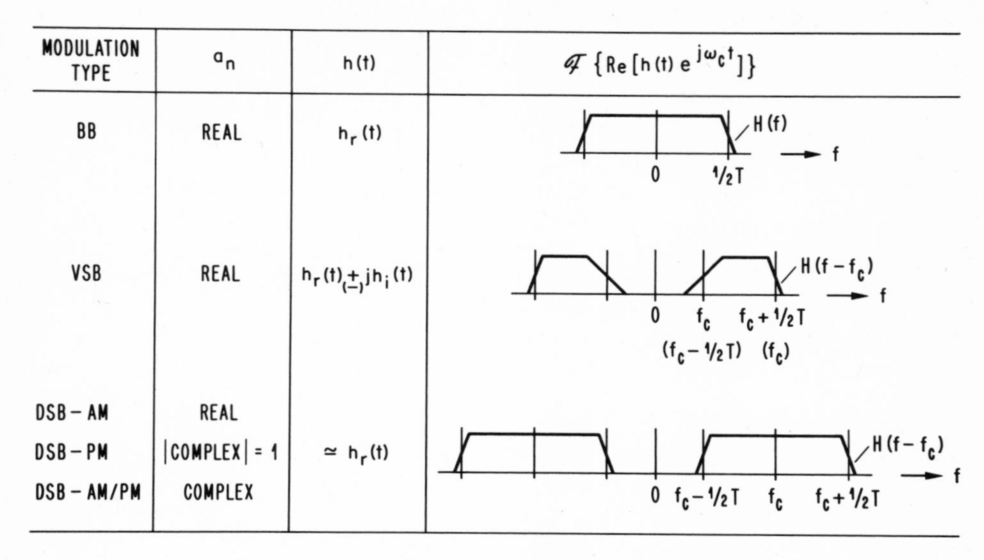

MODULATION TYPE	a_n	h(t)	$\mathcal{F}\{\mathrm{Re}[h(t)\,e^{j\omega_c t}]\}$
BB	REAL	$h_r(t)$	H(f); 0, 1/2T; f
VSB	REAL	$h_r(t) \underset{(-)}{+} jh_i(t)$	H(f − f_c); 0, f_c, f_c + 1/2T; (f_c − 1/2T), (f_c); f
DSB − AM	REAL		
DSB − PM	|COMPLEX| = 1	$\simeq h_r(t)$	H(f − f_c); 0, f_c − 1/2T, f_c, f_c + 1/2T; f
DSB − AM/PM	COMPLEX		

Fig. 1. Overview over modulation forms.

The first function to be performed by the receiver in the case of VSB or DSB, is transposing the signal into baseband* which leads from (1) to

$$x(t) \;=\; \sum_n a_n\, h(t - \tau_s - nT)\, e^{j\varphi_c} + w(t) \tag{2}$$

In the following we shall regard the received signal only in this form which exhibits the least differences between the various modulation forms. Note that for VSB and DSB, $x(t)$ is complex. In implementation this means that one deals concurrently with two real signals representing the real and imaginary parts of $x(t)$ (also called in-phase and quadrature components).

Ideally, τ_s and φ_c would be known, and depending on whether a_n is real-valued (BB, VSB, DSB-AM) or complex-valued (DSB-PM, DSB-AM/PM) $h(t)$ would satisfy

BB,VSB,DSB-AM:

$$h_r(\ell T) \;=\; h_0 \cdot \delta_\ell \;\Leftrightarrow\; \sum_\ell H_r\!\left(f + \frac{\ell}{T}\right) = h_0/T \tag{3a}$$

---

*E.g., by multiplying $x_c(t)$ by $e^{-j\omega_c t}$, and cancelling signal components around $f = -2f_c$ by low-pass filtering.

DSB-PM, DSB-AM/PM:

$$h(\ell T) = h_0 \cdot \delta_\ell \Leftrightarrow \sum_\ell H\left(f + \frac{\ell}{T}\right) = h_0/T \tag{3b}$$

(δ_ℓ ... Kronecker delta). This is the condition for zero *inter-symbol interference (ISI)* between consecutive pulses at sampling instants. The frequency-domain equivalent of this condition is well-known as the *first Nyquist criterion* and explains the two-sided bandwidth requirement of 1/T or 2/T Hz, plus spectral roll-offs, necessary for transmission of real or complex values a_n, respectively. In the absence of noise the values of a_n could simply be determined by sampling $x(t)\, e^{-j\varphi_c}$ at times $nT + \tau_s$:

$$a_n = (\mathrm{Re}) \frac{1}{h_0} x(t) e^{-j\varphi_c} \Big|_{t = nT + \tau_s} . \tag{4}$$

However, the fact that none of the above ideal suppositions is satisfied under real conditions, requires the following estimation functions to be performed in a receiver for synchronous data signals:

(a) Estimation of τ_s: "Timing-phase tracking"

(b) Estimation of φ_c: "Carrier-phase tracking"

(c) Estimation of $h(t)$ and corrective action in case of non-zero ISI: "Adaptive equalization"

(d) Estimation of $\{a_n\}$ in the presence of noise and ISI and/or redundant coding: "Signal detection".

These functions will be discussed in Sections III to V. In Section II we establish a basis for these discussions by indicating the general mathematical solution to the entire receiver problem.

II. GENERAL SOLUTION BY MEANS OF A SINGLE LIKELIHOOD FUNCTION

Suppose that the signal $x(t)$ is observed within the time interval $[0, NT]$, where N is sufficiently large, and let X_N denote a realization of $x(t)$ within this interval. Assuming $w(t)$ is stationary additive white Gaussian noise (AWGN) with real and, in the case of VSB and DSB, imaginary parts being independent and each having two-sided spectral noise power density N_0, one obtains the likelihood function (see, e.g., Refs. 1, 5, 6)

$$p(X_N|\{\hat{a}_n\}, \hat{\tau}_s, \hat{\varphi}_c) \sim \Lambda(X_N|\{\hat{a}_n\}, \hat{\tau}_s, \hat{\varphi}_c) =$$

$$= \exp\left\{-\frac{1}{2N_0}\int_0^{NT} |x(t)-\sum_n \hat{a}_n h(t-\hat{\tau}_s-nT)e^{j\hat{\varphi}_c}|^2 \, dt\right\}, \tag{5}$$

where $\{\hat{a}_n\}$, $\hat{\tau}_s$ and $\hat{\varphi}_c$ are estimates of these respective quantities. The condition of white noise is no restriction since it can always be induced by an appropriate whitening filter. The assumption of Gaussian noise is, however, somewhat artificial; it must be made in order to make the problem mathematically tractable. In (5), h(t) is assumed to be known, but it could also be replaced by an estimate $\hat{h}(t)$. Defining

$$y_n(\hat{\tau}_s) = \int \bar{h}(-t)\, x(nT + \hat{\tau}_s - t)\, dt \tag{6}$$

($\bar{h}$ ... conjugate complex of h) and

$$s_\ell = \int \bar{h}(-t)\, h(\ell T - t)\, dt = \bar{s}_{-\ell} \quad , \tag{7}$$

and taking the logarithm of (5), yields

$$\lambda(X_N|\{\hat{a}_n\},\hat{\tau}_s,\hat{\varphi}_c) = -\frac{1}{2N_0}\int_0^{NT} |x(t)|^2 \, dt \; +$$

$$+ \frac{1}{2N_0}\left\{\sum_{N=0}^{N} 2\mathrm{Re}\left[y_n(\hat{\tau}_s)e^{-j\hat{\varphi}_c}\cdot\bar{\hat{a}}_n\right] - \sum_{i=0}^{N}\sum_{k=0}^{N} \bar{\hat{a}}_i s_{i-k}\hat{a}_k\right\} \rightarrow \text{Max.} \tag{8}$$

The derivation of the sample values $y_n(\hat{\tau}_s)$ from x(t) is illustrated in Fig. 2.

These sample values represent, for $\hat{\tau}_s = \tau_s$, a *set of sufficient statistics*[1] for estimating $\{a_n\}$ from x(t). According to the con-

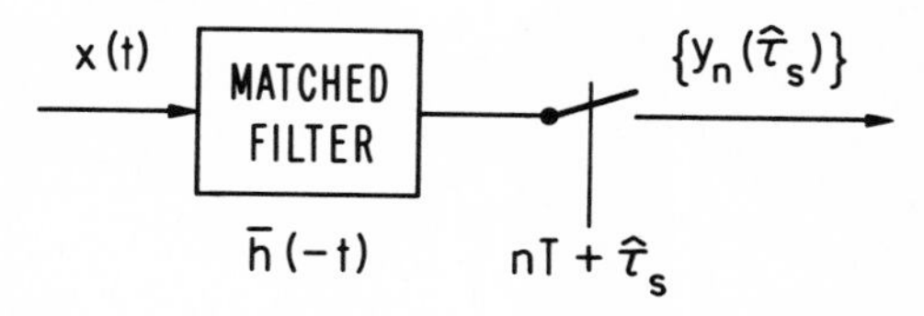

Fig. 2. Derivation of sample values $y_n(\hat{\tau}_s)$ by means of a matched filter and sampling.

cept of maximum likelihood (ML) an optimum receiver must jointly determine the quantities $\{\hat{a}_n\}$, $\hat{\tau}_s$ and $\hat{\varphi}_c$ [and perhaps also $\hat{h}(t)$] which maximize (8).

Formulation of the receiver problem in this general form is, however, impractical for solving in one step. It is more appropriate to approach the various problems individually. For timing- and carrier-phase estimation this will be discussed in the next section.

III. ESTIMATION OF TIMING PHASE AND CARRIER PHASE

Assume that the time-continuous output signal of the matched filter (MF) is given by

$$y(t) = \sum_n a_n \, s(t - \tau_s - nT) \, e^{j\varphi_c} + r(t) \quad , \tag{9}$$

and that $s(t)$ satisfies the condition for zero ISI: $s(\ell T) = 0$ for $\ell \neq 0$ (for VSB and DSB-AM $\mathrm{Re}[s(\ell T)] = 0, \ell \neq 0$ is sufficient). In the following one must distinguish between various degrees of knowledge about the transmitted values of a_n. The simplest case is when the values of a_n are already known (e.g., as a result of decisions made by the receiver).

a) $\{a_n\}$ *known*. According to (8), the best estimates of τ_s and φ_c are those which maximize

$$\lambda(X_N | \{a_n\}, \hat{\tau}_s, \hat{\varphi}_c) \sim \sum_{n=0}^{N} \mathrm{Re}\left[y_n(\hat{\tau}_s) \, e^{-j\hat{\varphi}_c} \, \bar{a}_n \right] \rightarrow \mathrm{Max} \quad . \tag{10}$$

Differentiating with respect to $\hat{\tau}_s$ and $\hat{\varphi}_c$ yields

$$\frac{\partial \lambda}{\partial \hat{\tau}_s} \sim \sum_{n=0}^{N} \mathrm{Re}\left[\dot{y}_n(\hat{\tau}_s) \, e^{-j\hat{\varphi}_c} \, \bar{a}_n \right] \rightarrow 0 \quad , \tag{11}$$

$$\frac{\partial \lambda}{\partial \hat{\varphi}_c} \sim \sum_{n=0}^{N} \mathrm{Im}\left[y_n(\hat{\tau}_s) \, e^{-j\hat{\varphi}_c} \, \bar{a}_n \right] \rightarrow 0 \quad , \tag{12}$$

and replacing summation over n by expectation indicates that the receiver must solve the regression problem[5]

$$E\left\{ \mathrm{Re}\left[\dot{y}_n(\hat{\tau}_s) \, e^{-j\hat{\varphi}_c} \, \bar{a}_n \right] \right\} \rightarrow 0 \quad , \tag{13}$$

$$E\left\{\mathrm{Im}\left[y_n(\hat{\tau}_s)\, e^{-j\hat{\varphi}_c}\, \bar{a}_n\right]\right\} \to 0\,, \qquad (14)$$

which may be accomplished by the simultaneous stochastic approximation[5,7] algorithms

$$\hat{\tau}_s^{(n+1)} = \hat{\tau}_s^{(n)} - \alpha^{(n)} \mathrm{Re}\left[\dot{y}_n(\hat{\tau}_s)\, e^{-j\hat{\varphi}_c}\, \bar{a}_n\right], \quad \alpha^{(n)} > 0, \qquad (15)$$

$$\hat{\varphi}_c^{(n+1)} = \hat{\varphi}_c^{(n)} - \beta^{(n)} \mathrm{Im}\left[y_n(\hat{\tau}_s)\, e^{-j\hat{\varphi}_c}\, \bar{a}_n\right], \quad \beta^{(n)} > 0. \qquad (16)$$

This approach is referred to as *decision-aided* timing- and carrier-phase tracking.

A slight modification of these algorithms is obtained if instead of maximizing (10) one minimizes the mean-square value of the decision error e_n:

$$E\left\{|e_n|^2\right\} = E\left\{\left|(\mathrm{Re})\, y_n(\hat{\tau}_s)\, e^{-j\hat{\varphi}_c} - a_n\right|^2\right\} \to \mathrm{Min}\,. \qquad (17)$$

It is easy to show that in (15) and (16), $\bar{a}_n$ must then be replaced by $-\bar{e}_n$. This has the effect of accelerating convergence considerately. The careful reader will observe that criterion (17) in fact is a combination of ML criterion (10) with ML criterion (20) or (21), discussed further below.

It is sometimes desirable to avoid the sampled time deviative in (15). Let $E\{\bar{a}_i a_k\} = \delta_{i-k}$. For estimating τ_s one may then exploit that $s(-T + \Delta\hat{\tau}_s) - s(T + \Delta\hat{\tau}_s) \gtrless 0$ for $\Delta\hat{\tau}_s = \hat{\tau}_s - \tau_s \gtrless 0$, which suggests the approximation algorithm[8]

$$\hat{\tau}_s^{(n+1)} = \hat{\tau}_s^{(n)} -$$

$$- \alpha^{(n)}\, \mathrm{Re}\left[e^{-j\hat{\varphi}_c}\left(y_{n-1}(\hat{\tau}_s)\bar{a}_n - y_n(\hat{\tau}_s)\bar{a}_{n-1}\right)\right], \quad \alpha^{(n)} > 0\,,$$

$$\downarrow$$

$$\text{Expectation:} \quad \mathrm{Re}\left[e^{j(\varphi_c - \hat{\varphi}_c)}\left(s(-T + \Delta\hat{\tau}_s) - s(T + \Delta\hat{\tau}_s)\right)\right]. \qquad (18)$$

The convergence properties of decision-aided algorithms for recovery of τ_s have been analyzed in Refs. 8 and 9.

b) $\{a_n\}$ *unknown, normally distributed.* Here one must refer back to (5) and integrate

$$\int p(X_N|\{a_n\},\hat{\tau}_s,\hat{\varphi}_c)\, dP(\{a_n\}) \tag{19}$$

in order to eliminate $\{a_n\}$, which for real a_n (BB: $\varphi_c = 0$, VSB, DSB-AM) yields

$$\lambda(X_N|\hat{\tau}_s,\hat{\varphi}_c) \sim \sum_{n=0}^{N} \left|\mathrm{Re}\left[y_n(\hat{\tau}_s)\, e^{-j\hat{\varphi}_c}\right]\right|^2 =$$

$$= \frac{1}{2}\sum_{n=0}^{N} |y_n(\hat{\tau}_s)|^2 + \frac{1}{2}\,\mathrm{Re}\left[\sum_{n=0}^{N} y_n^2(\hat{\tau}_s)\, e^{-j2\hat{\varphi}_c}\right] \rightarrow \mathrm{Max}, \tag{20}$$

and for complex a_n (DSB-AM/PM)

$$\lambda(X_N|\hat{\tau}_s,\hat{\varphi}_c) \sim \sum_{n=0}^{N} |y_n(\hat{\tau}_s)|^2 \rightarrow \mathrm{Max}\,. \tag{21}$$

The ML criterion (21) indicates that for DSB-AM/PM, with the assumption of $\{a_n\}$ being normally and identically distributed in its real and imaginary parts, carrier-phase estimation is not possible, whereas from (20) it follows that for VSB and DSB-AM, φ_c can be estimated only with an ambiguity of π. A further observation is that substituting in (20) the optimum value of $\hat{\varphi}_c$ would make (20) and (21) identical.

The traditional timing-phase tracking method without prior knowledge of $\{a_n\}$ has been squaring the received signal, i.e., computing $|y(t)|^2$, and extracting a spectral line at $f_T = 1/T$ Hz by means of a narrow bandpass filter ("timing tank"). The maxima or, with a phase shift of $\pm$ T/4, the zero crossings of the so obtained sinosoidal signal determine the sampling instants $nT + \hat{\tau}_s$. For BB modulation this direct method has been analyzed in Ref. 10, and for DSB-AM/PM the method is described in Ref. 11. Its resemblance to the ML approach suggested by (21) is rather apparent, if the bandpass-filter function is recognized as a sliding summation over T-spaced signal values.

Differentiating (20) with respect to $\hat{\varphi}_c$ yields

$$\frac{\partial\lambda}{\partial\hat{\varphi}_c} = \sum_{n=0}^{N} \mathrm{Re}\left(y_n(\hat{\tau}_s)\, e^{-j\hat{\varphi}_c}\right) \cdot \mathrm{Im}\left(y_n(\hat{\tau}_s)\, e^{-j\hat{\varphi}_c}\right) \rightarrow 0\,, \tag{22}$$

which in essence describes the control term used in Costa's loop for carrier-phase tracking in VSB and DSB-AM systems (Ref. 8, p. 184). A direct method for deriving a carrier signal with appropriate phase from the received-signal, when $f_c >> 1/T$, consists in squaring the received signal, extracting a sinosoidal component at $f = 2f_c$ by a narrow bandpass filter, and frequency-dividing this signal by two (with phase ambiguity $k\pi$), (Ref. 8, p. 182). The output signal of the bandpass filter can be written as

$$\mathrm{Re}\left[y^2(t)|_{LPF} \cdot e^{j2\omega_c t}\right] \sim \mathrm{Re}\left[e^{j2\omega_c t + j[\mathrm{arc}\, y^2(t)|_{LPF}]}\right].$$

Since the ML phase estimate from (20) is given by

$$2(\hat{\varphi} + k\pi) = \mathrm{arc}\left[\sum_{n=0}^{N} y_n^2(\hat{\tau}_s)\right] \simeq \mathrm{arc}\left[y^2(t)|_{LPF}\right] ,$$

if follows that the above direct method is essentially equivalent to ML estimation of φ_c.

It can be shown[12,13] that the variances of the maximum-likelihood estimates of τ_s and φ_c derived from (20) and (21) ($\{a_n\}$ unknown), depend critically on the spectral roll-off characteristics of $S(f) = \mathcal{F}\{s(t)\}$. Roughly, $\mathrm{Var}(\hat{\tau}_s)$ is proportional to $(N \cdot A)^{-1}$, whereas for VSB and DSB-AM, $\mathrm{Var}(\hat{\varphi}_c)$ is proportional to $(N \cdot B)^{-1}$, where A and B are defined in Fig. 3. With very steep roll-offs around $f = \pm 1/2T$, A approaches zero and hence timing-phase estimation becomes impossible. With respect to carrier-phase estimation one sees that VSB must be considerable inferior to DSB-AM since $B_{VSB} << B_{DSB}$. For single-sideband (SSB) modulation, B is zero and hence carrier-phase estimation with normally distributed values of $\{a_n\}$ cannot work. In SSB systems pilot tones have therefore often been used to provide carrier-phase information.

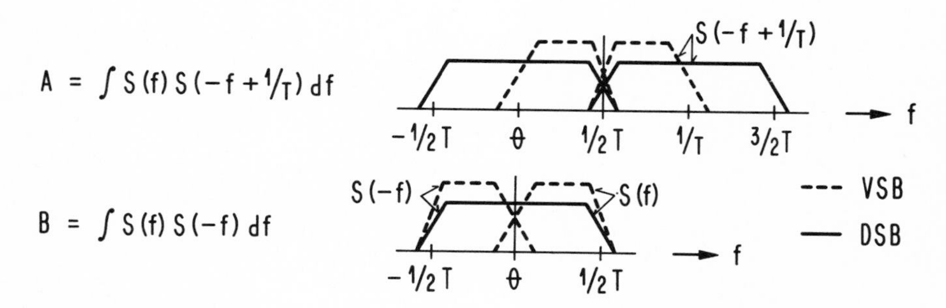

Fig. 3. Definition of values A and B.

In not yet published work[13] Cramer-Rao bounding techniques (Ref. 1, Part I) have been used to compute efficient lower bounds for the ML estimates of τ_s and φ_c for the various modulation schemes and for $\{a_n\}$ known or unknown (normally distributed). The results have been expressed in terms of the mean-square error, defined in (17), that results from noise and the uncertainities in estimating τ_s and φ_c. Thus bounds of the form

$$E\left\{|e_n(\hat{\tau}_s,\hat{\varphi}_c)|^2\right\} \geq E\left\{|e_n(\tau_s,\varphi_c)|^2\right\} \cdot (Q/N + 1) \ , \tag{23}$$

have been obtained, where $N.T$ is the length of the observation interval and Q represents an inverse quality factor. For VSB, DSB-AM and DSB-AM/PM this factor is shown in Fig. 4 as a function of the spectral roll-offs and for $\{a_n\}$ known and unknown. (For carrier-phase estimation in DSB-AM/PM $\{a_n\}$ must always be known, i.e., a decision-aided scheme must be used.) In (23) equality is well approached for large N, say $N > 10$. The small values of Q

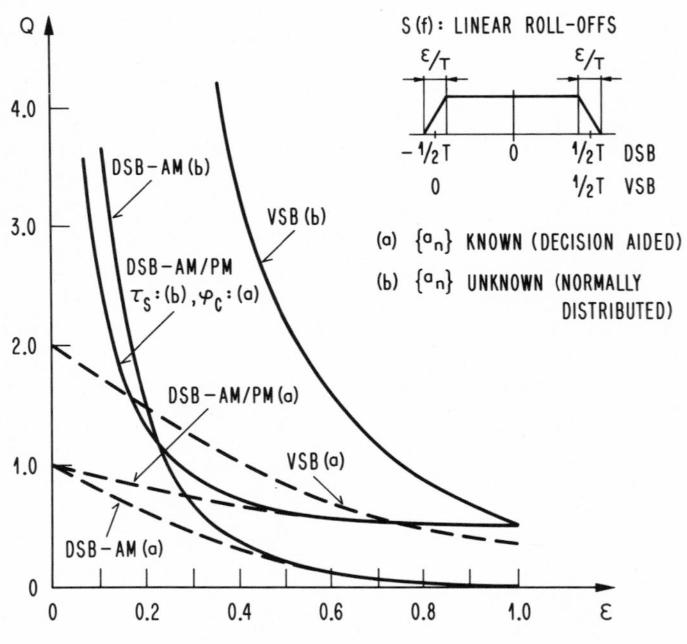

Fig. 4. Inverse quality factor Q as a function of the linear roll-off quantity ε for various modulation forms with τ_s and φ_c being known or unknown.

indicate then a remarkable capability for fast recovery, except when the roll-offs in DSB became smaller than 15% and in VSB smaller than 40%, and $\{a_n\}$ is not known. Clearly, this finding indicates a considerable weakness of VSB which explains why DSB-AM/PM and not VSB became the preferred carrier-modulation form for high-speed data-transmission systems.

Much more could be said about timing- and carrier-phase tracking. In particular, a performance comparison should be of interest between direct ML estimation of τ_s and φ_c by means of continuous-time processing, which in good approximation is realized in existing systems by signal squaring and bandpass filtering, and estimating τ_s and φ_c iteratively from T-spaced signal values by stochastic approximation. The latter technique becomes important, when digital realizations are envisaged. It also represents the more natural approach when knowledge about $\{a_n\}$ is available.

It has so far been assumed that $\{a_n\}$ is either known or unknown and normally distributed. In between are cases where $\{a_n\}$ is not known, but advantage can be taken from knowing its possible discrete values. For example, if $\{a_n\}$ is binary and BB modulation is used, from (19) the likelihood function[14]

$$\lambda(X_N|\hat{\tau}_s) \sim \sum_{n=0}^{N} \ln \cosh\left(\frac{1}{N_0}\, y_n(\hat{\tau}_s)\right) \simeq \sum_{n=0}^{N} \frac{1}{N_0}\, |y_n(\hat{\tau}_s)| \qquad (24)$$

is obtained. A full-wave linear rectifier instead of a square-law device, followed by bandpass filtering, will therefore in this case be more appropriate for timing-phase estimation. This subject is still open for further investigation and extension.

Tracking carrier phase in the presence of significant time variations of φ_c, termed carrier-phase jitter, which occurs on certain telephone channels, is another estimation problem which we shall not discuss here in further detail. Finally, the effect of signal distortion on the performance of the various timing- and carrier-phase tracking schemes must be considered as a further topic worth closer examination.

IV. ADAPTIVE EQUALIZATION

Suppose that τ_s and φ_c have already been determined, but signal distortion still causes a significant amount of intersymbol interference (ISI). This situation occurs especially in data transmission over telephone channels where, owing to a priori unknown and slowly changing channel characteristics, a fixed equalizer cannot adequately correct distortion. This led in the mid 1960's to the development of adaptive equalizers[4] which to some extent find now application also in data transmission over other channels. Let, after preconditioning the received signal by a fixed filter (possibly approximating a matched filter), the carrier-phase corrected signal at sampling instant $nT + \tau_s$ be

$$x_n = \sum_{\ell} h_\ell a_{n-\ell} + w_n , \tag{25}$$

where $h_\ell \neq 0$ for $\ell \neq 0$ causes ISI. A linear adaptive equalizer is usually realized in the form of a transversal filter with T-spaced taps and adjustable tap gains c_i, $0 \leq i \leq N$, as shown in Fig. 5. The output signal of the equalizer

$$z_n = \sum_{i=0}^{N} c_i x_{n-i} = \underline{c}^T \underline{x}_n \tag{26}$$

should approximate a_n as closely as possible. Using the mean-square decision-error criterion, one has

$$E\{|e_n|^2\} = E\{|z_n - a_n|^2\} =$$

$$= \underline{\bar{c}}^T \underline{R}\, \underline{c} - \mathrm{Re}(\underline{\bar{c}}^T \underline{b}) + E\{|a_n|^2\} \rightarrow \mathrm{Min}, \tag{27}$$

where $\underline{R}$ is the $(N+1) \times (N+1)$ autocorrelation matrix

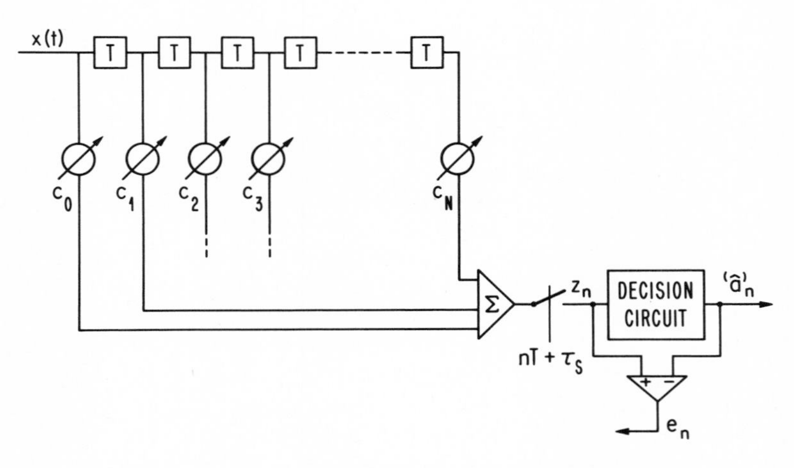

Fig. 5. Adaptive transversal equalizer.

$$\underline{R} = E(\bar{\underline{x}}_n \cdot \underline{x}_n^T) \tag{28}$$

and $\underline{b}$ is the (N+1) vector

$$\underline{b} = E(\bar{x}_n \cdot a_n) \ . \tag{29}$$

The tap gains that minimize (27) are[15]

$$\underline{c}_{opt} = \underline{R}^{-1}\, \underline{b} \ . \tag{30}$$

The estimation problem consists in estimating $\underline{c}_{opt}$ from $\{\underline{x}_n\}$ and $\{e_n\}$, assuming that $\{a_n\} = \{\hat{a}_n\}$ is known with already sufficiently low probability of error. The stochastic approximation algorithm[4,15,16]

$$\underline{c}^{(n+1)} = \underline{c}^{(n)} - \alpha^{(n)} \frac{1}{2} \frac{\partial |e_n|^2}{\partial \underline{c}}$$

$$= \underline{c}^{(n)} - \alpha^{(n)} e_n \bar{\underline{x}}_n \ , \quad \alpha^{(n)} > 0 \ , \tag{31}$$

solves this problem in most existing systems.

It has been shown that for the convergence of (31) the eigenvalues of $\underline{R}$ and the number of N play a significant role,[16,17] and that fastest initial convergence of (31) is generally obtained by using the step-size parameter[17]

$$\alpha_{opt} = 1/E\left\{\bar{\underline{x}}^T\, \underline{x}\right\} . \tag{32}$$

Therewith, $\approx 2.3\,N$ iterations are required to decrease $E\{|e_n|^2\}$ by one order of magnitude, provided the eigenvalue spread of $\underline{R}$ is not excessive.[17]

In half-duplex and multi-point systems, fast receiver set-up is very important, and many attempts have therefore been made to shorten equalization time. A significant new concept has been cyclic equalization,[18] where a known cyclic data sequence with period equal to the length of the equalizer delay line is used for equalizer training. This concept permits operation of the stochastic approximation algorithm at iteration rates above 1/T with precisely known values $\{a_n\}$ from the very beginning. Note that the cyclic nature of the problem also permits discrete Fourier transform techniques to be used for estimating $\underline{c}_{opt}$. Another new concept has been equalization with tap spacing somewhat smaller than T (fractional tap spacing) which makes equalization possible for arbitrary sample-timing phases.[19]

A theoretically interesting approach uses the Kalman filter algorithm for estimating the "states" $\underline{c}_{opt}$ from[20]

$$\underline{c}_{opt}^{(n+1)} = \underline{c}_{opt}^{(n)} \qquad \text{("state equation")} , \tag{33}$$

$$a_n = \underline{c}_{opt}^T \, \underline{x}_n - e_{opt\,n} \qquad \text{("measurement equation")} . \tag{34}$$

Convergence is considerably faster than with the usual stochastic approximation algorithm (31), but the algorithm is too complex for practical implementation. In effect, the Kalman filter algorithm can be regarded as a method for computing

$$\underline{c}^{(n)} = \left[\sum_{k=0}^{n} \bar{\underline{x}}_k \, \underline{x}_k^T \right]^{-1} \cdot \sum_{k=0}^{n} (\bar{\underline{x}}_k \, a_i) \quad [\text{cf. } (30)], \tag{35}$$

without ever inverting a matrix. The same can be achieved by computing

$$\left[\sum_{k=0}^{n} \bar{\underline{x}}_k \, \underline{x}_k^{-T} \right]^{-1}$$

iteratively by means of the matrix inversion lemma,[21,22] which leads to an algorithm still much more complicated than (31).

Further related topics are adaptive decision-feedback equalization[23] and adaptive matched filtering.[6] From an estimation point of view, these subjects are treated in much the same way as for linear equalization.

V. SIGNAL DETECTION

In the preceding two sections continuous-valued parameters had to be estimated which may change only slowly. The estimation of these parameters plays only an auxiliary role for the main function of the receiver: detecting the discrete signal values $\{a_n\}$. Let τ_s and φ_c again be already decided, but ISI not necessarily be removed from the signal,

$$x_n = \sum_{\ell=0}^{L} h_\ell a_{n-\ell} + w_n , \quad h_0, h_L \neq 0 , \tag{36}$$

where L is the ISI-memory order of the transmission system. Let $\{a_n\}$ be an aribitrary sequence of values from the finite alphabet $A_M = \{a^0, a^1, .. a^{M-1}\}$. If $L = 0$ (zero ISI), independent decisions can be made by determining $\hat{a}_n = a^i : |x_n - h_0 a^i|^2 \to \text{Min}$. This leads to the usual symbol-by-symbol detection approach. It is sometimes, however, very reasonable not to eliminate ISI completely, especially when the signal is strongly attenuated at the band edges and linear equalization would lead to severe noise enhancement.

In decision-feedback equalization only leading ISI is eliminated by linear equalization, whereas trailing ISI is removed without noise enhancement by using previous independent signal decisions to compute

$$x_n' = \sum_{\ell=0}^{L} h_\ell a_{n-\ell} + w_n - \sum_{\ell=1}^{L} \overset{(\wedge)}{h} \hat{a}_{n-\ell} . \tag{37}$$

Decision-feedback equalization leads to improved error performance, but still involves the problem of error propagation. To obtain the optimum ML receiver one must recognize that at the output of the linear section of an optimal decision-feedback equalizer[24] (= Forney's whitened matched filter[25]) the noise samples are independent.[26] Assuming Gaussian noise, from (36) the likelihood function

$$\lambda(X_N | \{\hat{a}_n\}) \sim \sum_n |x_n - \sum_{\ell=0}^{L} h_\ell \hat{a}_{n-\ell}|^2 \tag{38}$$

is obtained. Hence, the optimum receiver must find the discrete sequence $\{\hat{a}_n\}$ which minimizes (38) in an unbounded interval. This sequence can be determined efficiently by a dynamic programming algorithm, called the Viterbi algorithm.[25,27] To explain this algorithm let

$$\sigma_n \triangleq \left\{a_{n-L+1} \cdots a_n\right\} \tag{39}$$

be one of M^L states in a finite-state machine, and let (σ_{n-1}, σ_n) be a possible transition. Then (38) can be written as

$$\lambda(X_N | \{\hat{a}_n\}) \sim \sum_n |x_n - F(\hat{\sigma}_{n-1}, \hat{\sigma}_n)|^2 , \tag{40}$$

where

$$F(\hat{\sigma}_{n-1},\hat{\sigma}_n) = \sum_{\ell=0}^{L} h_\ell \, \hat{a}_{n-\ell} \quad , \tag{41}$$

and minimizing (38) is equivalent to determining a sequence of states $\{\hat{\sigma}_n\}$ in the state transition diagram ("trellis diagram") of the finite-state machine. Suppose now that up to time $n - 1$ the best paths through the trellis diagram are known that terminate individually in the M^L states $\hat{\sigma}_{n-1}$. Call

$$J(\hat{\sigma}_{n-1}) = \underset{\{..\,\hat{\sigma}_{n-2}\}}{\mathrm{Min}} \sum_{k=-\infty}^{n-1} |x_k - F(\hat{\sigma}_{k-1},\hat{\sigma}_k)|^2 \tag{42}$$

the survivor metric at $\hat{\sigma}_{n-1}$, and let the sequence $\{\ldots\,\hat{\sigma}_{n-2}\}$, which in (42) achieves minimum, be the associated path history. To extend from $n - 1$ to n, the following recursion must be used:

$$J(\hat{\sigma}_n) = \underset{(\hat{\sigma}_{n-1}\to\hat{\sigma}_n)}{\mathrm{Min}} \left\{ J(\hat{\sigma}_{n-1}) + |x_n - F(\hat{\sigma}_{n-1},\hat{\sigma}_n)|^2 \right\} \quad , \tag{43}$$

where minimization goes over all states $\hat{\sigma}_{n-1}$ that have a transition to $\hat{\sigma}_n$. Proceeding in this manner and updating path histories accordingly will leave, in the trellis diagram, a common path which represents the best estimate of $\{\hat{a}_n\} \triangleq \{\hat{\sigma}_n\}$. For $M = 2$ and $L = 1$, this is illustrated in Fig. 6.

An equivalent algorithm with different metric calculation can be obtained direct from (8), if one considers $s_\ell = \bar{s}_{-\ell}$ and $s_\ell = 0$ for $L > 0$, and therewith writes

$$\lambda(X_N|\{\hat{a}_n\}) \sim \sum_n \left[2\,\mathrm{Re}\,(y_n\,\bar{\hat{a}}_n) - G(\hat{\sigma}_{n-1},\hat{\sigma}_n) \right] \quad , \tag{44}$$

where

$$G(\hat{\sigma}_{n-1},\hat{\sigma}_n) = \bar{\hat{a}}_n\, s_0\, \hat{a}_n + 2\,\mathrm{Re}\,(\bar{\hat{a}}_n \sum_{\ell=1}^{L} s_\ell\, \hat{a}_{n-\ell}) \; . \tag{45}$$

The new recursion then becomes

$$J'(\hat{\sigma}_n) = 2\,\mathrm{Re}\,(y_n\,\bar{\hat{a}}_n) + \underset{(\hat{\sigma}_{n-1}\to\hat{\sigma}_n)}{\mathrm{Max}} \left\{ J'(\hat{\sigma}_{n-1}) - G(\hat{\sigma}_{n-1},\hat{\sigma}_n) \right\}. \tag{46}$$

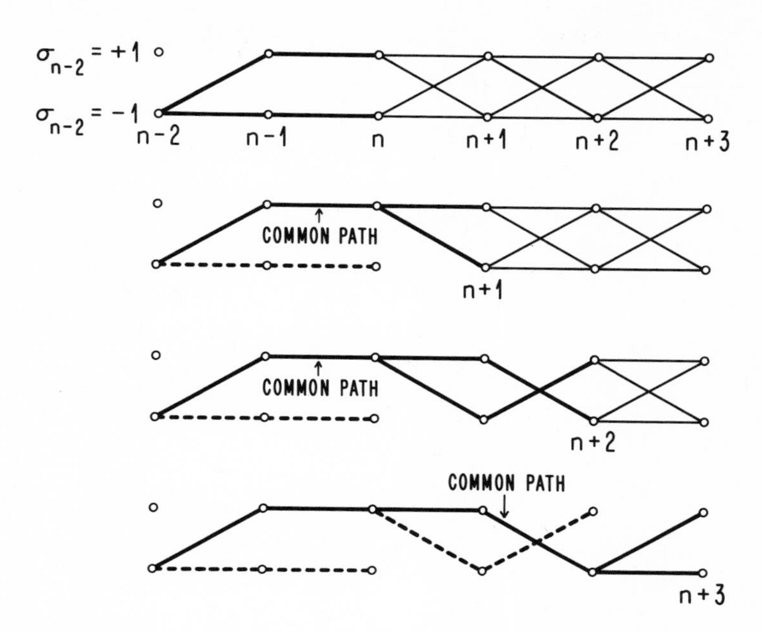

Fig. 6. Illustration of the Viterbi algorithm for $M = 2$ and $L = 1$.

The different metric calculation results from the fact that instead of the output x_n of a whitened matched filter or linear section of a decision-feedback equalizer, the output y_n of a matched filter is used direct, where in the case of non-zero ISI, the noise is correlated. Assuming that the quantities F and G are in both cases precomputed, the new metric calculation apparently offers computational advantages.

In Ref. 6 it has been shown that the algorithm is very robust against ISI in that ISI has essentially no influence on error probability if

$$\sum_{\ell \neq 0} |s_\ell| \leq s_0 \quad . \tag{47}$$

If (47) is satisfied with equality, there will normally be a spectral null in the signal spectrum. Equalization to remove ISI is then no longer possible. The Viterbi algorithm, however, still persuits signal detection with almost no degradation in error performance.

The Viterbi algorithm was originally developed for optimum decoding of convolutionally encoded signals,[28,29] where a different form of state-oriented dependence between successive signal values exists, but otherwise the same discrete optimization problem must be solved.

VI. CONCLUSION

In this paper the attempt has been made to demonstrate the various estimation and approximation problems arising in synchronous data-transmission systems. The presentation had necessarily to remain incomplete in many respects. It is hoped, however, that it may serve as a useful introduction into this field. For a more thorough understanding the reader is referred to the original literature.

VII. REFERENCES

[1]H.L. Van Trees, *Detection, Estimation and Modulation Theory*, Parts I, II and III, New York: Wiley 1968, 1971.

[2]W.R. Bennett and J.R. Davey, *Data Transmission*, New York: McGraw-Hill, 1965.

[3]J.M. Wozencraft and I.M. Jacobs, *Principles of Communications Engineering*, New York: Wiley, 1965.

[4]R.W. Lucky, J. Salz, and E.J. Weldon, Jr., *Principles of Data Communicaton*, New York: McGraw-Hill, 1968.

[5]H. Kobayashi, "Simultaneous adaptive estimation and decision algorithm for carrier-modulated data-transmission systems," *IEEE Trans. Commun. Technol.*, vol. COM-19, pp. 268-280, June 1971.

[6]G. Ungerboeck, "Adaptive maximum-likelihood receiver for carrier-modulated data-transmission systems," *IEEE Trans. Commun.*, vol. COM-22, pp. 624-636, May 1974.

[7]H. Robbins and S. Monro, "A stochastic approximation method," *Ann. Math. Stat.*, pp. 400-407, 1951.

[8]K.H. Mueller and M. Mueller, "Timing recovery in digital synchronous data receivers," *IEEE Trans. Commun.*, vol. COM-24, pp. 516-530, May 1976.

[9]D. Maiwald, "On the performance of decision-aided timing recovery," IBM Research Report, RZ 749, December 1975.

[10]L.E. Franks and J.P. Bubrouski, "Statistical properties of timing jitter in PAM timing recovery scheme," *IEEE Trans. Commun.*, vol. COM-22, pp. 913-920, July 1974.

[11]D.L. Lyon, "Timing recovery in synchronous equalized data communication," *IEEE Trans. Commun.*, vol. COM-23, pp. 269-274, February 1975.

[12]L.E. Franks, "Acquisition of carrier and timing data - I," presentation at NATO Advanced Study Institute on New Directions in Signal Processing, in Communications and Control, Darlington, U.K., August 1974.

[13]G. Ungerboeck, unpublished work.

[14]P.A. Wirtz and E.J. Luecke, "Performance of optimum and suboptimum synchronizers," *IEEE Trans. Commun. Technol.*, vol. COM-17, pp. 380-389, June 1969.

[15]A. Gersho, "Adaptive equalization of highly dispersive channels for data transmission," *Bell System Tech. J.*, vol. 48, pp. 55-70, January 1969.

[16]K. Moehrmann, "Einige Verfahren zur adaptiven Einstellung von Entzerrern für die schnelle Datenübertragung," *Nachrichtentechnische Zeitschrift*, vol. 24, pp. 18-24, January 1971.

[17]G. Ungerboeck, "Theory on the speed of convergence in adaptive equalizers for digital communications," *IBM J. Res. Develop.*, vol. 16, pp. 546-555, November 1972.

[18]K.H. Mueller and D.A. Spaudling, "Cyclic equalization - A new rapidly converging adaptive equalization technique for synchronous data communication," *Bell System Tech. J.*, vol. 54, pp. 369-406, February 1975.

[19]G. Ungerboeck, "Fractional tap-spacing equalizer and consequences for clock recovery in data modems," *IEEE Trans. Commun.*, vol. COM-24, pp. 856-864, August 1976.

[20]D. Godard, "Channel equalization using a Kalman filter for fast data transmission," *IBM J. Res. Develop.*, vol. 18, pp. 267-273, May 1974.

[21]A.P. Sage, *Optimum Systems Control*, Prentice-Hall, Englewood Cliffs, N.J., 1968.

[22]R.D. Githin and F.R. Magee, Jr., work to be published.

[23]P. Monsen, "Feedback equalization for fading dispersive channels," *IEEE Trans. Info. Theory,* vol. IT-17, pp. 56-64, January 1971.

[24]J. Salz, "Optimum mean-square decision-feedback equalization," *Bell System Tech. J.,* vol. 52, pp. 1341-1373, October 1973.

[25]G.D. Forney, "Maximum-likelihood sequence estimation of digital sequences in the presence of intersymbol interference," *IEEE Trans. Info. Theory,* vol. IT-18, pp. 363-378, May 1972.

[26]R. Price, "Nonlinearly feedback-equalized PAM vs. capacity for noisy filter channels," *Conference Record ICC 1972,* Philadelphia, pp. 22-12/16, June 1972.

[27]G.D. Forney, "The Viterbi algorithm," *Proc. IEEE,* vol. 61, pp. 268-278, March 1973.

[28]A.J. Viterbi, "Error bounds for convolutional codes and an asymptotically optimum decoding algorithm," *IEEE Trans. Info. Theory,* vol. IT-13, pp. 260-69, April 1967.

[29]A.J. Viterbi, "Convolutional codes and their performance in communication systems," *IEEE Trans. Commun. Technol.,* vol. COM-19, pp. 751-772, October 1971.

OPTIMAL APPROXIMATION IN AUTOMATED CARTOGRAPHY

Wigand Weber

Institut für Angewandte Geodäsie, Frankfurt a. M.

ABSTRACT

For several years, attempts are being made in many countries of the world to apply electronic data processing techniques in the production and updating of maps. Therein, cartographic generalization still is a vital problem.

Generalization in cartography is the process of transforming the contents of a map into a form appropriate for a map of smaller scale; it consists of several partial processes (as selection, geometric combination, qualitative summarization, simplification and displacement of map objects) which are interdependent in a non-hierarchic way.

It is shown that cartographic generalization is a problem of optimal approximation and that it may be described by a model of mathematical optimization. The basis of such a model is the calculation of semantic information content of the individual map objects in a way corresponding to the thesises of information theory and to the way people (most probably) read maps. Prediction techniques play an essential part in this respect.

The generalization model is subsequently used as a scheme for the classification and judgement of the most important contemporary partial solutions of automated map generalization.

Concluding, three other problems in automated cartography are described which are solved by using methods of approximation theory. These are the reduction of the quantity of digital cartographic data to a minimal amount allowing the 'exact' reconstruction of a line, the Helmert Transformation (which is a conformal transformation – optimal in the sense of the least-squares method) for the minimization of distortions in geometric data digitized from a distorted map sheet, and finally the calculation of splines for the reconstruction of lines from a limited number of given points for output on flatbed plotters.

1. INTRODUCTION

For several years many institutions all over the world try to apply techniques of electronic data processing also to the production and updating of topographic maps. The reason is not only curiosity as usual with scientists and engineers, but also and above all the necessity to offer a greater variety of different maps for a more and more specialized human community and economy. These maps must keep pace with the many changes in our surroundings in a way much quicker than it is usual up to now. There is no doubt, that such high demands cannot be met only by the usual manual methods, especially because a lack of highly specialized cartographic craftsmen has to be expected.

Automated cartography may be divided into four subtasks:

- Acquisition of graphic (analog) original data in digital form by suitable 'digitizers' from air photographs or already existing maps.

- Storage of these digital data (essentially coordinates and quality information) in a data base with a suitable data structure and with possibilities of direct access and detailed data selection.

- Manipulation of the stored data in batch or interactive mode due to the topographic changes in the map area or during the derivation of small-scale maps from large-scale maps.

- Output of the digital data from the data base in analog (graphic) form using computer-controlled plotters.

Experiences gained so far show that data output is the best known subtask in the whole process; data storage may also be considered optimistically. Data acquisiton, however, is actually the bottleneck in the system, and with data manipulation – especially with the derivation of small-scale maps from large-scale maps – human interaction via interactive graphic displays is still widely necessary.

At this point we have touched the problem of cartographic generalization which is central in this paper. Preceding a more precise definition of this term as given in section 3, 'generalization' may roughly be described as 'the reduction of a map's information content due to a reduction in scale'. Generalization, of course, takes already place when initially portraying the natural topography on a map. In this process the cartographer's task is to approximate the natural topography by his map 'as well as possible'. Thus it becomes evident that cartographic generalization belongs to the field of approximation theory: Cartographic generalization must be done in such a way that from the limited amount of information shown on the map, natural topography may 'optimally' be recovered.

Within generalization the following partial processes may be discerned; all of them are due to the reduction in the map surface caused by the reduction of scale:

– Selection of objects to be omitted.

– Simplification of lines (also area outlines).

– Combination of several objects into one object (mostly represented by enlarged symbolization).

– Summarization of several object qualities into one.

– Displacement of objects because the symbols of neighbouring objects require relatively more place at smaller scales.

The process of generalization is essentially complicated by the fact, that most of its partial processes depend on others (e. g. displacement depends on selection and vice versa). Although this dependence is not always hierarchic, attempts have been made to establish such a hierarchy for purposes of automated cartography [1]. This hierarchy corresponds to the sequence of the partial processes mentioned above.

The publications concerning automated generalization nearly exclusively deal either with the partial process of selection or simplification – mostly starting from quite different premises (see section 5). The lack of a comprehensive mathematical model of cartographic generalization is often deplored, and Steward is right, when he states [2] that the profit of such models is based on the conceptual clarification of existing relations which allows for a more reliable categorization and evaluation of preliminary partial solutions, – even though the model as a whole could not be converted into a software package under present conditions.

On the following pages a contribution is made to such a model, using the nomenclature of mathematical optimization, information theory, and statistics. Stimulations were also taken from publications [3], [4], [5], [19], [20] and [21].

2. DETERMINATION OF INFORMATION CONTENT

As during generalization a map's information content is changed in a systematic manner, the first task is to find a method of its determination. In information theory, 'information' is generally defined as the 'diminuation of uncertainty'; therefore, the information content of some character is assumed to be the bigger the greater the improbability of its occurrence is. According to this definition, the information content of a town shown on a map is not necessarily high, only because it has many inhabitants: rather depends its information content on whether it is situated in a region with many big towns (i. e. where one really 'expects' big towns), or in a lonely steppe, where the existence of a big town is very 'surprising'. Similarly a mountain rising abruptly from a plain must be depicted on a map, whereas the same mountain has only a low information content in a mountainous region and could therefore be omitted.

The information content I of an appearance which has the probability p is expressed by the formula

$$I = ld\ \frac{1}{p}\ \text{(bit)}. \tag{1}$$

In this formula 'I' means either the 'syntactic' or the 'semantic' information content depending on whether the probability p refers to the appearance of the characters used for display or to the meaning they transmit. Investigations published so far deal nearly exclusively with the syntactic information content of maps and mostly even with an average value only which refers to a whole map sheet and is called entropy. The entropy E is calculated from

$$E = \sum_i p_i \cdot ld\ \frac{1}{p_i}$$

for discrete appearances and for continuous appearances from

$$E = \int_{-\infty}^{+\infty} p_{(x)} \cdot ld\ \frac{1}{p_{(x)}} \cdot dx$$

supposing stochastic independence.

For generalization, however, the values calculated in this way are of little use, as only semantic information content is relevant to the map user – and this not for the whole map sheet, but for any individual cartographic object shown on it: only in this way can a decision be made, which object is to be canceled or combined with others during generalization without diminishing the map's total information content more than inevitable.

With a cartographic object (e. g. the image of a town, a forest or a road) several types of semantic information are to be distinguished.

Some of these types refer to the p r o p e r t i e s of the single objects. These are: their geometric information (position, length, area), their quality information (e. g. 'town', 'forest' etc.), and finally their numerical supplementary information (e. g. the number of inhabitants of a town).

Other types of information result from r e l a t i o n s between several objects, that is from sets of pairs or triples (etc.) of objects: The most important relations shown on a map are the connection relation in its traffic network, the neighbourhood relation of objects (as the 'over' or 'under' in the crossing of roads with railways or the adjacency or intersection of different objects), and finally the topological incidence relation in the planar graph describing a network of area-borderlines (e. g. of communities) or in the tree-graph describing a river-network.

Subsequently, a method of determining the semantic information content of individual cartographic objects is proposed, assuming the information content of an object to be a function of the difference between the data attached to it and the data to be predicted there from its surroundings according to some prediction strategy using statistical relations. (The data used for the prediction may be of equal or different kind as those to be predicted so far, as there is only a sufficiently strong interrelation between them.) This assumption corresponds both to the definition of 'information content' in information theory as cited above and imitates the behaviour of a map user, to which evidently only those appearances of a map are of higher interest which he is not able to deduce already from the general 'information trend' of the map or from his empirical knowledge about the interdependences of topographic facts.

The prediction techniques to be applied with properties are different, depending on whether data of equal or different type (compared to the data to be predicted) are used and whether these data are (quasi) continuous or discrete (see paragraphs a) through c)); prediction techniques with relations are dealt with in paragraph d):

a) As to the prediction of continuous data from data of equal type, its technique considers the fact, that the 'information trend' mentioned above is conveyed to the map user by the low frequencies of the data which can be imagined to be a surface above the map plane, whereas the information content is contained in the higher frequencies. Beside by a Fourier Transformation, the smoothed surface of information trend can also be determined by regression using polynoms of limited order, 'least-squares-interpolation' according to the theory of linear filters of Wiener-Kolmogoroff, or by applying the 'gliding weighted arithmetic mean'-technique (also called 'convolution', in which the points of the original surface are replaced by the mean of the values in their surroundings after weighting them according to their horizontal distance). The distances of data points from this trend-surface are finally the basis for the determination of the information content of the appertaining cartographic objects. – In this way the geometric information and the numerical supplementary information mentioned above may be treated*).

b) Another approach is to be taken, when the types of predicting and predicted data are different. E. g. there is a strong correlation between the position of rivers and the curvature of the earth's surface (as expressed in the contour lines) and mostly also vice versa. The strength of the (linear) statistical dependence between the two variables can be judged after the calculation of their covariance and the correlation coefficient to be deduced from it**). If its amount is not too far from 1, one variable may be used for the prediction of the other one, using a linear formula. The dependence of one variable of one or more other variables may also be found by regression.

---

*) As an example the geometric information of the rivers of a map is used: At first the map is subdivided by a raster of quadratic meshes of suitable width, and the total length of the rivers in each mesh is taken as z-coordinate of a control point to be placed above the center of the mesh. These control points are then used for the calculation of the trend-surface. The vertical distances of the control points from the trend-surface define the information content (using e. g. formula (2) to be given below), which is then distributed to the rivers (or river portions) of each mesh proportionally to their length. Similarly point and areal objects may be treated either with respect to their geometric information or to their numerical supplementary information as numbers of inhabitants, height values etc.

**) Correlation coefficients may also be calculated by 'factor analysis' as is shown e. g. in [7].

As in paragraph a) the differences between the calculated (predicted) value and the actual value are the basis for the determination of the information content of the appertaining cartographic object.

c) When trying to predict the quality (e. g. 'river', 'town', etc.) of a cartographic object on the background of its surroundings, one has to consider the fact that it is of discrete nature – in contrast to the information types dealt with in the paragraphs a) and b) above. Therefore, the conception of 'conditional probability' is used which is known from statistics. The conditional probability $p\,(B/A)$ of an event B with reference to event A means the probability of the occurrence of B, if A has occurred and can be expressed by the formula

$$p\,(B/A) = \frac{p\,(A \cap B)}{p\,(A)}\,.$$

E. g. there is a greater probability that a farm exists in a field than in a forest.

These conditional probabilities have to be determined empirically – and that separately for every distinct type of landscape.

d) As stated above, information in cartographic objects can also result from relations established by them, e. g. by roads in a traffic network. In such cases, predictions as to the existence of a connection (edge) between two nodes can be made. For this purpose, the methods of *Dijkstra* (for the calculation of the shortest path between two nodes in a labeled graph) or of *Ford* and *Fulkerson* (for the determination of maximal flows in networks) could be applied. – The differences between the path-length or the flows in the networks, containing and lacking the edge under consideration, is then used for the calculation of its probability.

In the above paragraphs (except c)) only differences between a predicted and an actual value of some information variable have been calculated instead of probabilities p as they are required for formula (1). For the transformation of differences into probabilities an empirical density distribution or in the lump the Gaussian law of errors can be used:

$$p_{(\epsilon)} = \frac{1}{m\sqrt{2\pi}} \cdot e^{\frac{-\epsilon^2}{2m^2}} \cdot d\epsilon. \qquad (2)$$

Herein m is the mean square error and $d\epsilon$ the tolerance of ϵ. From the formula it can be seen that the probability of some information about a cartographic object becomes the lower, the greater the difference between the predicted and the actual value is; consequently, its information content – to be calculated from formula (1) – will attain a high value.

A difficulty arises from the fact that (as already mentioned above) a cartographic object may have several types of semantic information for which predictions are made and probabilities are calculated separately. Theoretically, one could try to modify formula (2) by introducing a multi-dimensional density distribution or to have a separate object function for each of the different information types instead of only one for all together according to formula (3) (see below). However, for practice it seems to be sufficient, simply to add for each object the information contents of the different information types after appropriate weighting according to the multiplication law of statistics neglecting stochastic dependences.

The method of determination of information content described so far, still implies some subjective decisions to be made: among them are the choice of the degree of smoothing of the trend-surface, of the mesh width of the raster used with some of the predictions, and of the weights, when adding the different types of information contents. Also there is still some uncertainty about the question, which of the many interdependences in map information are really unconsciously perceived by map readers and used for their 'optimal estimation' of how the natural topography may look like.

Having determined the semantic information content, we dispose of an essential prerequisite for the generalization model.

3. A MODEL OF CARTOGRAPHIC GENERALIZATION

In the model of cartographic generalization described subsequently, the terminology of mathematical optimization calculus is used. As already mentioned in section 1, topography has to be approximated during generalization 'as well as possible' – that is o p t i m a l l y in a sense to be defined more exactly hereinafter.

Since the development of the simplex algorithm for linear optimization some 40 years ago, mathematical optimization has made considerable progress, especially with respect to nonlinear and integer optimization. The importance of mathematical optimization for geodesy and cartography was most recently emphasized by the 'International Union for Geodesy and Geophysics' in Resolution No. 22 of its General Assembly 1975 in Grenoble.

Within mathematical optimization the unknowns of a problem are to be determined in such a way that a so-called object-function is maximized or minimized. In this process the unknowns or functions of them have to fulfil supplementary conditions (restrictions), given either as equations or inequalities. Recent developments aim at a generalized optimization containing more than one object function.

During cartographic generalization, one tries to keep the total information content of the result as high as possible. Therefore, the object function of a map sheet is defined as follows:

$$\sum_i g_k \cdot e_j \cdot (z_{il} + z_{im}) \cdot I_i = \mathrm{Max} \tag{3}$$

I_i is the semantic information content of the ith cartographic object in the initial map as determined according to section 2; g_k is a constant weight with which the members of the kth category of cartographic objects of a pragmatic evaluation scale contribute to the total information content*). e_j, z_{il} and z_{im} belong to the unknowns of the optimization model: e_j is a 0–1 variable ('selection variable') and indicates, whether or not the jth ring of combinable objects of the initial map containing the object i will appear on the generalized map; z_{il} and z_{im} are the 'combination variables' of object i with reference to its neighbouring objects l and m**). The combination variables are e. g. allowed to vary between the values $z_{min} = 1/3$ (with full combination) and $z_{max} = 1/2$ (with no combination). The assumption of different degrees of combinations is in agreement with practice, where combined objects are symbolized by areas of more or less extent.

Now the restrictions of the optimization model will be specified. They are typical for cartographic generalization and responsible for the fact that normally the maximum of the object function will not reach the information content of the original map:

- Shift restrictions: In order to guarantee a certain positional accuracy, shifts of objects or parts of them caused by displacement (see above) may not exceed a certain maximum. If parts of objects are shifted differently, additional steadiness conditions have to be defined for the shift vectors.

- Distance restrictions: Caused by the dimensions of map symbols and resolving power of the human eye, the distance of adjacent map objects may not fall below certain minimal values D. The left side of a distance restriction, referring to the objects i and k, is represented by an expression for the distance of these objects after shifting them by still unknown shift vectors; their right sides are given by the following reference value R which depends on both the selection and combination variables of the objects i and k:

$$R = \frac{z_{ik} - z_{min}}{z_{max} - z_{min}} \cdot D + (1 - e_j \cdot e_l) \cdot N,$$

  where N is a negative constant of suitable value and e_j, e_l are the selection variables of the rings of combinable objects, in which the objects i and k respectively are contained.

- Density restrictions: Owing to the distance restrictions, it is already guaranteed to a certain degree that the generalized map remains readable after scale reduction. However, additional restrictions concerning maximal density of symbols per square unit are inevitable, if legibility of the map shall be maintained. Because symbol density will often change throughout a map sheet, several such restrictions have to be defined. Of course, these restrictions will contain both the selection and the combination variables of the map objects.

- Structure restrictions: By using these restrictions, it shall be assured that after displacement typical geometrical relations are maintained: e. g. that a straight row of houses remains straight.

---

*) Thus g_k transforms semantic information content into what information theory calls 'pragmatic' information content, i. e. the information content in which a special user is interested in.

**) In agreement with the 'assignment problem' of operations research, with each object exactly two references to (an) other object(s) have been presumed.

– Supplementary restrictions: They comprise the 0–1 condition for the selection variables and conditions forcing the values of combination variables into the range between z_{min} and z_{max}.

In the generalization model described above the partial processes of selection, combination, and displacement are taken into consideration, – summarization and simplification have not yet been mentioned. As to the latter, it can be treated in advance as a separate optimization problem, because the 'micro-information-content', exclusively changed by simplification, may be considered to be independent of the 'macro-information-content' used in our model. As to summarization, it has to be done in accordance with a rule which is uniform for a certain scale change and not only for the generalization of an individual map sheet, thus figuring as a constant in our model. Though these rules have been developed and standardized in a long period of cartographic practice, scientists try to find an objective basis also in this area using methods of operation research [8].

4. EVALUATION OF THE GENERALIZATION MODEL

Theoretically the model of section 3 refutes the thesis that the numerous and not only hierarchic interdependences of the partial processes of cartographic generalization could not be imitated by sequential algorithms as characteristic for EDP-machines. However, experts in mathematical optimization will notice that the numerical system, necessary for the generalization of only a rather small part of a map sheet, will (inspite of some simplifying neglections*)), be so large that optimization algorithms of today would have to fail with respect to time and numerical stability. This impression is intensified by the fact that 0–1 variables, non-linearities, and even non-convexities are implied.

In addition, it is to be taken into account that the different restrictions could presumably be written down by a human being looking at the map – that, however, for a computer the same task causes serious recognition problems: he must e. g. during the formation of the distance restrictions, be able to decide, which objects are adjacent. The same is true with respect to the definition of combination variables. Perhaps these problems could be overcome by using an adequate data structure in the data base**) or by solving the 'assignment problem' of operations research. However, when forming the structure restrictions, the very complex area of pattern recognition is already touched.

But what is the model then good for at all? – In my opinion, its use consists in the clarification of the relations between information content, degree of generalization, and 'optimality', and of the complex interdependence of the various partial processes of generalization. Additionally, such a model permits a more detailed judgement on

---

*) E. g. the assumption of the information content of map objects in the optimization model as being constant.

**) E. g. by representing the 'intersection graph' (which describes the neighbourhood relations and intersections of map objects by its 'cliques') in a data structure with n-array relations [9].

character, suitableness, and interrelations of partial solutions already developed. Some of these partial solutions will be described in the following section. So far, they have not yet been integrated into an integral mechanism of generalization.

5. SOME AVAILABLE PARTIAL SOLUTIONS OF AUTOMATED CARTOGRAPHIC GENERALIZATION

Selection: In automated cartography the so-called law of *Töpfer* is often used. It was found empirically and serves the calculation of the number n_F of objects on the generalized map from the number n_A of objects at the original scale:

$$n_F = n_A \cdot \sqrt{\frac{m_A}{m_F}} \,;$$

m_A, m_F are the reciprocal scales of the original and the generalized map. Supplementary coefficients are often introduced additionally to adapt the law of *Töpfer* to the special characteristics of individual object types. The law does not say anything about whether an individual object should be retained or cancelled. As to this question, until now 'rank descriptors' [10] are widely used which are at best determined by a weighted summation of different data types attached to an object and which, by their value, decide on the selection – independently of the information trend in the object's surroundings.

Simplification: Automatic methods of simplification are applied to lines, surfaces, and point agglomerations:

a) Lines: At present, lines are mostly smoothed by applying a 'gliding weighted arithmetic mean', independently of whether they are given in point- or raster representation. The method was taken from the filter theory and is known there as 'convolution' in the context of low-pass filtering; it was already described before. The diminuation of information content is achieved by eliminating the high frequency portions in the Fourier spectrum of the line – that is the very one equipped with the highest information content. As a measure of the amount of diminuation of the information content of the line, the increase of its 'sampling distance' (i. e. the smallest wavelength contained in the line) is taken. – Other smoothing methods apply regression polynoms (piecewise) or Fourier series and assume the line's information content to be proportional to their degree. – Still others assume the information content of a line to be expressed by the number of its relative curvature extrema and reduce them systematically during smoothing according to the law of *Töpfer**).

   Typical for all these methods is the fact that for the determination of the information content, the aspect of probability (according to (1)) is replaced by other more heuristic quantities, and that the demand for maximal information content after generalization is replaced by the principle of its reduction by a fixed percentage according to the law of *Töpfer*.

---

*) For more details of all these methods see [4].

b) Surfaces: Surfaces are simplified, i. e. smoothed, very similarly to lines. Additionally, 'multiquadratic functions' were introduced for smoothing interpolation [11], [12]. Also 'least squares interpolation' (sometimes called 'linear prediction') [5], [13] and 'gliding slant planes' as local regression functions [14] have been applied.

c) Point agglomerations: Point agglomerations (namely the houses of settlements) have been smoothed using the method of convolution already mentioned [15]. A measure for the diminuation of information content was not determined in this case. Points with a great information content (e. g. a lonely restaurant in a forest) are, of course, lost with this method.

Displacement: As far as I know, until now only the publications [16] and [17] deal with cartographic displacement. In the methods described there, maximal shifts and areas of shift influence are taken into account. However, conflicts in displacement, caused by lack of place, are not solved by selection or combination processes or displacement of more distant objects.

In this context, a remark of Boyle may be of interest, as it considers our problem from quite a different (rather Anglo-Saxon pragmatical) point of view: "Maps should be matched to the new capabilities. Generalization problems could, for instance, be simplified and map accuracy improved, if symbolization or variation of line thickness could be used in place of feature displacement".

6. OPTIMAL APPROXIMATION IN OTHER DOMAINS OF AUTOMATED CARTOGRAPHY

After having described cartographic generalization as a problem of optimal approximation, some of its other applications in different domains of automated cartography shall be mentioned here more briefly: especially, the topics of 'data reduction', 'Helmert transformation', and the construction of splines are concerned.

In data reduction, the problem is to find a minimal set of data sufficient for the 'exact' reconstruction of a line. This set may consist of point coordinates or coefficients of mathematical functions. Again the problem belongs to the domain of mathematical optimization. However, in contrast to generalization, here the amount of data, necessary for the coding of a line, is to be minimized (object function) with the restriction that either its information content remains constant or the reconstructed line differs from the original line only by predefined maximal amounts. The problem has been tackled until now using either the least-squares-method [4] or the Tschebyschef approximation [18].

The Helmert transformation is often used for the rectification of coordinates, digitized from distorted images, or as a quick approximative transformation of one type of planar representation of the earth's surface into another one. It is a conformal projection minimizing the sum of squares of the transformation residuals in the control points.

As to the calculation of splines please refer to other papers in this volume.

BIBLIOGRAPHY

[1] *Christ, F.; Schmidt, E.; Uhrig, H.:* Untersuchung der Generalisierung der topographischen Übersichtskarte 1 : 200 000 auf mögliche Gesetzmäßigkeiten. – Nachrichten aus dem Karten- und Vermessungswesen, Series I, Nr. 51, Frankfurt a. M. 1972.

[2] *Steward, H. J.:* Cartographic generalization, some concepts and explanation. – Cartographica, supplement No. 1 to the Canadian Cartographer, Vol. 11, 1974.

[3] *Hake, G.:* Der Informationsgehalt der Karte – Merkmale und Maße. – In: Grundsatzfragen der Kartographie, Österr. Geograph. Gesellschaft, Wien 1970, pp. 119–131.

[4] *Gottschalk, H.-J.:* Versuche zur Definition des Informationsgehaltes gekrümmter kartographischer Linienelemente und zur Generalisierung. – Deutsche Geodätische Kommission bei der Bayer. Akademie der Wissenschaften, Series B, No. 189, Frankfurt a. M. 1971.

[5] *Lauer, S.:* Anwendung der skalaren Prädiktion auf das Problem des digitalen Geländemodells. – Nachrichten aus dem Karten- und Vermessungswesen, Series I, No. 51, Frankfurt a. M. 1972.

[6] *Bollmann, J.:* Kartographische Zeichenverarbeitung – Grundlagen und Verfahren zur Quantifizierung syntaktischer Zeicheninformation. – Diplomarbeit, Freie Universität Berlin, 1976.

[7] *Mesenburg, K. P.:* Ein Beitrag zur Anwendung der Faktorenanalyse auf Generalisierungsprobleme topographischer Karten. – Dissertation, Universität Bonn, 1973.

[8] *Monmonier, M. S.:* Analogs between class-interval selection and location – allocation models. – The Canadian Cartographer, Vol. 10, 1973, pp. 123–131.

[9] *Weber, W.:* Datenstruktur eines digitalen Geländemodells. – CAD-Bericht KFK-CAD 11 'Das digitale Geländemodell', Gesellschaft für Kernforschung, Karlsruhe, September 1976.

[10] *Kadmon, N.:* Automated selection of settlements in map generalization. – Cartogr. Journal, 1972, pp. 93–98.

[11] *Hardy, R. L.:* Multiquadratic equations of topography and other irregular surfaces. – Journal of Geophys. Res., Vol. 76, No. 8, 1971, pp. 1905–1915.

[12] *Gottschalk, H.-J.:* Die Generalisierung von Isolinien als Ergebnis der Generalisierung von Flächen. – Zeitschrift für Vermessungswesen, 1972, pp. 489–494.

[13] *Kraus, K.:* Interpolation nach kleinsten Quadraten in der Photogrammetrie. – Bildmessung und Luftbildwesen 40, 1972, pp. 3–8.

[14] *Koch, K. R.:* Höheninterpolation mittels gleitender Schrägebenen und Prädiktion. – Vermessung, Photogrammetrie, Kulturtechnik, Mitteilungsblatt 71. ann. set, 1973, pp. 229–232.

[15] *Gottschalk, H.-J.:* Ein Rechnerprogramm zur Berechnung von Flächen aus flächenartig strukturierten Punktmengen mit Hilfe der Bearbeitung von Binärbildern. – Deutsche Geodätische Kommission, Series B, No. 205, Frankfurt a. M. 1974.

[16] *Gottschalk, H.-J.:* Ein Modell zur automatischen Durchführung der Verdrängung bei der Generalisierung. – Nachrichten aus dem Karten- und Vermessungswesen, Series I, No. 58, Frankfurt a. M. 1972.

[17] *Schittenhelm, R.:* The problem of displacement in cartographic generalization. – Attempting a computer-assisted solution. – Paper VIIIth International Cartographic Conference, Moscou 1976.

[18] *Tost, R.:* Mathematische Methoden zur Datenreduktion digitalisierter Linien. – Nachrichten aus dem Karten- und Vermessungswesen, Series 1, No. 56, Frankfurt a. M. 1972.

[19] *Pfaltz, J. L.:* Pattern recognition VI – MANS, a map analysis system. – University of Maryland, Techn. Report TR-67-42, 1967.

[20] *Freeman, J.:* The modelling of spatial relations. – University of Maryland, Techn. Report TR-281, 1973.

[21] *Kádár, I.; Karsay, F.; Lakos, L.; Ágfalvi, M.:* A practical method for estimation of map information content. – In: Automation the new trend in cartography, Final report on the ICA Commission III (Automation in cartography) scientific working session, August 1973 in Budapest, Budapest 1974.

RECONSTRUCTION FROM X-RAYS

K.T. Smith, S.L. Wagner and R.B. Guenther

Oregon State University, Corvallis, Oregon 97331
D.C. Solmon
State University of New York at Buffalo, New York 14240

Introduction. According to T.J. Rivlin (these Proc.) "The problem of optimal recovery is that of approximating as effectively as possible a given map of any function known to belong to a certain class from limited and possibly error-contaminated information about it." More precisely, in the scheme envisioned by Rivlin there are given spaces X of possible objects, Y of possible data, and Z of possible reconstructions of certain features of the objects. Also maps are given between the spaces as follows:

(0.1)

$$\begin{array}{ccc} \text{objects} \quad X & \xrightarrow{F} & Z \quad \text{reconstructed features} \\ D \downarrow & \nearrow R & \\ \text{data} \quad Y & & \end{array}$$

For any given object $\mathcal{O}$, $D(\mathcal{O})$, is the data coming from $\mathcal{O}$, and $F(\mathcal{O})$ is the selected feature of $\mathcal{O}$; for any given set of data d, R(d) is the reconstruction of the selected feature provided from the data d by the particular reconstruction method R in use. A metric ρ is given on the space Z, and the number

(0.2) $$e(R) = \sup_{\mathcal{O} \in X} \rho(F(\mathcal{O}), RD(\mathcal{O}))$$

is taken as a measure of the error in the reconstruction method R. The problem of optimal recovery is to minimize the error function e.

This research has been supported by the National Science Foundation under grant No. DCR72-03758 A01

In this article we shall discuss three medical reconstruction problems from the point of view of this abstract formulation. These problems, which involve reconstructions from x-ray data, are quite different in nature from most problems that have been considered in optimal recovery. Therefore, it can be expected that there will be open questions and new points of view in both domains.

In particular, it is useful to broaden the notion of optimal recovery a little. Suppose given, in addition to the above, a positive integer p and a norm on the space R^p satisfying:

$$\underline{\text{If}}\ 0 \le x_i \le y_i,\ i = 1,\dots,n,\ \underline{\text{then}}\ ||x|| \le ||y||.$$

Set $\tilde{X} = X^p, \tilde{Y} = Y^p, \tilde{Z} = Z^p$, and let $\tilde{F}$, $\tilde{D}$, and $\tilde{R}$ be the obvious maps. The norm on R^p and metric on Z induce a natural metric on Z^p: if $z = (z_1,\dots,z_p)$ and $w = (w_1,\dots,w_p)$, then

$$\tilde{\rho}(z,w) = ||\rho(z_1,w_1),\dots,\rho(z_p,w_p)|| .$$

Suppose given a subset $\tilde{X}_T \subset \tilde{X}$ for testing, and define

$$e(R) = \sup_{\tilde{\mathcal{O}} \in \tilde{X}_T} \tilde{\rho}(\tilde{F}(\mathcal{O}),\tilde{R}\tilde{D}(\mathcal{O})) . \tag{0.3}$$

In some problems $\tilde{X}_T$ may be equal to $\tilde{X}$, but in medical problems $\tilde{X}_T$ is always finite.

Of course this seemingly broader scheme fits into the original, but then objects become p-tuples of the actual objects, etc., which is rather unnatural.

Table of Contents

1. MATHEMATICAL GENERALITIES

Since dimensions 2 and 3 are both important in x-ray work, it is convenient to give the statements and formulas for the general dimension n. If $\mathcal{O}$ is an object in R^n and Θ is a direction (i.e. a point on the unit sphere S^{n-1}), then the information provided by an x-ray from the direction Θ is effectively the total mass of $\mathcal{O}$ along each of the lines with direction Θ. Thus, if f is the density function of $\mathcal{O}$, this information is the function

$$(1.1) \qquad P_\Theta f(x) = \int_{-\infty}^{\infty} f(x+t\Theta)dt \qquad \underline{\text{for}}\ x\in\Theta^\perp,$$

where $\Theta^\perp$ is the n-1 dimensional subspace orthogonal to Θ. $P_\Theta f$ is called the radiograph of $\mathcal{O}$(or of f) from the direction Θ. The reconstruction problem is that of recovering certain selected features of an unknown object $\mathcal{O}$ from (primarily) the knowledge of a certain finite number of radiographs. Since the radiographs are determined by the density function, the first basic question is the extent to which the density function itself can be recovered from the radiographs. (The density function does not tell the full story, however, even in radiology, for new radiographs can be taken after the object $\mathcal{O}$ has been modified in known ways, and the new radiographs can be compared with the old ones.) The answer to the question of the determination of the density function is contained in the simple formula relating the Fourier transforms of f and $P_\Theta f$ [3]:

$$(1.2) \qquad (P_\Theta f)^\wedge(\xi) = \sqrt{2\pi}\hat{f}(\xi) \qquad \underline{\text{for}}\ \xi\in\Theta^\perp,$$

where

$$(1.3) \qquad \hat{f}(\xi) = (2\pi)^{-\frac{n}{2}} \int e^{-i\langle x,\xi\rangle} f(x)\,dx\ .$$

This shows that $\hat{f}$, and therefore f itself, are uniquely determined by the $P_\Theta f$, and leads to two inversion formulas. [3]

<u>Fourier inversion formula</u>: <u>Given the</u> $P_\Theta f$, <u>set</u>

$$\hat{f}(\xi) = (1/\sqrt{2\pi})\ (P_\Theta f)^\wedge(\xi)\ \underline{\text{for}}\ \xi\in\Theta^\perp;$$

then

$$(1.4) \qquad f(x) = (2\pi)^{-\frac{n}{2}} \int e^{i\langle x,\xi\rangle}\hat{f}(\xi)\,d\xi$$

<u>Radon inversion formula</u>:

$$(1.5) \qquad f(x) = (2\pi)^{-1}|S^{n-1}|\,|S^{n-2}| \int_{S^{n-1}} \wedge P_\Theta f(E_\Theta x)\ d\Theta,$$

where $|S^k|$ is the k-dimensional area of the sphere S^k, E_Θ is the orthogonal projection on the subspace $\Theta^\perp$, and $\wedge$ is the operator defined by

$$(1.6) \qquad (\wedge g)^\wedge(\xi) = |\xi|\ \hat{g}(\xi).$$

The operator Λ can be expressed directly (i.e. without Fourier transforms) in the following way. If

$$(17) \qquad R_i g(x) = \int \frac{x_i - y_i}{|x-y|^{n+1}} g(y)\, dy,$$

the integral being taken in the sense of the Cauchy principal value, then

$$(1.8) \qquad \Lambda g = -\Sigma\, R_i(\partial g/\partial x_i).$$

Formula (1.5), with Λ defined by (1.8), is not quite the classical formula of Radon, for the Radon transform involves integration over the n-1 planes perpendicular to a direction Θ, rather than over the lines parallel to Θ, but it is the same sort of thing.

The Fourier and Radon inversion formulas presuppose that the radiographs $P_\Theta f$ are known for all directions Θ. Since all real objects have finite extent, i.e. the density functions have bounded support, the Fourier transforms are real analytic functions. This gives the following result. [3]

Theorem 1.9 *The density function of any object is completely determined by the radiographs from any infinite set of directions.*

Nevertheless, the practical case is the one in which only a finite number of radiographs are known, and in this case the theorem is rather unsettling. [3]

Theorem 1.10 *Suppose given any object $\mathcal{O}$ with density function f, its radiographs from any finite number of directions, a totally arbitrary density function f', and an arbitrary compact set K in the interior of the support of f. Then there is a new object $\mathcal{O}'$ with the same shape and the same radiographs as $\mathcal{O}$, and the density function f' on the compact set K.*

For technical reasons it is assumed in this theorem that the space X consists of all objects having infinitely differentiable density functions with compact support. It might appear more realistic to consider finite dimensional spaces of objects. In this case the theorem is as follows. [3]

Theorem 1.11 *Let X_o be a finite dimensional space of objects with density function basis $f_1,\ldots,f_N$. Let V be the set of directions Θ such that at least two objects in X_o have the same radiograph from the direction Θ. Then V is an algebraic variety in S^{n-1} and there are polynomials $q_1,\ldots,q_N$ such that if $\Theta \epsilon V$, then*

$$(1.12) \qquad \sum_{j=1}^{N} q_j(\Theta)\hat{f}_j(\xi) = 0 \qquad \underline{\text{for all}}\ \xi\in\Theta^{\perp}.$$

Since an algebraic variety is either the whole sphere or a very small subset, and since condition (1.12) almost never holds on the whole sphere, the theorem shows that for almost any finite dimensional space it is possible to distinguish between objects in the space by a single radiograph from almost any direction.

The most commonly used finite dimensional space X_o in reconstruction work is the space of step functions on a grid. Let $Q = \{x: |x_j| \leq 1\}$ be the unit cube in R^n, choose a number N, and partition Q by the planes $x_j = k_j/N$, where k_j is an integer and $-N \leq k_j \leq N$. Then X_o is the space of functions which are constant on each of the little cubes in the partition. Theorem 1.11 can be used to show easily that the only directions which do not distinguish between objects in X_o are those determined by the points k/N with $|k_j| \leq 2N$. The 2 dimensional case of this was established by Mercereau and Oppenheim [2].

This is a case where the infinite dimensional theorem reflects nature much more faithfully than the finite dimensional one. Indeed, it is evident that in practice an infinitesimal change in the x-ray direction cannot improve the information on a radiograph, while, on the other hand, experience shows that for practical purposes the infinite dimensional theorem is correct. Even with a great many x-ray directions, reconstruction methods do lead to medically feasible but wrong objects with the correct radiographs. This problem can be lessened by incorporation into the method of additional information about either the object or the reconstruction method - i.e. information not coming from the x-rays. A couple of examples are given in [3,5], but much work remains to be done.

The three reconstruction methods in current use for recovering the density function are the following (or modifications thereof).

1. Fourier Method. This is a numerical implementation of the Fourier Inversion Formula.

2. Convolution Method. This is a numerical implementation of the Radon Inversion Formula.

3. Kacmarz or ART Method. This is an iterative procedure which goes as follows. Fix a compact set K in R^n containing all objects under consideration, and regard the density functions as points in $L^2(K)$. Fix the x-ray directions $\Theta_1,\ldots,\Theta_M$, and let N_j be

the null space of P_{Θ_j} and Q_j be the orthogonal projection in $L^2(K)$ on $f+N_j$, where f is the unknown true object. It is easily seen that N_j consists of the functions in $L^{2(K)}$ which are constant on the lines with direction Θ_j, so that for any g in $L^2(K)$, $Q_j g$ is computable even though f is unknown. It can be shown that

$$(1.13) \qquad (Q_M \dots Q_1)^m g \to Qg \quad \underline{\text{in}}\ L^2(K),$$

Q being the orthogonal projection on f+N, where N is the intersection of the N_j. For any g, Qg is an object with the correct radiographs from all the given directions. The Kacmarz or ART method consists in choosing an initial guess g and in taking the left side of (1.13) with a suitable m as an approximation to the unknown f. A proof of the convergence in (1.13) is given in [3] and an explicit description of the rate of convergence is given in [7].

In the present situation where the objective is the recovery of an approximation to the density function, the optimal recovery scheme looks as follows. The space X of objects is the space $L^2(K)$, where K is a compact set in R^n containing all of the objects under consideration - i.e. X is the space of density functions of all objects under consideration. If M x-rays are taken, the space Y of data consists of all M-tuples of x-ray films - or, since each film is read at a finite number m of points, $Y = R^{mM}$. Normally, the space Z is the space of step functions on a grid, i.e. the space X_o described above, which means that the true density functions are approximated by step functions on the grid.

The fundamental initial problem in the program of minimizing the error function is that of defining a relevant metric on the space Z. In most significant medical problems, e.g. the distinction between normal and abnormal tissue, the density differences are extremely small. On the other hand, bone, for instance, is something like a honeycomb, so that the irrelevant difference between bone at one point and bone at another is likely to be very large. Consider, for example, the case of a head with a tumor, and two reconstruction methods, one of which finds the tumor but fuzzes up the skull a bit, while the other recovers a perfect skull, but does not find the tumor. Since the tumor is rather small and differs by very little from normal tissue, almost any natural metric on Z will show the second method to be very good - while in fact it is a disaster. Since the skull is rather large and very different in density from tissue, almost any natural metric on Z will show the first method to be a disaster - while in fact it is very good.

In this case some very subtle metric is needed, which first locates the skull and then gives very little weight to differences nearby.

2. COMPUTERIZED AXIAL TOMOGRAPHY

For reasons of computational simplicity, the recovery of the density function of an object is usually achieved two dimensionally with the recovery of cross sections. When the cross sections and x-ray directions are perpendicular to a fixed axis the procedure is called Computerized Axial Tomography, or CAT.

CAT came to medical prominence, although it had been in the air for some time, and had actually been used in other fields, about 1973 with the invention by G.N. Hounsfield of the EMI scanner, which reconstructed cross sections of the head by means of the Kacmarz method. The performance of the EMI scanner in pinpointing tumors, hemorrhages, blood clots, etc. was so dramatic as to revolutionize radiology. Indeed, this scanner was called the greatest advance in radiology since Roentgen's discovery of x-rays. In the interim scanners have proliferated. There are scanners using the Fourier and Convolution Methods in addition to those using the Kacmarz method - the trend seems to be toward the Convolution Method; and there are scanners designed for other specific parts of the body, such as breasts or forearms, and scanners designed for all parts of the body at once. Since the scanners are very expensive (with an average initial cost of about $500,000 and rather heavy maintenance charges), and since they have certain inherent disadvantages (some of which are discussed in [5]), it seemed worthwhile to see what could be done with standard hospital x-ray equipment commonly available even in towns and villages.

The results of our research and experiments are described in [1,3,5], and explicit quantative comparisons with the results of the EMI scanner are given in [5].

In this problem the optimization set up is that described at the end of §1, with n=2. The space X of objects is (or can be taken to be, since the recovery of the density function is what is desired) the space $L^2(K)$ of density functions on a compact set $K \subset R^2$ which contains all cross sections of the objects under consideration. The space Y of data is the space R^{mM}, where M is the number of x-ray directions and m is the number of points at which each x-ray is read. The space Z is the space of step functions on an N x N grid, or, equivalently, the space of N x N matrices, which are called reconstruction matrices. For a given object $\mathcal{O} \in X$, $F(\mathcal{O})$ (see (0.1)) is the step function whose value on a square of the grid is the average density of $\mathcal{O}$ on this square, and $D(\mathcal{O})$ is the mM tuple obtained by reading the M radiographs of $\mathcal{O}$ at the prescribed m points. At present, at least, the reconstruction map R is one of the three described: the Fourier Method, the Convolution Method, or the Kacmarz method - or a modification of one of these.

For the reason described at the end of §1, the optimization of recovery cannot even be studied, much less carried through: namely, no medically relevant metric on the space Z is known. Nevertheless, various quantitative limits to the accuracy of the reconstruction can be given in terms of the size N of the reconstruction matrix, the number M of tolerable x-rays, the number m of points at which they are read and the positive thickness of a real cross section as opposed to the zero thickness of a theoretical one. These limits and relationships between them are discussed in [5]. The other major factors affecting the accuracy of reconstructions are the Indeterminacy Theorem (1.10) and the noise in the data. These are discussed, and certain remedies are discussed, from a much less quantitative point of view in [5].

The lessons to be learned from this particular problem appear to be the following. The problem fits admirably into the set up of Rivlin. The spaces X, Y, and Z, as well as the maps F, D, and R, are known. However, despite considerable knowledge, both quantitative and qualitative, on the factors limited the accuracy of reconstructions, nothing whatever is known regarding a truly relevant metric on the space Z. For this reason the program of optimal recovery breaks down. This is not a defect in the program, but a defect in our present practical knowledge.

3. DISCRIMINATION BETWEEN CANCER AND FIBROCYSTIC DISEASE IN THE BREAST

Early and reliable diagnosis of breast cancer is one of the most pressing needs in medicine today. Each year, in the United States alone, some 450,000 women are discovered to have abnormalities in the breast, about 90,000 of these are found to have cancer, and about 33,000 women die of breast cancer.

The problem divides naturally into two parts: 1) The development of a large scale screening procedure for the early detection of abnormalities; 2) The development of a reliable and noninvasive procedure for discriminating between malignant and non-malignant abnormalities. (At the present time surgery is the only sufficiently reliable procedure for making this discrimination.)

We have begun research on the second problem and have developed a simple, cheap, and noninvasive procedure which, in the thirty-three cases studied so far, discriminates correctly between adenocarcinoma (the most common of the breast cancers) and fibrocystic disease (the most common non-malignant abnormality). Briefly it is as follows: details are given in [4,6,8].

On the mammogram (breast x-ray film) a line through the lesion is read in steps of .1 mm. Suppose that D(x) is the reading at the point x. The edges a and b of the lesion on this line are located by examination of the readings themselves, a graph of the readings, and the film. The linear function L with values D(a) at a and D(b) at b is taken as an approximation to the readings that would be produced by normal tissue, and the difference

$$A(x) = L(x) - D(x) \tag{3.1}$$

is called the abnormality function. A rectangle is circumscribed around the graph of A, and the ratio of the area of this rectangle to the area under the graph of A is called the linear mass ratio.

linear mass ratio = $\dfrac{\text{area of rectangle}}{\text{area under graph of A}}$

The original readings D are highly dependent on both the film and the exposure. The abnormality function A is independent of the exposure, but depends upon the film. The linear mass ratio has the interesting property of being independent of both.

The linear mass ratios from the thirty-three cases studied are as follows:

Cancer			Fibrocystic Disease		
1.29	1.4.	1.53	1.66	1.88	2.06
1.32	1.46	1.56	1.68	1.91	2.12
1.34	1.48	1.57	1.75	1.91	2.15
1.35	1.51	1.59	1.78	1.92	2.17
1.39	1.51	1.59	1.80	1.99	2.26
1.40			1.81	2.02	

The reason for the success of the linear mass ratio is still obscure. It is not (as might be expected) due to a greater average density in cancer than in fibrocystic disease. Indeed, the linear mass ratio is largely independent of the average density, for any round lesion with constant density would produce a ratio $4/\pi$. The linear mass ratio depends upon the internal distribution of density within the lesion, not upon the average density.

On the other hand, the average density seems to be of relatively low diagnostic importance in this problem. In the cases of this study, particularly in the typical cases where the lesion was not neatly circumscribed, but consisted of a primary lesion embedded in other abnormalities, no correlation could be found between the average density and the surgical diagnosis.

If it is correct that the distinction between cancer and fibrocystic disease lies in the internal density distribution rather than the average density, then the poor success of breast scanners is easy to explain. A 128 x 128 reconstruction matrix is being used to represent a cross section about 30 cm. in diameter, so that a fairly large 1 cm. lesion is represented by a 4 x 4 square in the matrix. A 4 x 4 square should give a fairly accurate picture of the average density, but no information whatever about the internal distribution.

In this problem the optimization set up as as follows. The space X of objects is the space of breasts, or of density functions of breasts. The space Y of data is R^m, where m is the number of points at which each mammogram is read; and the data map D takes each breast, or density function, into the m-tuple of readings. The space Z can be one of two: Z_1 = real numbers, or $Z_2 = \{0,1\}$. If $Z = Z_1$, then the reconstruction map R carries a given set of data to the linear mass ratio, while if $Z = Z_2$, then R carries a given set of data into 0 = cancer if the linear mass ratio is less than or equal to 1.63, and into 1 = fibrocystic disease if the ratio is larger than 1.63.

In the first case, where $Z = Z_1$ = real numbers, the feature map $F: X \to Z$ is, for practical purposes, uncomputable. Even with biopsies, autopsies, etc., it is not practical to compute total masses along lines spaced at .1 mm. intervals.

In the second case, where $Z = Z_2 = \{0,1\}$, the feature map is computable by biopsy. There is only one possible metric on Z_2 (up to a constant multiple), namely that with $d(0,1) = 1$. Thus $e(R) = 0$ if the reconstruction method R is perfect (on the class of test objects X_T, as indicated in §2), and otherwise $e(R) = 1$.

Obviously this is a case where the optimization (0.3) must be used rather than the original (0.2). If, for example, p = 100 and R^p is given the ℓ_1 norm, then e(R) is the maximum number of errors in any of the groups of 100 forming the test objects in $\tilde{X}_T$. This would seem to be a relevant measurement of error.

It is apparent that in most, if not perhaps all, cases where Z is a finite set the optimization (0.3) should be used.

4. NONINVASIVE ANGIOGRAPHY

The usual procedure in radiology for vizualizing blood vessels is called angiography. It uses the injection of a large amount of iodine contrast dye into the vessel in question directly at the site of the examination, which can be painful and dangerous and always calls for hospitalization. The reason for the injection of a large amount of dye directly into the vessel at the site of the examination is that unless the dye is present in high concentration, nothing is visible on the radiograph.

The scanning microdensitometer used in this work to read the x-ray films is capable of discriminating between 16,000 shades of gray. (The film is not capable of recording that many, but in some of the work about 1000 shades were needed.) With such sensitivity, iodine dye, even in very low concentrations is easily detected. However, the changes in the readings due to the dye cannot be interpreted, for they are lost in the changes due to changes in tissue from one point to another, changes in the thickness of the object, etc.

If one x-ray is taken before the dye injection, and others at various times afterwards, then, since the differences between the objects at these times are due solely to the dye, the differences between the densitometer readings should reveal the progress of the dye through the vessels and tissues.

This procedure was carried out in an examination of the blood vessels of the neck. A 50 cc. bolus of 28% iodine solution was injected into the principal (antecubital) vein of the arm. One x-ray was taken before the injection. The injection itself took 10 seconds. A second x-ray was taken just at the end, a third 5 seconds after the end, and a fourth 10 seconds after the end.

It turned out that the blood circulated much more rapidly than we had anticipated. By the time of x-ray 2 the dye was already widespread throughout the tissues and veins. At the time of x-ray 4 it remained in the tissues and veins, but had left the arteries. As might be expected, the difference between x-ray 1 and x-ray 4 revealed the veins and the thyroid gland, which has a very heavy blood supply. The difference between x-ray 2 and x-ray 4 revealed the arteries. The veins and tissues effectively cancelled because the dye was present at both times. (The difference between x-ray 1 and x-ray 2 did not reveal much of anything except widespread dye, which, before we appreciated how fast the blood had circulated, led us to believe the experiment had been a failure. X-ray 3 was not used.)

Although this is certainly a reconstruction problem, namely, to reconstruct the blood vessels in the neck, we do not see how to fit it into the optimal recovery scheme. In this case, the space X must be the space of actual necks, not of density functions. If there are two films, one before and one after, and if m lines are read on each film with n readings per line, then $Y = M_{mn} \times M_{mn}$, where M_{mn} is the space of m x n matrices. Finally, $Z = M_{mn}$. The data map D is obvious, and the reconstruction map R is, as described, the difference. The feature map F, however, is unreachable. Even in theory there is no way to pass from a given neck to a matrix of numbers representing total masses along lines before and after dye injection.

In such cases a less refined feature space must be selected, i.e. a space Z' with a map r: Z→Z', for which a computable feature map F':X→Z' is available. Thus

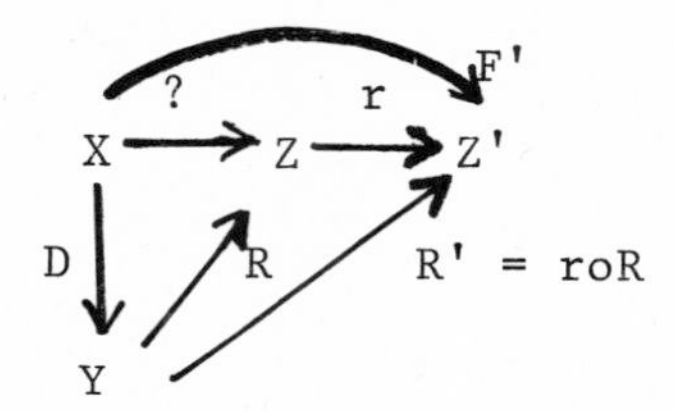

For example, if the medically relevant question is whether the carotid artery is obstructed, then Z' = {0,1}. The map F' is computable either by standard angiography or by surgery, and the map R' is the diagnosis obtained by looking at the reconstruction.

If, however, the problem is to visualize the blood flow through the various vessels, we have no idea whether a suitable Z' can be found. Radioactive scans are used for this purpose, but they are much too crude to be of any use in measuring the error in this procedure.

REFERENCES

1. R. Guenther, C. Kerber, E. Killian, K. Smith, and S. Wagner, "Reconstruction of Objects from Radiographs and the Location of Brain Tumors", Proc. Natl. Acad. Sci. USA 71:4884-4886 (1974).

2. R. Mersereau and A. Oppenheim, "Digital Reconstruction of Multidimensional Signals from their Projections", Proc. IEEE 62:1319-1338 (1974).

3. K. Smith, D. Solmon, and S. Wagner, "Practical and Mathematical Aspects of the Problem of Reconstructing Objects from Radiographs",

Bull. Amer. Math. Soc. (To appear).

4. K. Smith, S. Wagner, R. Guenther, and D. Solmon, "The Diagnosis of Breast Cancer in Mammograms by the Evaluation of Density Patterns" (To appear).

5. K. Smith, S. Wagner, R. Guenther, and D. Solmon, "Computerized Axial Tomography from Ordinary Radiographs - An Alternative to Scanners" (To appear).

6. D. Solmon, K. Smith, S. Wagner, and R. Guenther, "Breast Cancer Diagnosis from Density Studies of Mamograms", International Conference on Cybernetics and Society, IEEE, November 1-3, 1976, Washington, D. C.

7. D. Solmon and C. Hamaker, "The Rate of Convergence of the Kacmarz Method", J. Math. Anal. & Appl. (To appear).

8. S. Wagner, K. Smith, R. Guenther, and D. Solmon, "Computer Assisted Densitometric Detection of Breast Cancer", Application of Optical Instrumentation in Medicine V, SPIE 96: 418-422 (1976).

PLANNING OF RADIATION TREATMENT

Udo Ebert

Computing Center, University of Münster

West Germany

ABSTRACT

The aim of radiotherapy is to destroy a malignant tumour. In accomplishing this goal one has to spare the healthy tissue and organs as much as possible since otherwise they could perish. Thus, the total irradiation time should be short.

These considerations lead to a mathematical model in which constraints reflect the restrictions on the dosage - lower bounds in the tumour and upper bounds in the healthy tissue. The time of irradiation becomes the objective function of the optimization problem. By fixing some of the possible parameters of the treatment one gets a model in which the velocity of the source of radiation can be determined. This solution is approximated, solving a linear and quadratic or parametric programming problem. The model is implemented as a programming system for radiation-treatment planning. An example is given showing its application to a kidney tumour.

O. INTRODUCTION

In this paper a mathematical model of radiation-treatment planning is presented. It is possible to determine within this model optimal treatment plans: The source of radiation can move along certain orbits around the patient and, as a result, an approximation of the optimal velocity can be computed, taking into account additional restrictions.

Although the model discussed here is different from the F-model [2] (in which fixed positions for the source of radiation are determined) it is possible to extend the usual programming system for radiation-treatment planning to this optimizing model too [1,3].

1. FOUNDATIONS OF RADIOTHERAPY

The aim of every radiation treatment program is to destroy malignant tissue within a patient's body. When a tumour is irradiated, the healthy tissue which is confronted with the high-energy rays may also be damaged. The task of radiotherapy is to determine a treatment plan which destroys the tumour and injures the healthy tissue as little as possible. Moreover, one has to take into account that in some parts of the body and some organs (e.g. the liver or spinal card) the dosage must not exceed a certain amount since otherwise these parts may perish. Therefore, a treatment plan has to satisfy the following conditions: In some parts of the patient's body minimal doses must be achieved (e.g. in the tumour) and in some parts (e.g. in the liver) maximal doses must not be exceeded.

These restrictions are reflected in a mathematical model by constraints on the dosage in the body. The goal of minimal injury of healthy tissue is reached by minimization of the total irradiation time.

The dosage at a point of the body depends on the source of radiation, the condition at the patient, and the time of irradiation. Essential points are the type of the source of radiation, its position, its shape and its alignment. In the patient's body different tissues absorb radiation in different manners. Thus the dosage at a point P is described by

$$(1.1) \qquad D(P,p_1,\ldots,p_s)\cdot t \quad .$$

t denotes the irradiation time and $p_1,\ldots,p_s$ designate the remaining parameters. We may assume that D is continuous.

The central ray of the cone of radiation is named the central axis.

If the source irradiates the patient for a fixed time from a fixed location with constant parameters, one calls it a fixed field, or simply <u>field</u>.

In the following, a model with a moving source of radiation is developed.

2. MODEL FOR SPATIAL DISTRIBUTION OF FIELDS

2.1 DESCRIPTION

The model for radiation treatment which shall be examined here tries to distribute the directions in space from which the patient is irradiated in as uniform a manner as possible. It is called the S-model (model for spatial distribution of fields).

The underlying idea is easily explained: if one tries to destroy a tumour with one field then certainly the tissue lying between the tumour and the treatment unit will perish, too. Because of the absorption of radiation by the patient's body the dose there has to be even higher than in the tumour. As the doses of separate fields add up in every place it is more advantageous to use more than one field. These fields should come from very different directions and overlap only in the tumour. This concept is generalized in the S-model: it is assumed that the source of radiation moves along one or more orbits around the patient. For each orbit the central axis of the field has to go through a fixed point within the patient, the target point, which is situated in the tumour. The remaining parameters of the source are also constant and are prescribed in advance. The orbits are chosen according to anatomical and clinical factors, for instance by providing a plane and a constant distance between the source and the point of entry of the central axis into the patient.

Then the S-model determines the velocity with which the source moves along the orbit: v [cm/sec] or its inverse $w=1/v$ [sec/cm], the marginal "staying" rate at a point of the orbit. The objective function of the optimization problem is the total irradiation time, and it is to be minimized. As a result, advantageous parts of the orbits are used, i.e. the source of radiation will move slowly there. Disadvantageous parts will be avoided, i.e. the source will move very rapidly or not at all over these parts.

2.2 MATHEMATICAL MODEL

For an analytical representation of the model it is assumed that there are n orbits O_i. Each orbit O_i can be characterized by the continuously differentiable functions

$$x_i(u),\ y_i(u),\ z_i(u). \tag{2.1}$$

These describe the coordinates of the orbit within a fixed global coordinate system (the patient coordinate system). We can choose the parametric representation so that $u \in [0,1]$ for all orbits. c_i denotes the proposed constant parameters of the i-th orbit. Let $w_i(u)$ be the marginal staying rate on O_i; $w_i(u)$ is assumed to be Lebesgue-integrable on $[0,1]$. Let P be a point of the body B, a compact connected subset of 3-dimensional space, and define $w=(w_1,\dots,w_n)$. Then we have the total dose

$$(2.2)\quad D_s(P,w)=\sum_{i=1}^{n}\int_0^1 D(P,x_i(u),y_i(u),z_i(u),c_i)w_i(u)\cdot \sqrt{x_i'(u)^2+y_i'(u)^2+z_i'(u)^2}\,du$$

and the total time of irradiation

$$(2.3)\quad F_s(w)=\sum_{i=1}^{n}\int_0^1 w_i(u)\sqrt{x_i'(u)^2+y_i'(u)^2+z_i'(u)^2}\,du.$$

For $P \in B$ there exist constraints: the minimal dose $M_l(P)$ and the maximal dose $M_u(P)$. We may suppose the M_l and M_u are continuous functions. For medical reasons it is convenient to have a maximum marginal staying rate M, i.e. 1/M is a minimum velocity of the source; there should be no quasi-fixed fields.

So the system for the S-model can be expressed in the form

$$(2.4)\quad M_l(P) \le D_s(P,w) \le M_u(P) \qquad P \in B$$

$$(2.5)\quad 0 \le w_i(u) \le M \qquad u \in [0,1], \qquad i=1,\dots,n$$

$$(2.6)\quad w \in L[0,1] = \{(w_1(u),\dots,w_n(u)) \mid \text{for } i=1,\dots,n \;\; w_i(u) \text{ Lebesgue-integrable on } [0,1]\}$$

$$(2.7)\quad \text{minimize } F_s(w).$$

We have

Theorem 1: If there exists at least one w satisfying (2.4)-(2.6) then there is an optimal solution w^* to the problem (2.4)-(2.7).

Proof: For the exact details about the underlying (topological) assumptions and the proofs of this and the following theorems see [1].

2.3 DISCRETIZATION OF THE MODEL

The relationships (2.4)-(2.7) describe the S-model as an ideal one. In practice we do not have complete information about the patient's body, only an approximation. Moreover, the weight function, w, should lend itself to practical application. This leads to the following considerations:

1) The constraint (2.4) is weakened. It is replaced by a finite number of inequalities

$$(2.4') \quad M_l(P_j) \leq D_s(P_j,w) \leq M_u(P_j) \qquad j=1,\dots,m.$$

If we choose P_j suitably, (2.4) and (2.4') are almost equivalent:

<u>Theorem 2</u>: For every $\varepsilon>0$ we can find a natural number $m=m(\varepsilon)$ and m points $P_1,\dots,P_m$ such that from (2.4'), (2.5), (2.6) follows

$$M_l(P)-\varepsilon \leq D_s(P,w) \leq M_u(P)+\varepsilon \quad \text{for all } P\in B \text{ and } w\in L[0,1], \text{ satisfying (2.5)}.$$

2) The function class (2.6) is restricted to a finite dimensional subspace of $L[0,1]$. Here polygonal functions are considered (analogous results can be shown for step functions).

Let $\mathfrak{U}$ be a partition of $[0,1]$ into $r=r(\mathfrak{U})$ subintervals with $0=u_0<u_1<\dots<u_r=1$. Then we have the weight functions

$$(2.6') \quad w_i(u)=w_i^{k-1}\frac{u_k-u}{u_k-u_{k-1}}+w_i^k\frac{u-u_{k-1}}{u_k-u_{k-1}} \qquad u\in[u_{k-1},u_k] \quad k=1,\dots,r;\ i=1,\dots,n.$$

In the case of a closed orbit we suppose $w_i^0 = w_i^r$ for $i=1,\dots,n$. For this restricted problem we obtain an analogous result:

<u>Theorem 1'</u>: If there exists at least one $w_{\mathfrak{U}}$ satisfying (2.4), (2.5), (2.6') then there is an optimal solution $w_{\mathfrak{U}}^*$ to the problem (2.4), (2.5), (2.6'), (2.7).

The restriction to polygonal functions does not lead to a considerably worse solution; indeed, one can show

<u>Theorem 3</u>: Let $\mathfrak{U}_\nu$ be a sequence of partitions of the interval $[0,1]$ satisfying $\mathfrak{U}_\nu \subset \mathfrak{U}_{\nu+1}$ and

$$\lim_{\nu\to\infty}\ \max_{i=1,\dots,r(\mathfrak{U}_\nu)} |u_i-u_{i-1}| = 0 \text{ then}$$

$$\inf_{\nu\in\mathbb{N}} F_s(w_{\mathfrak{U}_\nu}) = F_s(w^*).$$

3) The above reflections prove that the S-model can be considered in the discretized form (2.4'), (2.5), (2.6'), (2.7). Combining (2.4') and (2.6') we have

$$(2.4'') \quad M_l(P_j) \leq \sum_{i=1}^{n} \Big(a_{io}^{+}(P_j)w_i^{o} + \sum_{k=1}^{r-1} (a_{ik}^{+}(P_j)+a_{ik}^{-}(P_j))w_i^{k} + a_{ir}^{-}(P_j)w_i^{r}\Big) \leq M_u(P_j) \qquad j=1,\ldots,m$$

where

$$a_{ik}^{+}(P) = \int_{u_k}^{u_{k+1}} D(P,x_i(u),y_i(u),z_i(u),c_i)\,\frac{u_{k+1}-u}{u_{k+1}-u_k}\sqrt{x_i'(u)^2+y_i'(u)^2+z_i'(u)^2}\,du$$

and

$$a_{ik}^{-}(P) = \int_{u_{k-1}}^{u_k} D(P,x_i(u),y_i(u),z_i(u),c_i)\,\frac{u-u_{k-1}}{u_k-u_{k-1}}\sqrt{x_i'(u)^2+y_i'(u)^2+z_i'(u)^2}\,du.$$

These integrals can be computed without knowing the w_i. The objective function (2.7) also turns out to be linear in the w_i^j. So the S-model can be reduced to a problem in linear programming.

2.4 SOLUTION

To solve this problem we can use any of the known methods of linear programming [11]. Applying one of these methods we find a basic solution (a vertex or extremal point of the polytope of the feasible solutions). If, because of symmetry, there is more than one optimal solution there will exist optimal solutions which are not basic solutions. Some of these will distribute the directions of radiation better than most other optimal solutions (e.g. the average of all optimal basic solutions).

2.5 IMPROVING THE SOLUTION

In such cases we can improve the solution with the help of one of the following two methods. For the sake of an easier description a new and simpler notation is introduced. The S-model can be represented in the following form

$$(2.8) \qquad Ax \leq b$$

$$(2.9) \qquad 0 \leq x_i \leq M \quad i=1,\ldots,r$$

$$(2.10) \qquad \text{minimize } c'x$$

where A is a matrix of dimensions $q \times r$, $b=(b_1,\ldots,b_q)'$, $x=(x_1,\ldots,x_r)' \geq 0$ are column-vectors, and $c=(c_1,\ldots,c_r)$ is a row-vector. Let x^* be an optimal solution.
We want to find among all optimal solutions a most appropriate one. Therefore we add to (2.8), (2.9) the constraint

$$(2.11) \qquad c'x = c'x^*$$

and replace (2.10) by a different objective.

1) Modifying (2.10) we get

$$(2.10') \qquad \text{minimize } \sum_{i=1}^{r} c_i x_i^2 \, .$$

Substantially this means seeking a time-optimal solution of such a kind that its functional values have as little variation as possible. Then this solution will distribute the directions of radiation more uniformly than x^*.
It is difficult to solve this quadratic problem for large q,r. A slight variation leads to a simpler problem:

In the first step we seek all basic solutions $x^1,\ldots,x^l$ on the hyperplane (2.11) (cf. [11]). Generally there will be less than r. In the second step we find a linear convex combination of $x^1,\ldots,x^l$,

i.e. $x = \sum_{i=1}^{l} \lambda_i x^i$ satisfying (2.11).

Thus we must solve the essentially simpler quadratic problem:

$$(2.12) \qquad \sum_{i=1}^{r} \lambda_i = 1$$

$$(2.13) \qquad \lambda_i \geq 0 \quad i=1,\ldots,l$$

$$(2.14) \quad \text{minimize} \sum_{i=1}^{n} c_i x_i^2 = \sum_{j=1}^{l} \sum_{k=1}^{l} q_{jk} \lambda_j \lambda_k$$

where $Q = (q_{jk})$ is a symmetric positive definite matrix.
2) The second method requires less computation. To get a uniform distribution here we try to force down the maximum of the weight function as much as possible. We choose as the objective

(2.10'') minimize M.

The problem (2.8), (2.9), (2.10''), (2.11) is solved by parametric programming.
Remarks: In general the results of both methods do not agree. Also it is convenient to replace (2.11) by

$$(2.11') \quad c'x \leq c'x^* + \varepsilon$$

with small $\varepsilon > 0$ and to similarly apply both methods.

3) Example
The model described above has been implemented. The dose function is approximated with the help of Sterling's field equations [12,13,14,15].
In the optimizing part the S-model uses the MPS (Mathematical Programming System [4,5]) for the linear programming and a PL/I-version of Lemke's algorithm [7,10] for the quadratic programming. The output part makes use of a program written by Pudlatz [9] for the graphic representation of the result. All programs are written in PL/I [6].

The following example describes the irradiation of a kidney tumour. It has already grown very extensively and the lymphatic ganglions situated in front of the spinal cord must also be irradiated (fig. 1).

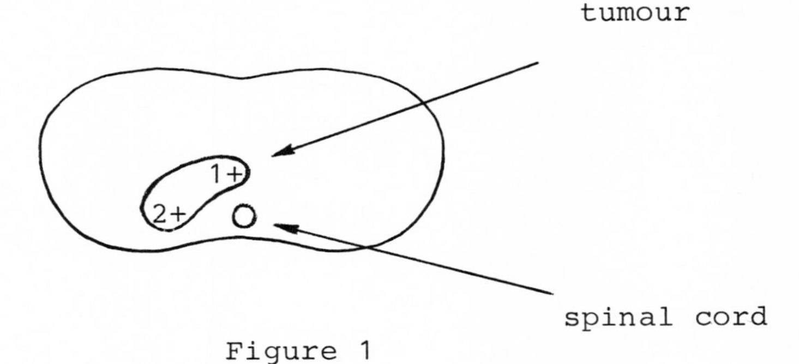

Figure 1

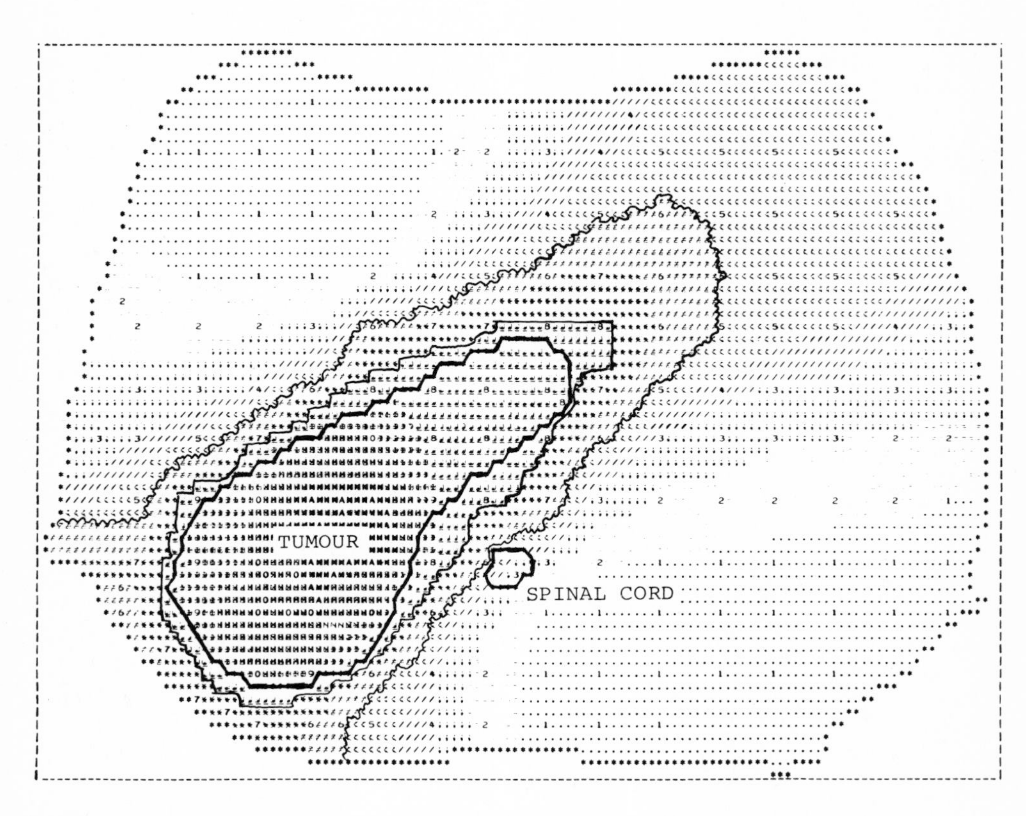

Figure 2. Result of the S-model
——5000 rd isodose ∿∿ 4000 rd isodose

In addition to the tumour the spinal cord has been drawn because it contains relatively sensitive nerve fibres. It should receive not more than 4000 rd since otherwise it could be irreparably damaged. The tumour itself must get as a minimum 5000 rd to be destroyed. Moreover, there are restrictions on the dosage at the surface (4500 rd).

There are two admissible orbits for the source of radiation; each has a fixed target point (crosses 1 and 2 in fig. 1). The orbits are selected so that the central axis of the fields always penetrates the target point, and so that the resulting source-skin-distance is constant. The number of orbits and the choice of target points suggest themselves by the shape of the target area. Moreover, a fixed field size on each orbit was determined in advance (for the orbit around target point 1 it is smaller than for orbit 2). The weight function (marginal staying-rate) has the form of a polygonal function. An improvement of the solution was obtained using parametric programming. Figure 2 shows the result.

ACKNOWLEDGEMENT

The auther wishes to thank Prof. Dr. H. Werner for suggesting the problem and for his advice, and Prof. Dr. E. Schnepper and Dr. H. Terwort for explaining the problems of radiation-treatment planning.

REFERENCES

[1] U.Ebert: Optimale Auslegung von Bestrahlungsplänen, Schriftenreihe des Rechenzentrums der Universität Münster, Nr. 19, 1976

[2] U.Ebert: Computation of Optimal Radiation Treatment plans, to appear

[3] U.Ebert: A System for Calculating Optimal Radiation Treatment Plans, to appear

[4] IBM Application Program, Mathematical Programming System /360: Control Language, User's Manual, IBM-Form H20-0290-1

[5] IBM Application Program, Mathematical Programming System /360 Version 2: Linear and Separable Programming - User's Manual, IBM-Form GH20-0476-2

[6] IBM System /360 Operating System, PL/I(F): Language Reference Manual, IBM-Form GC28-8201-3

[7] C.E.Lemke: Bimatrix Equilibrium Points and Mathematical Programming, Management Science 11 (1965), 681-689

[8] S.Matschke, J.Richter, K.Welker: Physikalische und technische Grundlagen der Bestrahlungsplanung, Leipzig 1968

[9] H.Pudlatz: GEOMAP - ein FORTRAN-Programm zur Erzeugung von Choroplethen- und Isolinienkarten auf dem Schnelldrucker, Schriftenreihe des Rechenzentrums der Universität Münster, Nr. 16, 1976

[10] A.Ravindran: A Computer Routine for Quadratic and Linear Programming Problems (alg.431), Communications of the ACM 15 (1972), 818-820

[11] M.Simmonard: Linear Programming, Englewoods Cliffs, 1966

[12] T.D.Sterling, H.Perry, L.Katz: Automation of Radiation Treatment Planning, IV. Derivation of a mathematical expression for the per cent depth dose surface of cobalt 60 beams and visualisation of multiple field dose distributions, British Journal of Radiology 37 (1964), 554-550

[13] T.D.Sterling, H.Perry, J.J.Weinkam: Automation of Radiation Treatment Planning, V. Calculation and visualisation of the total treatment volume, British Journal of Radiology 38 (1965), 906-913

[14] T.D.Sterling, H.Perry, J.J.Weinkam: Automation of Radiation Treatment Planning, VI. A general field equation to calculate per cent depth dose in the irradiated volume of a cobalt 60 beam, British Journal of Radiology 40 (1967), 463-468

[15] J.J.Weinkam, A.Kolde: Radiation Treatment Planning System Manual, Saint Louis

SOME ASPECTS OF THE MATHEMATICS OF LIMULUS

K.P.Hadeler

Lehrstuhl Biomathematik, Univ. Tübingen

Auf der Morgenstelle 28, 74 Tübingen,W.Germany

Abstract

The compound eyes of the horse-shoe crab exhibit one of the few nervous networks which are well understood, mainly because of its simple and repetitive structure. The theory of this network poses various mathematical questions such as existence and uniqueness of stationary solutions, stability problems and, at present of greatest interest, the existence of periodic solutions of differential equations with retarded argument.

1. Facts from biology

The horseshoe crab Limulus is a marine arthropode from the coastal waters of the north american east coast. It is one of the five recent species of the ancient order Xiphosura, a group more closely related to spiders than to true crabs.

The animal is up to 60 cm long. The main exterior feature is a huge shield covering the entire body and almost the legs, and a tail-rod. The shield consists of two parts, connected by a hinge. Inserted in the anterior part of the shield are two rather large compound eyes.

These eyes have been introduced as an object of investigation by Hartline in 1928 [16] and have since then proved to be very useful, so much that at present "Limulus" is a field of nervous physiology as well as "Drosophila" is a field in genetics.

It is somewhat astonishing that up to now the role the eyes play in the life of Limulus has not been completely clarified, their purpose seems somewhat doubtful at least in adult animals.

The nervous system of the Limulus eye represents a rather regular network, with each ommatidium connected to its neighbors. These connections can be schematically depicted as in fig.1. Connections are not restricted to nearest neighbors.

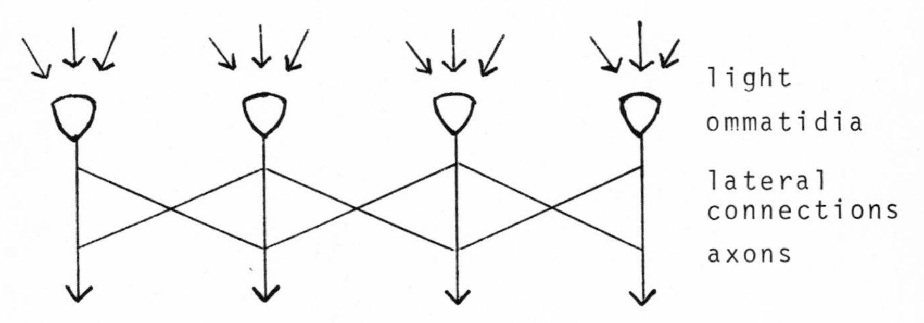

Fig.1: Scheme of retina

First we consider a single ommatidium. For simplicity we neglect that fact that each ommatidium contains several retinal cells. Light of intensity x is falling from the exterior into the retinal cell and generates a potential (generator potential) y. The potential y is a monotone function of x in the form of a saturation curve. In the following we shall forget about x and consider y as a measure of the excitation of the ommatidium.

If the excitation is below a certain threshold $\bar{y}$ nothing will happen. If y exceeds $\bar{y}$ then the nerve cell releases nervous impulses (spikes) with a frequency z proportional to the excess potential. Thus, under stationary conditions, the function of the ommatidium can be described as

$$z = b\,\vartheta(y-\bar{y}), \qquad (1)$$

where ϑ is the function defined by

$$\vartheta(x)=\max(x,0)=\frac{1}{2}(x+|x|). \qquad (2)$$

Experiments have shown that ommatidia act inhibiting upon each other, i.e. a large excitation in one ommatidium tends to decrease the excitation in the neighbors. The inhibition acts through the nervous connections exhibited in fig. 1. The inhibitory effects decrease with distance. We assume that the inhibition acts linear. The relations of x,y, and z are represented in fig.2.

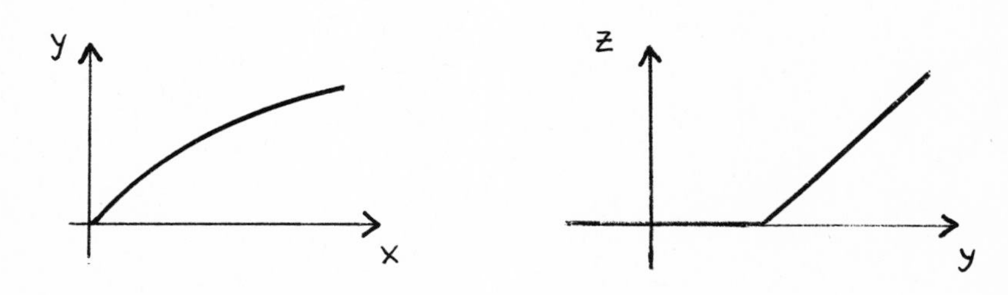

Fig.2: Transition $x \longrightarrow y \longrightarrow z$

The inhibitory effect of the k-th ommatidium on the j-th ommatidium is measured by the inhibition coefficient $\beta_{jk} \geq 0$. Self-inhibition may be present, thus β_{jj} need not be zero. Then a compound eye with n ommatidia is described by the equations

$$z_j = b_j \vartheta(y_j - \bar{y}_j - \sum_{k=1}^{n} \beta_{jk} z_k), \ j=1,\ldots,n. \qquad (3)$$

Equations (3) are known as the Hartline-Ratliff model. The model describes a stationary situation. It has been discussed by Reichardt and McGinitie [23], Varjú [24], Walter [25].

A possible "purpose" of lateral inhibition is compensation of a loss in acuity of image during perception. Suppose the animal is shown a spatial step function (in fig.3 we consider one space dimension).

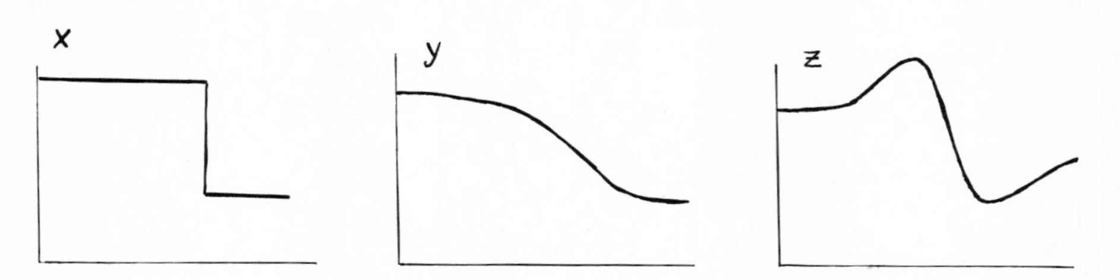

Fig.3: Regain of contrast

Since each ommatidium will perceive light from various points of the exposed pattern, the excitation, as a function of space, is some smoothened step function, and the result of lateral inhibition is a regain in contrast, leading to the formation of Mach bands. (Mach bands are a famous effect in human psychophysics: In a pattern of black and white stripes the white adjacent to black appears very white such that in the middle of white stripes some greyish stripes seem to appear).

2. The model

If the eye is exposed to patterns varying in time then the y_j are functions of time, and so are the z_j. As appropriate state variables we choose the reduced excitations

$$v_j = b_j(y_j - \bar{y}_j - \sum_{k=1}^{n} \beta_{jk} z_k). \qquad (4)$$

The quantity v_j is the actual excitation which generates the nervous impulses, taking all inhibitory effects into account, multiplied by the factor b_j.

We remark that in the work of Coleman and Renninger ([3]- [8]) the term "reduced excitation" denotes the excitation that would occur in an ommatidium disconnected from all others, i.e.

$$b_j(y_j-\bar{y}_j-\beta_{jj}z_j) \ . \tag{5}$$

From (3), (4) follows

$$z_j=\vartheta(v_j), \qquad j=1,\ldots,n. \tag{6}$$

Under non-stationary conditions the reduced excitation will not immediately respond to a change of the input (y_j), but adapt to a change with a time constant γ,

$$\gamma\dot{v}_j+v_j=b_j(y_j-\bar{y}_j-\sum_{k=1}^{n}\beta_{jk}z_k) \ .$$

In this equation we can replace z_k by $\vartheta(v_k)$, thus

$$\gamma\dot{v}_j+v_j=b_j(y_j-\bar{y}_j-\sum_{k=1}^{n}\beta_{jk}\vartheta(v_k)). \tag{7}$$

Such models have been considered by Morishita and Yajima [20] . Recently it has been found that inhibiton acts with a delay τ which is independent of the distance of the ommatidia. There is no such delay in self-inhibition. If this delay is incorporated into the model then the equations read

$$\gamma\dot{v}_j(t)+v_j(t)=b_j(y_j(t)-\bar{y}_j-\beta_{jj}\,\vartheta(v_j(t)) \tag{8}$$

$$-\sum_{\substack{k=1\\k\neq j}}^{n}\beta_{jk}\,\vartheta(v_k(t-\tau)), \ j=1,\ldots,n.$$

The model can be adapted more closely to reality by assuming that the coefficient β_{jk} depends explicitely on z_k and that inhibition acts with a threshold, too.

Indeed there is some indication that inhibition increases with excitation, such that β_{jk} should be a non-decreasing function of z_k.

With such modifications the model becomes

$$\gamma\dot{v}_j(t)+v_j(t)=b_j(y_j(t)-y_j(t)-\bar{y}_j-\beta_{jj}(\vartheta(v_j(t))\vartheta(v_j(t)-\bar{z}_{jj}) \\ -\sum_{\substack{k=1\\k\neq j}}^{n}\beta_{jk}(\vartheta(v_k(t-\tau)(\vartheta(v_k(t-\tau)-\bar{z}_{jk}), \quad j=1,\ldots,n. \tag{9}$$

However, to obtain mathematically tractable equations we shall stay with the model (8). By introducing new variables

$$v_j:=b_j(y_j-\bar{y}_j), \quad \beta_{jk}:=b_j\beta_{jk} \tag{10}$$

we arrive at

$$\gamma\dot{v}_j(t)+v_j(t)=y_j(t)-\beta_{jj}\vartheta(v_j(t))-\sum_{\substack{k=1\\k\neq j}}\beta_{jk}\vartheta(v_k(t-\tau)), \quad j=1,\ldots,n. \tag{11}$$

With vector notation

$$v=(v_j), \qquad y=(y_j),$$

$$D=(\beta_{jj}\delta_{jk}), \quad B=(\beta_{jk}),$$

equations (11) take the form

$$\gamma\dot{v}(t)+v(t)=y(t)-D\vartheta(v(t))-(B-D)\vartheta(v(t-\tau)). \tag{12}$$

3. The equilibrium states

Consider equation (12) with a constant input $y(t)\equiv y$. A stationary solution v satisfies the equation

$$v=y-B\vartheta(v) . \tag{13}$$

We show that this equation is equivalent to the stationary Hartline-Ratliff model

$$z = \vartheta(y - Bz) \tag{14}$$

in the sense that either both equations have no solution or there is a one-to-one correspondence between solution sets.

If v is a solution of (13), then $z=\vartheta(v)$ satisfies $z=\vartheta(y-B\vartheta(v))=\vartheta(y-Bz)$. On the other hand, if z solves (14) then $v=y-Bz$ fulfils $\vartheta(v)=\vartheta(y-Bz)=z$, thus $v=v-B\vartheta(v)$.

If $v^{(1)}$, $v^{(2)}$ are two solutions of (1) and $z^{(1)}=\vartheta(v^{(1)})$, $z^{(2)}=\vartheta(v^{(2)})$ then from $z^{(1)}=z^{(2)}$ follows that $v^{(1)}, v^{(2)}$ coincide in their positive components, and from (13) follows $v^{(1)}=v^{(2)}$. On the other hand, if $z^{(1)}$, $z^{(2)}$ are two solutions of (14) and $v^{(1)}=y-Bz^{(1)}$, $v^{(2)}=y-Bz^{(2)}$ then from $v^{(1)}=v^{(2)}$ follows $Bz^{(1)}=Bz^{(2)}$ and thus $z^{(1)}=z^{(2)}$ from (14).

To equation (14) we relate the mapping $T: \mathbb{R}^n \longrightarrow \mathbb{R}^n$,

$$Tz = \vartheta(y - Bz), \tag{15}$$

solutions of equation (14) correspond to fixed points of T. T maps $\mathbb{R}^n$ into $\mathbb{R}^n_+$, and T maps $\mathbb{R}^n_+$ into the compact convex domain

$$M = \left\{ z:\ 0 \leq z \leq \vartheta(y) \right\}. \tag{16}$$

By Brouwer's fixed point theorem the set M contains at least one fixed point; and all fixed points are contained in M.

The corresponding stationary states of equation (8), i.e. solutions of equation (13), are contained in

$$M' = \left\{ v : y - B\vartheta(y) \leq v \leq y \right\}. \tag{17}$$

If the spectral radius $\varrho(B)$ of the matrix B is less than 1 then T is a contraction with respect to an appropriate norm

$$\|x\| = \sum_{j=1}^{n} \alpha_j |x_j| ,$$

where the α_j are the components of a positive left eigenvector of the matrix $\tilde{B}=(\beta_{jk}+\varepsilon)$, $\varepsilon>0$ sufficiently small. Thus $\varrho(B)<1$ implies uniqueness of the stationary state ([11]).

In general (i.e. for $\varrho(B)\geq 1$) the stationary state will not be unique. This is important insofar as in the Limulus eye normally inhibition is as strong as $\varrho(B)>1$.

4. Oscillating solutions

Comparably recent results (Barlow, Adler, see [3] - [8]) indicate that even with a constant excitation the nervous impulses do not appear equally spaced in time, but in "bursts". The impulse frequency is oscillating or periodic. Apparently Coleman and Renninger ([3] - [8]) were the first explaining these bursts as a consequence of the delay with which the inhibition acts.

To investigate these effects we simplify the model by assuming that the compound eye is homogeneous and that it is exposed to a spatially homogeneous excitation. To avoid at present any complications related to the geometry of the eye we assume that the inhibition coefficients β_{jk}, $j\neq k$, do not depend on j,k, furthermore we neglect self-inhibition. Then, if only spatially homogeneous solutions are considered, each component of such a solution satisfies a scalar difference-differential equation

$$\gamma\dot{v}(t)+v(t)=y(t)-\beta\vartheta(v(t-\tau)) . \qquad (18)$$

By rescaling the time variable we achieve

$$\gamma\dot{v}(t)+v(t)=y-\beta\vartheta(v(t-1)).$$

The stationary state is $\bar{v}=y/(\beta+1)$, and by way of the substitution $v=\bar{v}+x$ we shift the stationary state into zero

$$\dot{x}+\frac{\tau}{\gamma}x = -\vartheta\left(\frac{\tau}{\gamma}\beta\bar{v}+\frac{\tau}{\gamma}\beta x\right)+\frac{\tau}{\gamma}\beta\bar{v}$$

or

$$\dot{x}(t)+\nu x(t)=-\left[\vartheta(a+bx(t-1))-a\right] \tag{19}$$

where

$$\nu=\frac{\tau}{\gamma}, \quad a=\frac{\tau}{\gamma}\beta v, \quad b=\frac{\tau}{\gamma}\beta \quad . \tag{20}$$

For some special cases Coleman and Renninger [7] were able to exhibit the existence of periodic solutions. We shall prove the existence of such solutions for equation (19) for all cases where the zero solution is unstable.

This equation is a special case of the equation

$$\dot{x}(t)+\nu x(t)=-f(x(t-1)) \tag{21}$$

where ν is a non-negative parameter and the continuous function $f: \mathbb{R}\longrightarrow\mathbb{R}$ has the properties

$$f(x)x > 0 \quad \text{für } x \neq 0, \tag{22}$$

f is differentiable at $x = 0$ and $f'(0)=\alpha > 0$. (23)

The linearized equation corresponding to (21) is

$$\dot{x}(t)+\nu x(t)+\alpha x(t-1)=0, \tag{24}$$

and the characteristic equation is

$$\nu+\lambda+\alpha e^{-\lambda} = 0 \quad . \tag{25}$$

For $\nu \geq 0$ let α_ν be the smallest positive solution of the equation

$$\nu + \alpha\cos\sqrt{\alpha^2-\nu^2} = 0 \quad . \tag{26}$$

One can show (see [13]): For $\alpha<\alpha_\nu$ the characteristic equation has no roots in the right half-plane, and the zero solution of (24) is stable. For $\alpha>\alpha_\nu$ the characteristic equation (25) has a root λ with $\mathrm{Re}\lambda > 0$, and the zero solution of (24) is unstable.

If α increases from 0 to infinity then the solutions of equation (21) in the neighborhood of 0 become oscillatory before the stationary solution 0 looses its stability property, namely (see [13]):

Suppose there is $\delta > 0$ and $c > \nu/(e^{\nu}-1)$ ($c > 1$ if $\nu=0$) such that

$$|f(x)| \geq c\ |x| \qquad \text{for} \quad |x| \leq \delta. \tag{27}$$

Let $\mathcal{K}$ be the cone of all functions $\varphi \in C[-1,0]$ with the properties

1) $e^{\nu t}\varphi(t)$ is a non-decreasing function of t, where $-1 \leq t \leq 0$.

2) $\varphi(-1)=0$.

Let $\mathcal{K}_0 = \mathcal{K} - \{0\}$.

Let $x(t)=x(t,\varphi)$ be the solution of equation (21) generated from the initial function $\varphi \in K_0$. Then:

1) x has denumerably many zeros $z_1 < z_2 < z_3 < \ldots$, and

2) $z_1 > 0$, $z_{n+1}-z_n > 1$ for $n=1,2,3,\ldots$

3) $\dot{x}(z_{2k-1}) < 0$, $\dot{x}(z_{2k}) > 0$ for $k=1,2,\ldots$

4) The function $e^{\nu t}\ x(t)$ is non-increasing on each interval $(z_{2k-1}, z_{2k-1}+1)$ and non-increasing on each interval $(z_{2k}, z_{2k}+1)$, $k=1,2,3,\ldots$.

5) For every $M > 0$ there is a constant $C_M > 0$ such that from $\|\varphi\| \leq M$ follows $z_2 \leq C_M$.

These properties of the solution enable one to define a mapping $\mathcal{F}: \mathcal{K} \rightarrow C[0,1]$ by

$$\mathcal{F}(\varphi)=x(z_2(\varphi)+1+t,\varphi),\ -1\leq t\leq 0, \quad \text{for } \varphi \neq 0 \tag{28}$$

$$\mathcal{F}(0)=0;$$

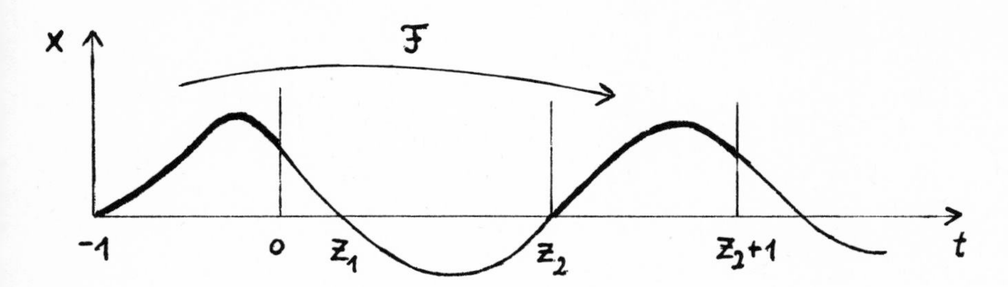

Fig.4: The operator $\mathfrak{F}$.

Fig.4 shows how $\mathfrak{F}$ acts. One can show $\mathfrak{F}\mathcal{K} \subset \mathcal{K}$, $\mathfrak{F}$ is continuous and compact. If, in addition, f satisfies the condition

There is a number $\varkappa > 0$ such that $f(x) \geq -\varkappa$ for all x (29)

then $\mathfrak{F}$ maps the closed, bounded convex set of infinite dimension $\mathfrak{D} = \{\varphi \in \mathcal{K} : \|\varphi\| \leq \varkappa\}$ into itself. Therefore Schauder's theorem would immediately yield the existence of a fixed point of $\mathfrak{F}$ and thus to a periodic solution, if it were not for $\mathfrak{F}(0)=0$.
Following the proof of R. Nussbaum [21] for the case $\nu=0$, we apply the Browder-Horn theorem on ejective fixed points. A fixed point $\bar{\varphi}$ of $\mathfrak{F}$ is ejective, if there is an open neighborhood $U \in \mathfrak{D}$, $\bar{\varphi} \in U$, such that for every $\varphi \in U$, $\varphi \neq \bar{\varphi}$, there is an integer $n=n(\varphi)$ such that $\mathfrak{F}^{(n)} \varphi \notin U$. The Browder-Horn theorem [1] states that a continuous compact mapping of a closed bounded convex set of infinite dimension has a non-ejective fixed point. This theorem yields the following result (see [13]):

If $\nu=0$ is given and the function f has properties (22), (23), where $\alpha > \alpha_\nu$, and (29), then equation (21) has a non-constant periodic solution.

Similar but weaker results have been shown by Pesin [22] by an immediate application of Schauder's theorem, and Kaplan and Yorke [18] have proved the existence of periodic solutions for the general equation

$$\dot{x}(t)=-f(x(t),x(t-1))$$

by a phase plane method. Again, when restricted to equation (21), their result is somewhat weaker than the above. See also Chow [2] and the earlier work by Jones [17] and Grafton [10].

The equation (19) satisfies all conditions required for a periodic solution if $b > \alpha_\nu$.

5. The vector system

Consider again equation (12) with constant input

$$\gamma\dot{v}(t)+v(t)=y-D\vartheta(v(t))-(B-D)\vartheta(v(t-\tau)). \tag{12}$$

A stationary solution $\bar{v}$ satisfies

$$(B+1)\bar{v}=y . \tag{30}$$

We assume that equation (30) has a positive solution $\bar{v}$ and that the non-negative matrix B is non-singular. In (12) we substitute

$$v=\bar{v}+Bx$$

and obtain

$$\gamma\dot{x}(t)+x(t)=\bar{v}-B^{-1}D\vartheta(v+Bx(t))-(I-B^{-1}D)\vartheta(\bar{v}+Bx(t-\tau)). \tag{31}$$

If we exclude self-inhibition (D=0) then equation (31) simplifies to

$$\gamma\,\dot{x}(t)+x(t)=-[\vartheta(\bar{v}+Bx(t-\tau))-\bar{v}]$$

or, with $a=\tau\bar{v}/\gamma$, $\bar{B}=\tau B/\gamma$, $\nu=\tau/\gamma$

$$\dot{x}(t)+\nu x(t)=-[\vartheta(a+Bx(t-1))-a] \tag{32}$$

similarly to equation (19).

For x sufficiently small equation (32) is linear

$$\dot{x}(t)+\nu x(t)+Bx(t-1)=0 . \tag{33}$$

With $x(t)=\xi\exp(\lambda t)$ we obtain the characteristic eigenvalue problem

$$B\xi + (\lambda+\nu)e^{\lambda}\xi = 0 \quad .$$

Now the possible complex or negative eigenvalues considerably effect the stability of equation (33). If $\lambda=\beta+i\gamma$ is a characteristic root, then $-(\lambda+\nu)e^{\lambda}=\mu$ is an eigenvalue of the matrix B. In general nothing will be known about μ except $\mu=\alpha\exp(i\psi)$ with $0\leq\alpha\leq\varrho(B)$. Then

$$\nu+\beta+\alpha e^{-\beta}\cos(\gamma-\psi)=0, \quad \gamma-\alpha e^{-\beta}\sin(\gamma-\psi)=0 \quad .$$

These equations have a solution with $\beta>0$ for an appropriate ψ iff $\nu<\alpha$.

Therefore we have the following situation: Let ν be fixed, for $\varrho(B)<\nu$ all characteristic exponents have negative real parts, for $\varrho(B)>\alpha_{\nu}$ there is at least one characteristic exponent with positive real part. Thus the eigenvalues of B which are not real and positive tend to cause instabilities.

6. Related problems from ecology

One of the classical ecological models is the Verhulst equation, in normalized form

$$\dot{u}(t)=\alpha u(t)(1-u(t)). \qquad (34)$$

For any initial data $u(0)>0$ the solution $u(t)$ approximates the carrying capacity $u=1$. All solutions are monotone.
Equation (34) is equivalent with

$$\dot{x}(t)=-\alpha x(t)(1+x(t)). \qquad (35)$$

If the growth rate depends on history we arrive at Hutchinsons's equation [15] (see [9],[24] for biological details)

$$\dot{u}(t)=\alpha u(t)(1-u(t-1)) \qquad (36)$$

or, equivalently,

$$\dot{x}(t)=-\alpha x(t-1)(1+x(t)). \qquad (37)$$

These equations have oscillating solutions for sufficiently large α. The characteristic equation

$$\lambda + \alpha e^{-\lambda} = 0$$

has a zero λ with $\mathrm{Re}\lambda > 0$ iff $\alpha > \pi/2$, and it is known that equations (36) and (37) have periodic solutions if $\alpha > \pi/2$.
Dunkel [9] , Halbach [14] proposed the more general equation

$$\dot{x}(t) = -\alpha \int_{-1}^{0} x(t+s)\, d\sigma(s)(1+x(t)), \tag{38}$$

where $\alpha > 0$ and σ is a function, monotone and continuous from the left, with $\sigma(-1-)=0$, $\sigma(0+)=1$. The corresponding characteristic equation is

$$\lambda + \alpha \int_{-1}^{0} e^{\lambda s}\, d\sigma(s) = 0. \tag{39}$$

A serious obstacle on the way to the proof of existence of periodic solutions is the location of the zeros of the characteristic equation. For which σ, α are there zeros in the right half-plane? Obviously there are no zeros in the right half-plane if σ has a step of height 1 at s=0. Hence we conjecture that there is a zero in the right half-plane at least for large α if the "mass" of σ is mainly concentrated in the past. The two results so far known support this conjecture: H.O.Walther [27] showed that with

$$\int_{-1}^{0} |\sin \pi s|\, d\sigma(s) > 1/\pi \tag{40}$$

there is a zero for α sufficiently large. The author [12] showed that if $\sigma(s)=1$ for $s > -\varepsilon$ then there is a zero in the right half-plane for

$$\alpha \sin \frac{\pi}{2\varepsilon} > \frac{\pi}{2\varepsilon} \quad . \tag{41}$$

Walther [27] succeeded to prove the existence of periodic solutions of

$$\dot{x}(t)=-\alpha\int_{-T}^{-1} x(t+s)d\sigma(s)(1+x(t)) \qquad (42)$$

for $\alpha > \pi/2$ and T-1 sufficiently small. It would be desirable to treat problems like

$$\dot{x}(t)+\gamma x(t)=-\alpha\int_{-T}^{-1} x(t+s)d\sigma(s)(1+x(t)) \qquad (43)$$

but approaches so far known lead to tremendous difficulties.

References

[1] Browder,F.E., A further generalization of the Schauder fixed point theorem, Duke Math.J.32, 575-578 (1965)

[2] Chow,S.-N., Existence of periodic solutions of autonomous functional differential equations, J.Diff. Eq. 15, 350-378 (1974)

[3] Coleman,B.D., and Renninger,G.H., Theory of delayed lateral inhibition in the compound eye of Limulus, Proceedings Nat.Acad.Sc. 71, 2887-2891 (1974)

[4] Coleman,B.D., and Renninger, G.H., Consequences of delayed lateral inhibition in the retina of Limulus I. Elementary theory of spatially uniform fields, J. Theor. Biol. 51, 243-265 (1975)

[5] Coleman,B.D., and Renninger,G.H., Consequences of delayed lateral inhibition in the retina of Limulus. II. Theory of spatially uniform fields, assuming the 4-point property, J. Theor. Biol. 51,267-291 (1975)

[6] Coleman,B.D., and Renninger,G.H., Periodic solutions of certain nonlinear integral equations with a time-lag, SIAM J.Appl.Math. 31, 111-120 (1976)

[7] Coleman,B.D., and Renninger,G.H., Periodic solutions of a nonlinear functional equation describing neural action, Istituto Lombardo Acad.Sci.Lett.Rend.A, 109, 91-111 (1975)

[8] Coleman,B.D., and Renninger,G.H., Theory of the response of the Limulus retina to periodic excitation, J.of Math.Biol.3, 103-119 (1976)

[9] Dunkel,G., Single species model for population growth depending on past history in: Seminar on Differential euqations and dynamical systems, Lecture Notes in Mathematics Vol. 60, Berlin Springer 1968.

[10] Grafton,R.B., A periodicity theorem for autonomous functional differential equations, J.Diff.Equ.6, 87-109 (1969)

[11] Hadeler,K.P., On the theory of lateral inhibition, Kybernetik 14, 161-165 (1974)

[12] Hadeler,K.P., On the stability of the stationary state of a population growth equation with time-lag, J.Math. Biol. 3, 197-201 (1976)

[13] Hadeler,K.P., and Tomiuk,J., Periodic solutions of difference-differential equations, Arch.Rat.Mech.Anal. to appear.

[14] Halbach,K. und Burkhard,H.J., Sind einfache Zeitverzögerungen die Ursachen für periodische Populationsschwankungen? Oecologia (Berlin) 9, 215-222 (1972).

[15] Hutchinson,G.E., Circular causal systems in ecology, Ann.N.Y. Acad.Sci. 50, 221-246 (1948)

[16] Hartline,H.K., Studies on excitation and inhibition in the retina, Chapman and Hall, London (1974)

[17] Jones,G.S., The existence of periodic solutions of $f'(x)=-2f(x-1)(1+f(x))$, J.Math.Anal.Appl.5,435-450 (1962).

[18] Kaplan,J.L. and Yorke,J.A., Existence and stability of periodic solutions of $x'(t)=-f(x(t),x(t-1))$, in:

Cesari Lamberto (ed.) Dynamical Systems I,II Providence (1974)

[19] May,R., Stability and complexity in model ecosystems, Pop.Biol.6, Princeton Univ.Press (1974)

[20] Morishita,I., and Yajima,A., Analysis and simulation of networks of mutually inhibiting neurons, Kybernetik 11, 154-165 (1972)

[21] Nussbaum,R.D., Periodic solutions of some nonlinear autonomous functional differential equations, Annali di Matematica Ser.4, 51, 263-306(1974)

[22] Pesin,J.B., On the behavior of a strongly nonlinear differential equation with retarded argument. Differentsial'nye uravnenija 10, 1o25-1o36 (1974)

[23] Reichardt,W., and MacGinitie,G., Zur Theorie der lateralen Inhibition, Kybernetik 1, 155-165(1962)

[24] Varjú,D., Vergleich zweier Modelle für laterale Inhibition, Kybernetik 1, 200-208 (1962)

[25] Walter,H., Habilitationsschrift, Saarbrücken 1971.

[26] Walther,H.-O, Existence of a non-constant periodic solution of a nonlinear autonomous functional differential equation representing the growth of a single species population, J.Math.Biol.1, 227-240 (1975)

[27] Walther,H.-O., On a transcendental equation in the stability analysis of a population growth model, J.of Math.Biol.3, 187-195(1976)

ANALYSIS OF DECAY PROCESSES AND APPROXIMATION BY EXPONENTIALS

Dietrich Braess

Institut für Mathematik, Ruhr-Universität

4630 Bochum, F.R. Germany

In several sciences, e.g. in physics, chemistry, biology, and medicine, often decay processes have to be analyzed. Then an empirical function f(t) is given, its domain may be an interval or a discrete point set; and a sum of exponentials of the form

$$\sum_{\nu=1}^{n} \alpha_\nu e^{-\lambda_\nu t}$$

is to be fitted to the function f. Here, n is the given number of matters, while the α_ν's are the concentrations and the λ_ν's are the decay constants. The latter 2n parameters are to be determined by the fit.

During the last decade an involved theory for the mathematical treatment of this problem has been developped, see e.g. [1,4,6]. In this talk, however, we will restrict our attention to those problems which arise when the theory is applied. Then the following question seems to be crucial: How sensitive are the different methods to the noise (pollution) of the empirical data? Although there is a certain connection with the numerical stability [3], this problem must be considered seperately.

Let us start our discussion with the "peeling-off-method" (German: Abschälmethode). It is one of the first effective methods. We do not know who has developped it. But it will not be discussed for historical reasons. At first glance the peeling-off-method seems to be a poor algorithm from the mathematical point of view. Yet it shows the limitation of any method, and therefore it is probably superior to a highly developped mathematical procedure which is used only in the black-box-manner.

The peeling-off-method proceeds in an inductive way. Note that the parameters of a single exponential term $\alpha \cdot e^{-\lambda t}$ may be determined by a simple interpolation at two points t_1 and t_2:

$$\lambda = \frac{1}{t_2 - t_1} \cdot \log \frac{f(t_1)}{f(t_2)}, \qquad \alpha = f(t_1) e^{\lambda t_1}. \tag{1}$$

Moreover, if the decay constants are ordered: $\lambda_1 < \lambda_2 < \ldots < \lambda_m$, then all terms die away faster than the first one does. Therefore one gets a reasonable estimate for the first term by interpolating $f(t)$ at two large t's. The result $\alpha_1 e^{-\lambda_1 t}$ is peeled of the curve, i.e. it is subtracted from $f(t)$. Then the process is repeated.

Now, let us look for the conditions under which the peeling-off-method makes sense. They are obvious in the case when there is only one term and when f is given at m points $t_1 < t_2 < \ldots < t_m$. Then we require

$$\lambda_i \gtrsim \frac{1}{5} \frac{1}{t_m - t_1}, \tag{2}$$

$$\lambda_i \lesssim \frac{1}{t_3 - t_1}. \tag{3}$$

The first condition says that the time must be so large that the decay causes an observable decrease of the amplitude. The second one says that the curve must not die out, before the proper measuring starts. According to my experience the second requirement seems to be a hard one. Often there are transients because of the switching-on effects of the apparatus. Then the first measured data have to be abandoned.

If there are two or more terms, there are even more restrictive conditions. The decay constants must not be too close. Instead of (2) we have

$$\lambda_{i+1}-\lambda_i \gtrsim \frac{1}{5} \cdot \frac{1}{t_m - t_1} .$$

Otherwise the terms cannot be distinguished within the observation time. Moreover, to be more precise we should insert only the time interval which is used for determining the i-th term, instead of referring to the total time $(t_m - t_1)$. In any case, the conditions above seem to be intrinsic ones and they are independent of the applied numerical method.

The main drawback of the peeling-off-method results from the fact that generally the data are noisy. The interpolation (1) is very sensitive to a small change of the data. Since the formula refers only to two values, there is no smoothing, no cancelling of the statistical errors. Some methods, which make use of the solution of certain difference equations, are even worse, c.f. [5, p. 278].

Therefore the problem has to be considered and solved in the framework of approximation theory. Given f the sum of exponentials F is to be determined which minimizes the norm of the error function

$$\| f - F \| .$$

It is not à priori clear which norm has to be chosen. If the function would depend linearly on the unknown parameters then the least-squares fit is appropriate unter the assumption that the errors are normally distributed. But we do not get a serious hint from statistics, when the parameters enter into the ansatz in a non-linear manner. Here, both the ℓ_2-norm and the uniform norm have to be considered. Which is the better one for eliminating the pollution of the data?

Let us discuss the ℓ_2-approximation first. Then the (square of the) norm of the error function

$$\|f-F\|^2 = \sum_{i=1}^{m} [f(t_i)-F(t_i)]^2$$

depends explicitly on all the data $f(t_1), f(t_2), \ldots, f(t_m)$. There are no distinguished points. Consequently, the pollution of the data is not as dangerous as the interpolation. This is an advantage of ℓ_2-norm. Moreover, the best approximation is unique, if the distance of the given function to the approximating one is small [7].

But that means that the error of the data must be small. In many actual situations the analysis has to be performed, though the data are noisy. Then the uniqueness result above is of no use. Moreover, another complication seems to be a more severe drawback. It is not sufficient to regard only the best approximations! When computing a solution, one has to apply iterative procedures: Newton-like algorithms or gradient methods. Starting with a reasonable fit, the approximating function is successivily improved. Then the local best approximations cause trouble. No descent algorithm can distinguish between local and global minima. Therefore, each local best approximation acts like a mouse trap at least as long as you have not available a very good guess to start the iteration. From this viewpoint Wolfe's result [7] on the approximation by exponentials on an interval is discouraging.

THEOREM. Given an integer m there is a function $f_o \in L_2$ such that each f in a neighborhood of f_o has at least m local best approximations.

Now we turn to uniform approximation by sums of exponentials. Then no trouble is caused by the fact that there may be local best approximations which are not global minima. The number of local solutions is bounded [2] and the bound is independent of the function f. Moreover,

n	c_n
1	1
2	2
3	3
4	≤ 9

Table 1. Maximal number of local best approximations in the family of sums of exponentials with order n.

there is a numerical algorithm for computing all the local best approximations (see below).

How sensitive is the best approximation to pollution? The solution is characterized by an alternant of length j with $j \leq 2n+1$, i.e., there are j points $\tilde{t}_1 < \tilde{t}_2 < \ldots < \tilde{t}_j$ such that

$$(-1)^i [f(t_i) - F(t_i)] = \pm \|f-F\|, \quad i = 1,2,\ldots,j.$$

These m points are distinguished. At first glance the norm and therefore also the solution seem to depend only on the data at these points. But have the other points really no influence?

Indeed, small changes of those data which do not enter into the alternant have no influence to the result. Nevertheless they prevent that a large change of the solution occurs when the distinguished data are perturbed. Therefore the situation is not comparable with interpolation. Moreover, small perturbations of data cannot cause that local solutions vanish or that new ones appear. There is only one exception. The number of solutions is not (locally) constant for a neighborhood of f, if a local best approximation of f is degenerate, i.e., when one of the factors α_ν vanishes.

As a conclusion we proceed as follows when a reasonable fit by a sum of exponentials is to be determined:

1. Compute the best uniform approximation to the given function by sums with 1,2,...,n terms. Here the local solutions with k-1 terms, k=2,3,...,n, are used as the starting points for the iteration at the k-th stage.
2. Use the best uniform approximation to start an iteration for computing a (local) best ℓ_2-approximation.
3. Check the solution with respect to the conditions for the peeling-off-method. Moreover, check whether the degree of approximation with n terms is significantly better when the comparison with n-1 terms is performed. Otherwise return to the result for n-1.

We conclude with a numerical example, where the data stem for a reactor experiment.

Here, approximations with nonnegative factors α_ν were required. The results for the uniform approximation are presented in Table 3. The restriction $\alpha_\nu \geq 0$ implies, that the approximation cannot be improved even when 4 or more terms are admitted.

Table 2. Data from a reactor experiment.

t	f(t)	t	f(t)
o	86o	9	148
1	6o3	1o	128
2	491	11	1o9
3	4o7	12	96
4	341	13	82
5	288	14	71
6	242	15	62
7	2o5		
8	174		

Table 3. Results for data in Table 2.

n	best approximation	degree
1	$814.7\ e^{-o.229x}$	45.2
2	$2oo.5\ e^{-1.36x} + 654.6\ e^{-o.162x}$	4.87
3	$152.8\ e^{-2.327} + 556.7\ e^{-o.216x}$	o.994
	$+151.4\ e^{-o.o868x}$	

REFERENCES

1. D.Braess, Über Approximation mit Exponentialsummen. Computing 2, 3o9-321 (1967)

2. -, Chebyshev approximation by γ-polynomials, III: On the number of local solutions (submitted)

3. -, Zur numerischen Stabilität des Newton-Verfahrens bei der nichtlinearen Tschebyscheff-Approximation. Proceedings of the "Kolloquium über Approximationstheory", Bonn, Juni 1976

4. D.W.Kammler, Existence of best approximations by sums of exponentials. J.Approximation Theory 9, 173-191 (1973)

5. C.Lanczos, Applied Analysis. Prentice Hall, Englewood Cliffs 1956.

6. E.Schmidt, Zur Kompaktheit bei Exponentialsummen. J.Approximation Theory 3, 445-454 (197o)

7. J.M. Wolfe, On the unicity of nonlinear approximation in smooth spaces. J.Approximation Theory 12, 165-181 (1974)

OPTIMAL STATE ESTIMATION AND ITS APPLICATION TO INERTIAL NAVIGATION SYSTEM CONTROL

W. Hofmann

Deutsche Forschungs- und Versuchsanstalt für Luft- und Raumfahrt e.V. (DFVLR)

D-8031 Oberpfaffenhofen / West Germany

ABSTRACT

The behaviour of most functions in practical situations can always -- at least approximately -- be described by a finite-dimensional system of differential equations. The theory of optimal state estimation deals with the minimum error-variance reconstruction of the functions from a limited information in a noisy environment. The solution approach, algorithm and stability properties for the linear Gaussian estimation problem are reviewed. A general and typical way for the design of a suboptimal estimation algorithm with special regard to system properties and realization requirements is shown for the technical problem of the error estimation of an aided inertial navigation system. According to the closed-loop system requirement of high accuracy in a noisy environment the considerations about the estimator design are extended to the design of simple, (sub-) optimal controllers. A comparison to the optimal estimation accuracy shows the efficiency of proposed output feedback controller design procedure in this technical example.

1. INTRODUCTION

In a practical situation a control engineer is faced with the problem to find control inputs for a physical system such that the performance of the system behaves in a desired manner. Information about the system is available from measurement outputs. The behaviour of the output and input functions can always -- at least approximately -- be described mathematically by a finite dimensional system of differential equations, the state equations of the physical system and the controller. The modelling of physical and random disturbing states is discussed in chapter 2. For a broad class of controllers it is necessary to reconstruct the complete system state vector from the limited information of the output functions. The theory of optimal state estimation deals with the optimal e.g. minimum-error-variance reconstruction of the state in a noisy environment. The solution approach, algorithm and stability properties for the linear Gaussian estimation problem are reviewed in chapter 3. The technical problem of state estimation and platform error angle control of a Doppler-aided inertial navigation system is considered in chapter 4. A general and typical way for the design of a suboptimal estimation algorithm is shown with special regard to system properties and realization requirements. According to the closed-loop system requirement of a most accurate performance a design procedure for simple, minimum-variance controllers is derived. The direct equivalence to the optimal estimation problem is shown, such that all results of estimation theory can be used directly for the proposed output feedback controller design procedure. The procedure is applied to the design of simple feedback networks for the control of the platform error angles. The suboptimal accuracy results are compared with the performance of the optimal (Kalman-Bucy) filter, which gives the lowest reachable bound of accuracy for state vector feedback controllers.

2. MODELLING OF PHYSICAL AND RANDOM DISTURBING STATES

Since in general physical systems have the capability of storing energy, they have to be treated as dynamical systems. To describe the dynamical behaviour mathematically, differential equations have to be used. The mathematical model, which is at least only an approximation for the dynamical behaviour of the physical reality, has to be chosen to provide for an as good as necessary fitting in the input-output-behaviour. If the parameters of the physical system are lumped, the dynamical behaviour can be described by nonlinear time-varying differential equations [1]. Since the general treatment of nonlinear differential equations is rather difficult and solutions can often be obtained for a distinct class of equations only, it is advantageous to split the general control problem in two parts (fig. 2.1):

1. The design of a guidance, i.e. steering commands to keep desired nominal trajectories.
2. The design of a controller, i.e. control commands to keep the deviations from the nominal values small.

Whereas the first part involves the treatment of nonlinear differential equations, it is admissible in many practical problems to linearize the nonlinear model about the nominal trajectories by a first-order Taylor expansion [2]. The dynamical behaviour of the

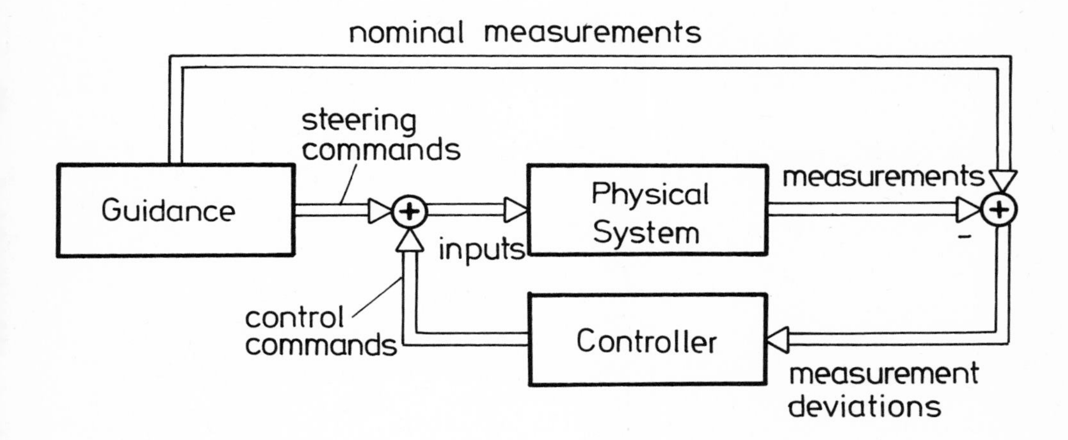

Fig. 2.1 Splitting of the General Control Problem

deviations from the nominal values can be described by linear time-varying differential equations, the so-called state equations [2]. The deviation measurements are linear algebraic functions of the deviations.

The linear model of a dynamical system can always be presented by a linear time-varying first-order vector differential equation [2]

$$\dot{\underline{x}}(t) = F(t)\ \underline{x}(t) + G(t)\ \underline{w}(t) + D(t)\ \underline{u}(t) \tag{2.1}$$

$\underline{x}(t)$: n-dimensional state vector with Gaussian initial conditions $\underline{x}(t_o) \sim N(\underline{0}, P_o)$ +) ,

$\underline{w}(t)$: s-dimensional Gaussian white noise input vector with $\underline{w}(t) \sim N(\underline{0}, Q(t))$,

$\underline{u}(t)$: r-dimensional (control) input vector

and a linear measurement equation

$$\underline{z}(t) = H(t)\ \underline{x}(t) + \underline{v}(t) \tag{2.2}$$

$\underline{z}(t)$: m-dimensional (measurement) output vector,

$\underline{v}(t)$: m-dimensional Gaussian white noise output vector with $\underline{v}(t) \sim N(\underline{0}, R(t))$.

The system matrices are of appropriate dimension. The structure of the mathematical model is illustrated in (fig. 2.2). The presentation by vector equations forms the basis for the application of state-space methods, a powerful tool in modern control theory. Especially with regard to the use of digital computers state-space methods provide for advantageous system analysis and control design procedures [2, 3].

In many practical applications it is possible to set up differential equations for the physical state vector $\underline{x}_1(t)$ by the insight in the physical relations of the considered system, yielding a linear vector differential equation of order n_1

$$\dot{\underline{x}}_1(t) = F_{11}(t)\ \underline{x}_1 + D_1(t)\ \underline{u}(t)\ ; \qquad \underline{x}_1(t_o) = \underline{x}_{10} \tag{2.3a}$$

and a measurement equation

$$\underline{z}(t) = H_1(t)\ \underline{x}(t)\ . \tag{2.3b}$$

+) This notation describes a Gaussian (normal) distribution with mean vector $\underline{0}$ and covariance matrix P_o [4].

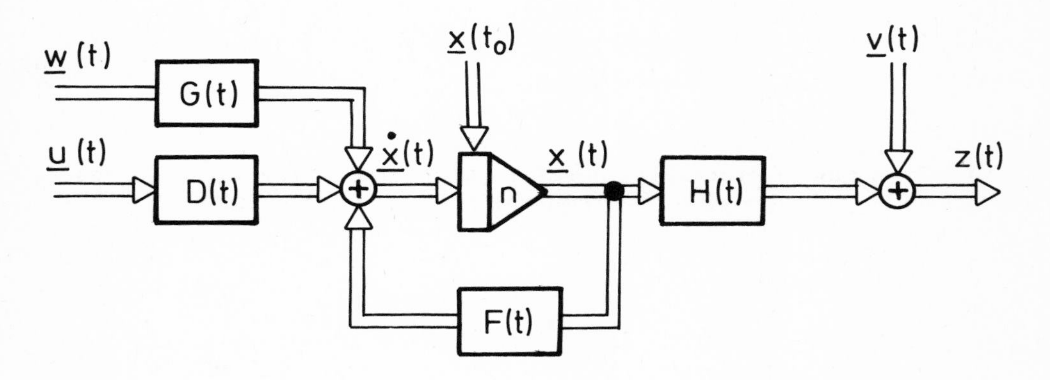

Fig. 2.2 General Structure of a Linear Dynamical System

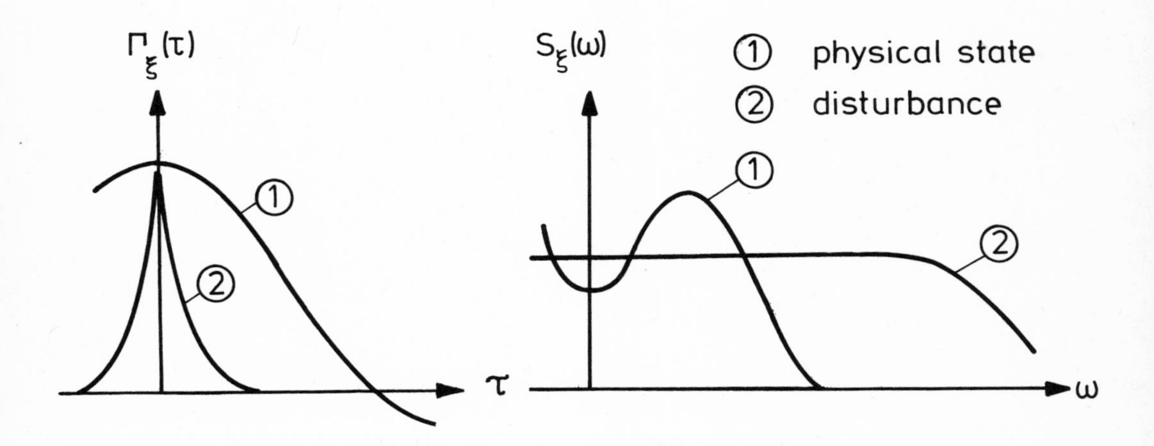

Fig. 2.3 Autocorrelation and Spectral Density Functions

Sometimes the deterministic viewpoint is sufficient to describe the dynamical behaviour of the physical system. But especially in the case of high accuracy requirements in a noisy environment it is necessary to include random disturbances in the model

$$\underline{\dot{x}}_1(t) = F_{11}(t)\ \underline{x}_1(t) + D_1\ \underline{u}(t) + \underline{w}^*(t)\ ;\quad \underline{x}_1(t) \sim N(\underline{0}, P_{10}) \tag{2.4a}$$

$$\underline{z}(t) = H_1(t)\ \underline{x}_1(t) + \underline{v}^*(t) \tag{2.4b}$$

with the random disturbance vectors $\underline{w}^*(t)$ and $\underline{v}^*(t)$ which in general cannot be influenced by the control. The modelling of stochastic disturbances often requires extensive experimental investigations. Due to the linear point of view it is often possible to treat the modelling problem within the framework of Gauss-Markov processes [4].

Complete statistics for a Gauss-Markov process $\underline{\xi}(t)$ are given by the mean-vector $\underline{m}_{\underline{\xi}}(t)$ and the covariance matrix $C_{\underline{\xi}}(t)$

$$\underline{m}_{\underline{\xi}}(t) = E\{\underline{\xi}(t)\} \tag{2.5}$$

$$C_{\underline{\xi}}(t) = E\{(\underline{\xi}(t) - \underline{m}_{\underline{\xi}}(t)(\underline{\xi}(t) - \underline{m}_{\underline{\xi}}(t))^T\} \quad . \tag{2.6}$$

Treating a single stationary [4] random process $\xi(t)$ two definitions related to the second-order moment are introduced [5]:

the autocorrelation function

$$\Gamma_{\xi}(\tau) = E\{\xi(t+\tau)\ \xi(t)\} \tag{2.7a}$$

and the spectral density function, which is the Fourier transform of the autocorrelation function

$$S_{\xi}(\omega) = \int_{-\infty}^{+\infty} e^{-j\omega\tau}\ \Gamma_{\xi}(\tau)\ d\tau \quad .$$

Some examples are given in (fig. 2.3, 2.4). The definitions (eq. 2.7) form the theoretical basis for several random disturbance analysers.

It is convenient [4] to distinguish between time-uncorrelated (white) and time-correlated (coloured) disturbances, since the modelling of the latter requires an augmentation of the state vector. Disturbances, whose spectral densities are broadband relative to the system power spectrum (fig. 2.3), are usually modelled by Gaussian white noise, a fictitious mathematical model with frequency independent spectral densitiy and no time correlation, which is expressed by the delta-impulse autocorrelation function (fig. 4.3.).

If the time constants of the physical system cannot be neglected, the disturbances can be modelled by the outputs of linear and Gaussian white noise driven shaping filters (fig. 2.4), whose dynamical behaviour can be described by linear vector differential equations.

Following the previous discussion the disturbance vectors $\underline{w}^*(t)$ and $\underline{v}^*(t)$ in (eq. 2.4) can be modelled by a linear superposition of Gaussian white and coloured noise

$$\underline{w}^*(t) = F_{12}(t)\ \underline{x}_2(t) + G_1(t)\ \underline{w}(t) \tag{2.8a}$$

$$\underline{v}^*(t) = H_2(t)\ \underline{x}_2(t) + \underline{v}(t) \quad . \tag{2.8b}$$

The dynamical behaviour of the coloured noise state vector $\underline{x}_2(t)$ of the n_2-order shaping filters is given by the linear vector differential equation

$$\dot{\underline{x}}_2(t) = F_{22}(t)\ \underline{x}_2(t) + G_2(t)\ \underline{w}(t) \quad . \tag{2.8c}$$

Combining (eqs. 2.4, 2.8) leads to a system description according to (eqs. 2.1, 2.2) for the augmented state vector of order $n = n_1 + n_2$ and a special structure of the system matrices

$$\underline{x}^T(t) = [\underline{x}_1^T(t) \quad \underline{x}_2^T(t)] \tag{2.9a}$$

$$F(t) = \begin{bmatrix} F_{11}(t) & F_{12}(t) \\ 0 & F_{22}(t) \end{bmatrix} ; \qquad G(t) = \begin{bmatrix} G_1(t) \\ G_2(t) \end{bmatrix} \tag{2.9b}$$

$$D(t) = \begin{bmatrix} D_1(t) \\ 0 \end{bmatrix} , \qquad H(t) = [H_1(t) \quad H_2(t)] \ .$$

The structure of the system including random disturbance modelling is pointed out in (fig. 2.5).

3. THE LINEAR, GAUSSIAN ESTIMATION PROBLEM

The application of several control schemes requires the exact or at least estimated knowledge about the complete state vector [2, 3]. The problem of estimating the state vector of a linear system with Gaussian disturbances (eq. 2.1) from noisy measurements (eq. 2.2) such that the estimation error-variance is minimized can be formulated as follows [4]:

Given a set $Z(t_f)$, $t_o \leq t \leq t_f$, of measurements $\underline{z}(t)$ of the state vector $\underline{x}(t)$ according to (eq. 2.2), where the dy-

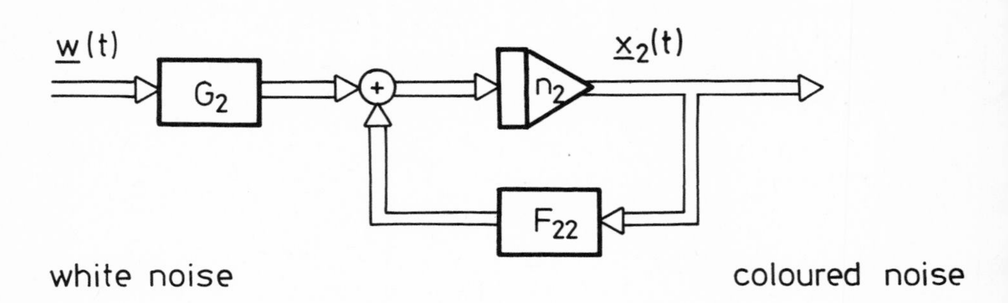

Fig. 2.4 Structure of a Shaping Filter

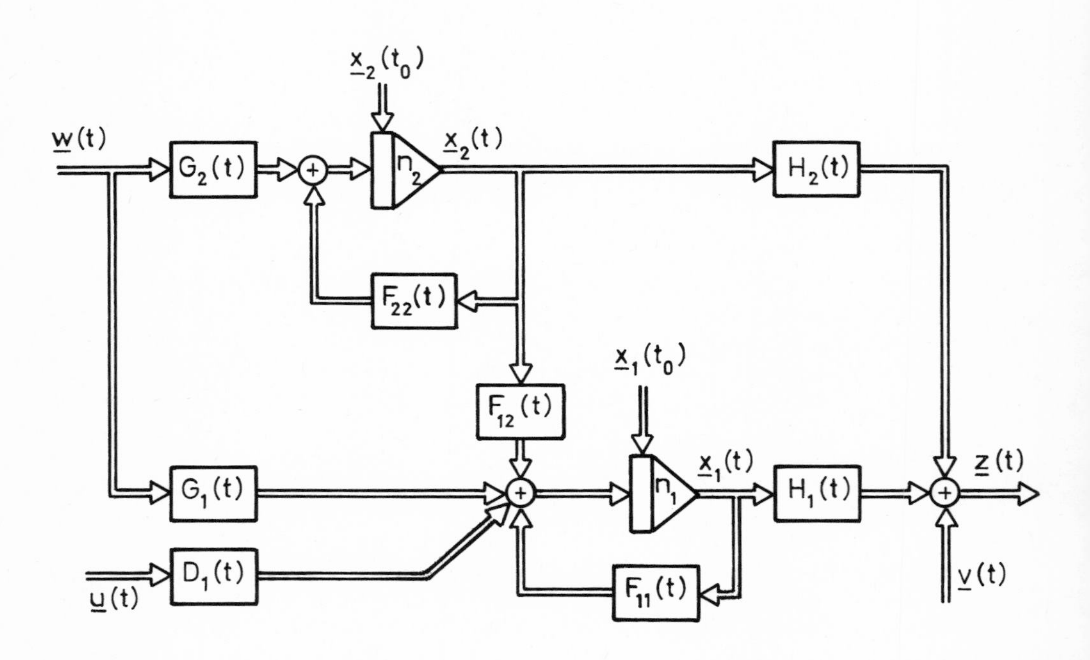

Fig. 2.5 System Structure including Random Disturbance

namical behaviour of $\underline{x}(t)$ is described by the linear vector differential equation (eq. 2.1). Find an estimate $\hat{\underline{x}}(t_f)$, such that the cost functional

$$J = E\{\tilde{\underline{x}}^T(t_f)\, X(t_f)\, \tilde{\underline{x}}(t_f) \mid Z(t_f)\}\,; \qquad X(t_f) > 0 \tag{3.1}$$

is minimized with the estimation error vector

$$\tilde{\underline{x}}(t_f) = \underline{x}(t_f) - \hat{\underline{x}}(t_f) \quad . \tag{3.2}$$

Since the exact derivation of the optimal state estimation procedure -- the famous Kalman-Bucy filter -- is well-known from the literature [4, 6], a more heuristic approach [7] is taken, which gives a deaper insight in the action of the optimal filter.

The structure of the state estimator (fig. 3.1) is assumed to be described by a linear vector differential equation

$$\dot{\hat{\underline{x}}}(t) = F(t)\,\hat{\underline{x}}(t) + K(t)(\underline{z}(t) - H(t)\,\hat{\underline{x}}(t)) + D(t)\,\underline{u}(t) \tag{3.3}$$

$$\hat{\underline{x}}(t_o) = \underline{0} \quad .$$

The choice of the structure can be motivated by the following considerations:

1. The assumptions of linearity of system dynamics, measurements and state estimator and of the Gaussian distribution of the random disturbances implies a Gaussian conditional density

$$p(\underline{x}(t_f) \mid Z(t_f)) \sim N(\hat{\underline{x}}(t_f),\, S(t_f)) \quad , \tag{3.4}$$

which contains all information about the state $\underline{x}(t_f)$ given the set of measurements $Z(t_f)$. The conditional density (eq. 3.4) is completely characterized by its first two moments [4], the conditional mean $\hat{\underline{x}}(t_f)$ and the conditional error-covariance matrix $S(t_f)$

$$\hat{\underline{x}}(t_f) = E\{\underline{x}(t_f) \mid Z(t_f)\} \tag{3.5a}$$

$$S(t_f) = E\{\tilde{\underline{x}}(t_f)\,\tilde{\underline{x}}^T(t_f) \mid Z(t_f)\} \quad . \tag{3.5b}$$

In contrary to an arbitrary conditional density (eq. 3.4), e.g. in the nonlinear filtering case, the linear filtering problem involves the solution of a finite dimensional problem only.

2. The structure involves the simulation of the system dynamics, which are driven by the estimated output error (innovation [4]) $\underline{\nu}(t) = \underline{z}(t) - H(t)\,\hat{\underline{x}}(t)$ and the known deterministic inputs. This assumption is meaningful, since the only information to compare the estimate $\hat{\underline{x}}(t)$ is given by the measurements $\underline{z}(t)$.

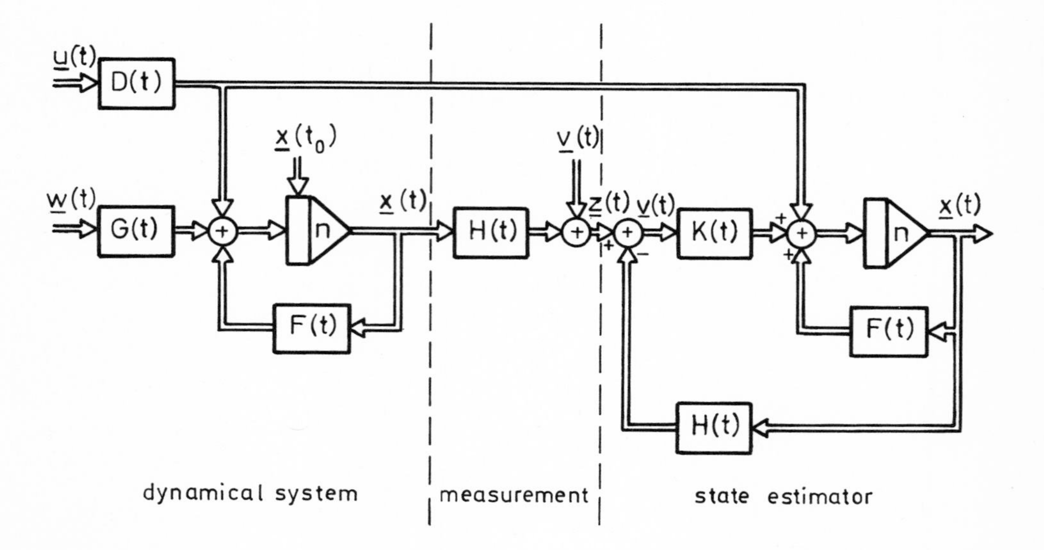

Fig. 3.1 Structure of State Estimators for Linear Systems

Imaging the noise free case and exact state knowledge, the estimator would use the system dynamics and deterministic inputs only to produce exact estimates. The structure of the state estimator is the same as the structure of observers in the deterministic observation problem [2, 3]. The state estimator works on the instantaneous measurements $\underline{z}(t)$ only, such that no storage of measurements is necessary.

The gain matrix K(t) of the state estimator still needs to be determined. In contrary to the goal of observer design -- a desired stability behaviour -- the optimal filter gain matrix K(t) is found to minimize the cost functional (eq. 3.1). The error behaviour of the linear state estimator (eq. 3.3) can be described by the error-covariance matrix S(t), which is solution of the matrix differential equation

$$\begin{aligned}\dot{S}(t) = &(F(t) - K(t)\,H(t))\,S(t)\\ &+ S(t)\,(F(t) - K(t)\,H(t))^T\\ &+ K(t)\,R(t)\,K^T(t)\\ &+ G(t)\,Q(t)\,G^T(t)\;; \qquad S(t_o) = P_o\,. \end{aligned} \tag{3.6}$$

Using S(t) the cost functional (eq. 3.1) can be reformulated as

$$J = \text{trace}\{\,X(t_f)\;S(t_f)\,\} \tag{3.7}$$

which is no longer a conditional cost functional, since the error-covariance matrix S(t) is independent of a realization $\underline{z}(t)$, an important property of the linear optimal filter. Using the matrix minimum principle [8], the gain matrix K*(t) minimizing (eq. 3.7) can be found as [4]

$$K^*(t) = P(t)\,H^T(t)\,R^{-1}(t)\quad, \tag{3.8}$$

where the optimal error-covariance matrix P(t) is solution of the nonlinear matrix Riccati differential equation

$$\begin{aligned}\dot{P}(t) = &F(t)\,P(t) + P(t)\,F^T(t)\\ &- P(t)\,H^T(t)\,R^{-1}(t)\,H(t)\,P(t)\\ &+ G(t)\,Q(t)\,G^T(t)\;; \qquad P(t_o) = P_o\quad. \end{aligned} \tag{3.9}$$

From (eqs. 3.8, 3.9) it is to be seen that a necessary condition for the existence of the optimal filter is:

$$R^{-1}(t) \quad \text{exists or} \quad R(t) > 0\quad. \tag{3.10}$$

This requirement of the positive definiteness of the output noise covariance matrix R(t) is equivalent to the assumption that all measurements contain white noise. In cases where measurements contain no or only coloured noise -- the singular problem of filtering -- a solution can be found by reducing the order of the system [9].

Some properties of the Kalman-Bucy filter are summarized [4]:

1. The optimal filter (eq. 3.3) is linear. No other estimation structure -- even nonlinear -- minimizes (eq. 3.1).
2. The optimal filter is an unbiased estimator

$$E\{(\underline{x}(t) - \hat{\underline{x}}(t)\} = 0 \quad \text{or} \quad E\{\underline{x}(t)\} = E\{\hat{\underline{x}}(t)\}. \tag{3.11}$$

3. The optimal filter provides for optimality uniformly in the time t, i.e. it is independent of the terminal-time t_f and the weighting matrix $X(t_f)$.
4. The optimal gain matrix $K^*(t)$ and the minimum error-covariance matrix $P(t)$ (eqs. 3.8, 3.9) are independent of the measurements so that the filter parameters and the expected filter accuracy can be computed off-line.

One important question still remains to be discussed: the stability of the state estimator (eq. 3.3), i.e. the stability behaviour of the filter dynamics $(F(t) - K(t)\,H(t))$. It is well-known that optimality does not imply stability. Moreover problems about the existence of a bounded, positive semidefinite solution for the error-covariance matrix $P(t)$ arise, since it is solution of the nonlinear matrix Riccati differential equation (eq. 3.9).

T1: If the system (eqs. 2.1, 2.2) is completely observable and completely controllable by the noise input vector $\underline{w}(t)$ (a more adequate expression seems to be completely disturbable) and if $P_o \geq 0$, then the optimal filter dynamics $(F(t) - K(t)\,H(t))$ are asymptotically stable and the positive definite solution $P(t)$ is bounded from above.

If these conditions are met, the solution $P(t)$ is numerically stable, i.e. the numerical solution $P_c(t)$ tends to the real solution $P(t)$ in the presence of disturbances as e.g. round-off errors [11].

The previous summarized results can be specialized to the practically important case of the constant gain filter design for systems with time-invariant system matrices:

Assuming the system (eqs. 2.1, 2.2) to be time-invariant and the observation interval to be infinite, $-\infty \leq t \leq t_f$, the optimal filter is given by

$$\dot{\hat{\underline{x}}}(t) = F\,\hat{\underline{x}}(t) + K_\infty\,(\underline{z}(t) - H\,\hat{\underline{x}}(t)) + D\,\underline{u}(t) \tag{3.12a}$$

with the constant filter gain

$$K_\infty = P_\infty H^T R^{-1} \quad . \tag{3.12b}$$

The stationary error-covariance matrix P_∞ is solution of the algebraic matrix Riccati equation

$$0 = F P_\infty + P_\infty F^T - P_\infty H^T R^{-1} H P_\infty + G Q G^T \quad . \tag{3.12c}$$

For this case there are necessary and sufficient conditions for stability and the uniqueness of $P_\infty \geq 0$ [12].

T2: The optimal filter is asymptotically stable, i.e. Re $\lambda(F - K_\infty H) < 0$, and the positive semidefinite solution P_∞ of (eq. 3.12c) is unique, iff the system (eq. 2.1, 2.2) is detectable and stabilizable by the input noise $\underline{w}(t)$.

There is a final remark necessary on the computational solution of the Kalman-Bucy filter. Except in simple examples the solution of the Kalman-Bucy filter cannot be given analytically. It is necessary to compute it pointwise on a digital computer. Efficient algorithms for the solution of the matrix Riccati differential or algebraic equation are presented in [13].

4. STATE ESTIMATION AND PLATFORM ERROR ANGLE CONTROL OF A DOPPLER-AIDED INERTIAL NAVIGATION SYSTEM

The considerations about physical and random disturbing state modelling (chapter 2) are now applied to the error modelling of inertial navigation systems. To get an imagination of the at most reachable accuracy of state vector feedback based control schemes the behaviour of the Kalman-Bucy filter is investigated. Using system properties (observability and controllability) and regarding realization requirements (simple and reliable realization) a general and typical way for the design of a suboptimal estimation algorithm is shown. Finally an output feedback controller for the error angle control is proposed, which owns the properties of simple and reliable implementation and as accurate as possible behaviour of the closed-loop system. The results of optimal filtering can be used directly for this type of optimal control.

Inertial navigation systems (INS) are technical measuring units supporting [14] information about velocity and position of an arbitrarily moving carrier (e.g. satellite, rocket, aircraft, ship) relative to a distinct -- inertially fixed or rotating -- coordinate system. They consist of (fig. 4.1a, 4.1c) gyros to keep the stable platform fixed in directions of the desired coordinate system, accelerometers, mounted on the platform to measure the accelerations of the carrier along the axis of the desired coordinate system, and a navigational computer producing velocity and position information by appropriate integration schemes and -- if necessary -- control commands to rotate the stable platform. A conventional coordinate system for earth-fixed carriers is represented by the local, vertical coordinate system (fig. 4.1b) with the three orthogonal axis in east, north and vertical direction [14]. In this case it is necessary to keep the platform "horizontally aligned" by commands depending on the earth's rotation and the carrier's motion. A very simple realization for one horizontal axis of a Schuler-tuned INS is indicated in (fig. 4.1c), where also the general structure of an INS is pointed out.

Due to error signals arising from the nonideal properties of the gyros and of the accelerometers, e.g. nonorthogonal adjustment, the alignment of the stable platform no longer coincides with the desired coordinate system. The velocity and position information contains slowly varying (low-frequency) erros. The error behaviour of the platform error angles is shown in (fig. 4.2). It is governed by a drift due to constant (random) inputs and by an oscillation -- the Schuler oscillation -- due to the Schuler-tuned feedback for "horizontal alignment". It is obvious that it is necessary to use redundant, external information about velocity and/or position to get a better performance of the INS, especially for the long-time behaviour. In [15] the application of an external velocity information, supplied by a Doppler-radar, is investigated for the on-line INS-alignment in moving aircrafts. The Doppler-velocity information contains rapidly varying (broad-band) erros besides the true information. Defining a difference measurement (fig. 4.1c) between INS-velocity and Doppler-velocity the true velocity is eliminated. The difference measurement contains the errors of INS and Doppler-radar only. The control design problem can now be formulated:

Find a controller using the difference measurement only to compensate the random disturbances, such that the accuracy of the physical states of the closed-loop system behaves "best in the sense of their variances" and such that the realization of the controller is "as simple as possible" for reasons of reliable implementation.

Assuming only small deviations from the nominal values, i.e. from the "horizontal alignment" of the stable platform, and exact integration schemes of the accelerations, it is possible to set up a linear error model for the physical error states and the stochastic disturbances. It should be emphasized that this assumption is not severe because the goal of a satisfying controller design is just to keep the deviations as small as possible.

The physical error states of the Doppler-aided INS are the platform error angles of the east, north and vertical axis α, β and γ,

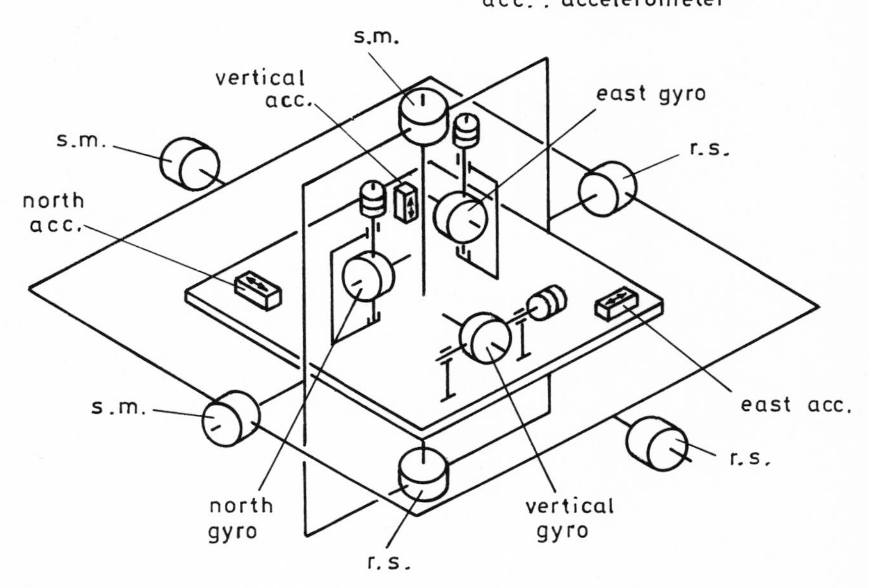

Fig. 4.1a

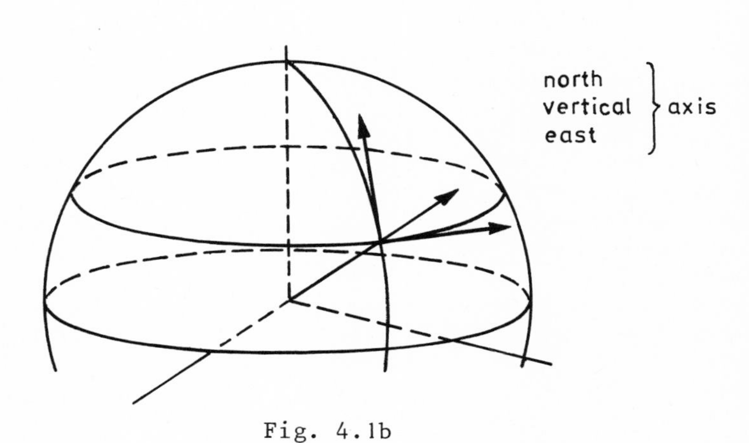

Fig. 4.1b

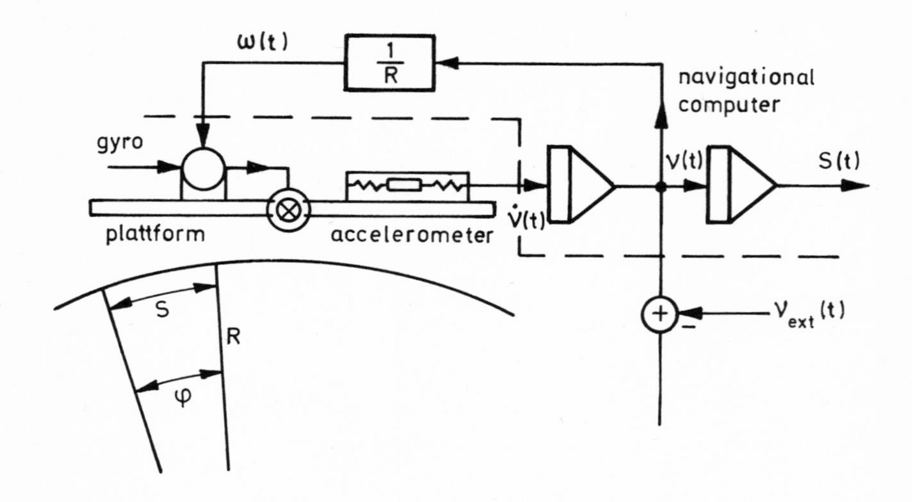

Fig. 4.1c INS: Stable Platform (a), Coordinate System (b), and One-Axis Structure

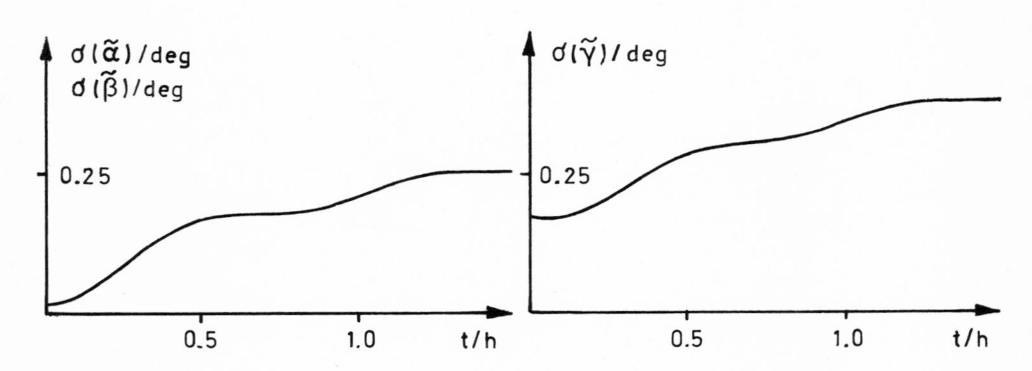

Fig. 4.2 Standard Derivations of Platform Error Angles

the east and north velocity error Δv_e and Δv_n and the east and north position error Δx_e and Δx_n, summarized in the physical state vector

$$\underline{x}_1^T(t) = [\alpha, \beta, \gamma, \Delta v_e, \Delta v_n, \Delta x_e, \Delta x_n] \ . \tag{4.1}$$

Using physical relationships as linearized gyro-dynamics and rotational kinematics the dynamical behaviour can be described by a linear vector differential equation according to (eq. 2.4a)

$$\dot{\underline{x}}_1(t) = F_{11}(t)\, \underline{x}_1(t) + \underline{u}(t) + \underline{w}^*(t);$$

$$\underline{x}_1(t_o) \sim N(\underline{0}, P_{10}) \tag{4.2}$$ +)

The dynamic matrix F_{11} depends on the nominal flight velocity and the geographic latitude of the aircraft such that $F_{11} = F_{11}(t)$ is a function of the independent variable t. Another important property is given by the equal number of independent control input components and of physical state variables, such that every state component can be influence directly either by control moments applied to the stable platform or control signals added in the navigational computer.

The velocity difference measurements for the east and north components $z_e = \Delta v_e + v_1$ and $z_n = \Delta v_n + v_2$ with Doppler-radar velocity errors v_1 and v_2 can be summarized in a linear output vector equation according to (eq. 2.4b)

$$\underline{z}(t) = H\, \underline{x}_1(t) + \underline{v}^*(t) \ . \tag{4.3}$$

Besides the modelling of the physical states $\underline{x}_1(t)$ there still remains the more elaborate analysis of the random system and measurement disturbances $\underline{w}^*(t)$ and $\underline{v}^*(t)$. In many practical problems the modelling of stochastic disturbances cannot be performed by a physical - insight - based way, but is done by an experimental correlation function or spectral density analysis. The underlying basic disturbance property of these analysis methods is the assumption of stationary Gauss-Markov processes, the mathematical description of which is already discussed in chapter 2. The reasons for this assumption are on one hand the necessity of formulating the disturbance identification problem as a -- at least approximately -- finite dimensional problem. Moreover it is often more promising to use simple, low-order disturbance models (shaping filters) than sophisticated, high-order models. There is not only the greater effort for analysis, but also a higher bound of confidence in the identified model parameters. On the other hand it is known

---

+) For parameter values of the complete INS-error model see [15].

by the central limit theorem [4] that the superposition of a large number of small independent random effects, regardless of their individual distribution, causes the distribution of the sum of these effects to be -- at least approximately -- Gaussian. In addition, if a dynamical system with low-pass filter characteristic is driven by a disturbance with nearly arbitrary distribution, the distribution of the output of the dynamical system is -- at least approximately -- Gaussian [5].

Returning to the INS-disturbance modelling it is necessary to investigate the random disturbance inputs to the gyros, to the accelerometers and the Doppler-radar velocity error. Extensive analysis and a comparison of the influence of different effects show that the random disturbances can be modelled using the following (most simple) mathematical models.

1. Gaussian white noise $\xi_w(t)$:

   A random disturbance with negligible time-correlation (relative to the other time-constants of the system) can be modelled by Gaussian white noise, whose spectral density and correlation function are illustrated in (fig. 4.3)

$$\xi_w(t) \sim N(0,\ q) \quad . \tag{4.4a}$$

2. Random constant (bias) $\xi_b(t)$:

   A random disturbance, which is constant -- at least nearly constant during the considered time-intervall -- can be modelled by a random constant (bias) (fig. 4.3)

$$\dot{\xi}_b(t) = 0 \ ; \qquad \xi_b(0) \sim N(0,\ \sigma_b^2) \quad . \tag{4.4b}$$

3. Exponentially correlated noise $\xi_c(t)$:

   If the time-correlation of a random disturbance is not negligible, it can often be modelled by exponentially correlated noise (fig. 4.3)

$$\dot{\xi}_c(t) = -\ 1/T_c\ \xi_c(t) + w(t) \tag{4.4c}$$

$$\xi_c(0) \sim N(0,\ \sigma_c^2)$$

$$w(t) \sim N(0,\ q_c) \quad .$$

Since stationary processes are assumed it is necessary to choose the white noise variance $q_c = 2 \cdot \sigma_c^2/T_c$ such that the correlated noise variance $p_c(t) = E\{\sigma_c^2(t)\}$ as solution of

$$\dot{p}_c(t) = -\ 2/T_c\ p_c(t) + q_c \ ; \qquad p_c(0) = \sigma_c^2 \tag{4.5}$$

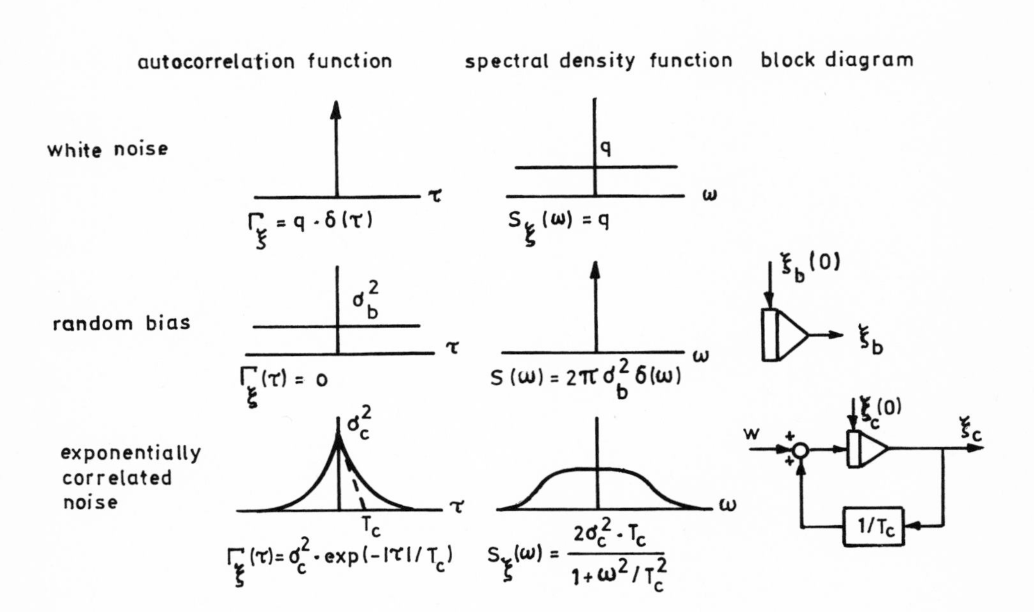

Fig. 4.3 Description of Simple Random Processes

is stationary by $p_c(t) = p_c(0) = \text{const.}$

Summarizing the disturbing states resulting from the shaping filters (eqs. 4.4b, 4.4c) in the disturbance state vector $\underline{x}_2(t)$ of order $n_2 = 5$ the findings of [15] can be expressed according to (eq. 2.5)

$$\begin{aligned} \dot{\underline{x}}_2(t) &= F_{22}\, \underline{x}_2(t) + G_2\, \underline{w}(t) \\ \underline{w}^*(t) &= F_{12}\, \underline{x}_2(t) + G_1\, \underline{w}(t) \\ \underline{v}^*(t) &= \underline{v}(t) \end{aligned} \tag{4.6a}$$

with the statistics

$$\begin{aligned} \underline{x}_2(t_o) &\sim N(\underline{0},\, P_{20}) \\ \underline{w}(t) &\sim N(\underline{0},\, Q) \\ \underline{w}(t) &\sim N(\underline{0},\, R) \quad . \end{aligned} \tag{4.6b}$$

Augmenting the physical state model (eq. 4.2) by the disturbance state model (eq. 4.6) the complete error model of the Doppler-aided INS of order n = 12 is given by

$$\begin{bmatrix} \dot{\underline{x}}_1(t) \\ \dot{\underline{x}}_2(t) \end{bmatrix} = \begin{bmatrix} F_{11}(t) & F_{12} \\ 0 & F_{22} \end{bmatrix} \begin{bmatrix} \underline{x}_1(t) \\ \underline{x}_2(t) \end{bmatrix} + \begin{bmatrix} G_1 \\ G_2 \end{bmatrix} \underline{w}(t) + \begin{bmatrix} \underline{u}(t) \\ \underline{0} \end{bmatrix}$$

$$\underline{z}(t) = [H_1 \quad 0] \begin{bmatrix} \underline{x}_1(t) \\ \underline{x}_2(t) \end{bmatrix} + \underline{v}(t) \tag{4.7}$$

which can be treated within the general framework of chapter 2.

Since several control schemes require the knowledge about the complete state or at least an optimal state estimate, the Kalman-Bucy filter (eq. 3.9) for the optimal error state estimation of the Doppler-aided INS (eq. 3.7) is investigated for a 1.5 h, east-west-east flight. The estimation accuracy is a lower bound for the reachable closed-loop system accuracy [16]. Since especially the accuracy of the stable platform alignment is of importance, the standard deviations of the estimation errors of the platform error angles are plotted in (fig. 4.4). A comparison to the error behaviour of the non-aided INS shows that the final uncertainty is reduced by about 75% for the horizontal and by about 50% for the vertical error angles.

The solution of the matrix Riccati differential equation (3.9)

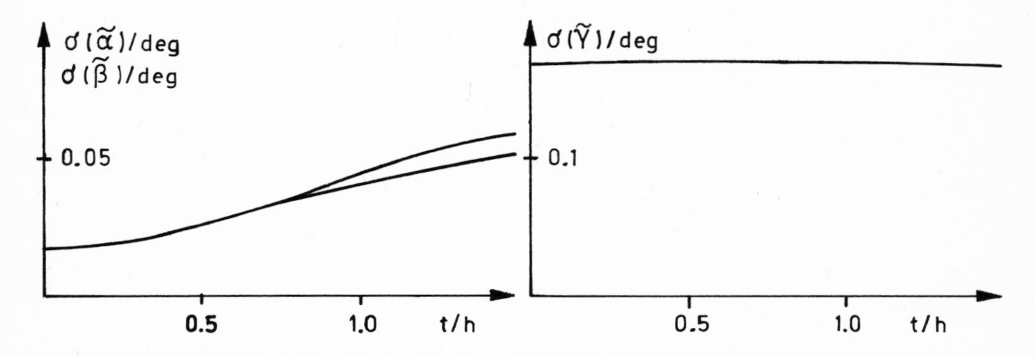

Fig. 4.4 Optimally Platform Error Angle Standard Deviations

for the optimal error-covariance matrix P(t), which is of dimension (12,12), requires the simultaneous solution of 78 nonlinear differential equations. The (optimal) Kalman-Bucy filter cannot be suggested to be realized on an onboard-computer. A suboptimal, but much simpler estimation algorithm has to be developed. The main steps of a general and typical design procedure with special regard to system properties and realization requirements are summarized below. An extensive discussion can be found in [15].

The finally aspired structure of the suboptimal state estimator is given by

$$\dot{\hat{\underline{\xi}}}(t) = A\,\hat{\underline{\xi}}(t) + K(\underline{z}(t) - C\,\hat{\underline{\xi}}(t)) + D_{\xi}\,\underline{u}(t)\ , \tag{4.8}$$

an as low-order as possible (<< 12) filter with constant filter matrices. It requires the solution of a low-order time-invariant vector differential equation on the onboard-computer only. (Eq. 4.8) represents a realization rule for analogue computers directly. If there is available a digital computer, (eq. 4.8) can be solved by means of the transition matrix [2] yielding a vector difference equation.

To get the desired algorithm (eq. 4.8) the following steps are executed:

1. Assumptions on the system:

   For the given INS-activity there is no prefered flight direction leading to the assumption of zero flight velocity. This implies the system matrix $F_{11}(t)$ in the INS-error model (eq. 4.7) to be independent of the time and so the complete system to time-invariant.

2. Observability and (noise) controllability analysis:

   From (chapter 3, T2) it is known that an asymptotically stable, constant gain filter can be designed by means of the algebraic Riccati equation only for the detectable and (noise) stabilizable subsystem of a given system. An observability and controllability analysis, which can be performed either analytically or numerically [2], shows the order of the detectable and (noise) stabilizable subsystem to be five. The transformed state variables of this subsystem are the first five physical states in (eq. 4.1), i.e. the three platform error angles α, β, γ and the two INS-velocity errors Δv_e and Δv_n, which are linearly combined with the disturbing state $\underline{x}_2(t)$.

   The dynamics of the detectable and (noise) stabilizable subsystem can be described by a time-invariant 5th-order vector differential equation

$$\dot{\underline{\xi}}(t) = A\,\underline{\xi}(t) + B\,\underline{\xi}(t) + D_{\xi}\,\underline{u}(t) \tag{4.9a}$$

together with a linear measurement vector equation

$$\underline{z}(t) = C\,\underline{\xi}(t) + \underline{v}(t) \quad . \tag{4.9b}$$

3. Constant gain filter design via Riccati equation:

   Since the subsystem (eq. 4.9) is timeinvariant, detectable and (noise) stabilizable, the stationary Kalman-Bucy filter algorithm (eq. 3.12) can be applied to find a filter gain matrix K for the filter algorithm (eq. 4.8). It should be emphasized that the filter gain matrix K is optimal only under the assumptions of zero flight velocity and an infinite measurement intervall. In all other cases it works suboptimal. The result of one typical simulation of the filter algorithm (eq. 4.8), estimating the platform error angles from noisy measurements for the above mentioned flight, is shown in (fig. 4.5).

4. Q-stabilization for better filter performance:

   From (fig. 4.5) it is to be seen that the filter designed in step 3 using the nominal statistics of the INS-error model (eqs. 4.2, 4.6) is missing the capability of reducing the initial platform error angles within the finite measurement intervall of 1,5 h.

   This is a typical result for filters designed by the algebraic Riccati equation (eq. 3.12c) for systems with small system to measurement noise ratio and system eigenvalues near to the imaginary axis. The disadvantageous property can be explained by the assumed infinite measurement intervall. Loosely spoken the filter gain matrix K is such big that it keeps the filter to be "just stable", i.e. to reduce initial errors very slowly. On the other hand it is as small as possible to increase the error variances not by to much measurement noise driving the filter (eq. 4.8). Such designed filter tend to possess another undesirable property: divergence. This means that the true estimation error behaviour is not asymptotically stable in the realistic environment due to e.g. modelling errors [4].

   To prevent these disadvantages the Q-stabilizing method [4] can be used. By an artificial increasing of the system noise covariance matrix Q the filter (eq. 4.8) tends to be more stable, i.e. to reduce initial errors faster. A more systematic but more sophisticated procedure for the inclusion of stability constraints in the filter design is derived in [17].

   Applying the Q-stabilizing method to the constant gain INS-filter design -- a suitable Q-matrix was found by trial and error -- the results for the standard deviations of the estimation errors of the horizontal platform error angles are shown in (fig. 4.6).

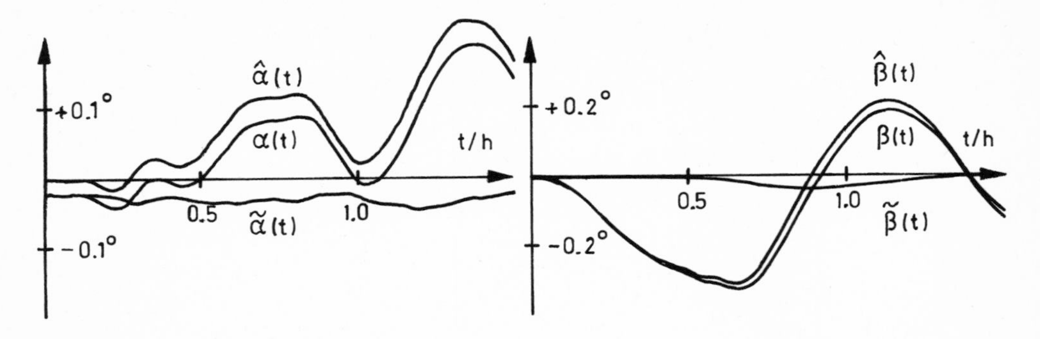

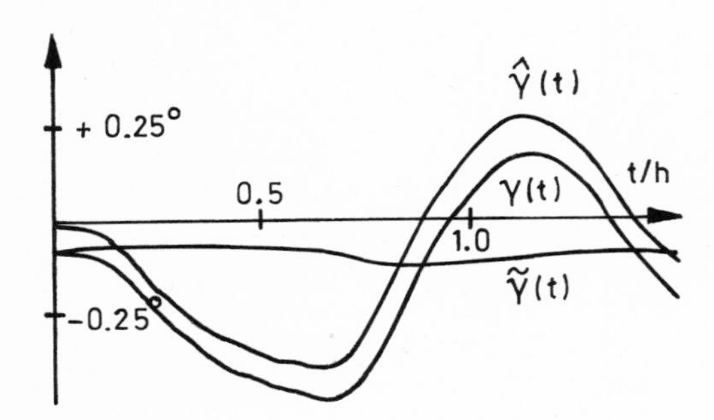

Fig. 4.5 Simulation of Platform Angle Error Behaviour (Suboptimal Estimation)

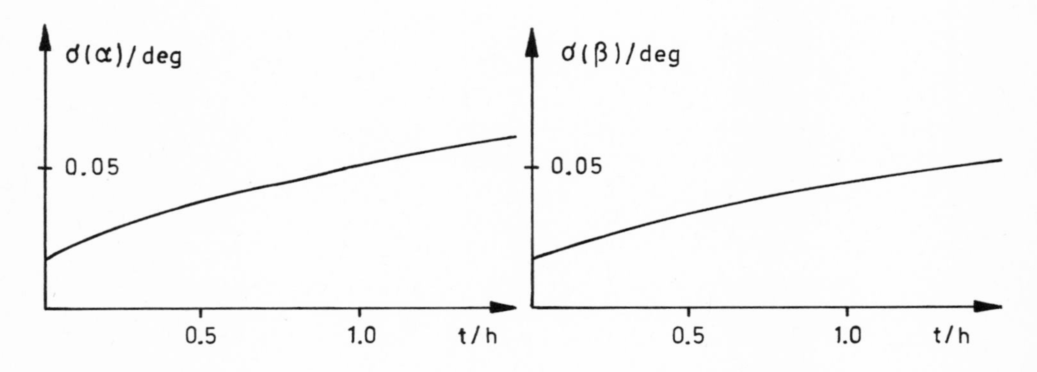

Fig. 4.6 Suboptimally (Q-stabilized) Platform Error Angle Standard Deviations

The vertical error accuracy does not differ from the optimal one (fig. 4.4) within the illustration accuracy. In comparison to the optimally reachable accuracy of the Kalman-Bucy filter (fig. 4.4) the loss of accuracy is negligible compared to the simple implementation of the filter algorithm (eq. 4.8).

If the Kalman-Bucy filter (eq. 3.3, 3.8, 3.9) or the suboptimal estimation algorithm (eq. 4.8) is used as a state estimator together with some control law [2, 3, 16] in a closed-loop system, the presented results (fig. 4.4, 4.6) are at most lower bounds for the reachable accuracy. It seems to be more adequate to attack the control problem with the requirement of most accurate closed-loop system behaviour directly by minimizing a cost functional, which gives regard to the closed-loop system variances. The following optimal control problem is considered:

A dynamical system is described by the linear vector differential equation (eq. 2.1) and a linear output vector algebraic equation (eq. 2.2), where the state vector $\underline{x}(t)$ is partitioned according to (eq. 2.6a) and the system matrices are of the structure (eq. 2.6b). It will be assumed that the output noise covariance matrix $R(t)$ and the control input matrix $D_1(t)$ are invertible.

Let an output vector feedback controller (fig. 4.7) be given by

$$\underline{u}(t) = - K_1(t)\ \underline{z}(t) + K_3(t)\ \hat{\underline{x}}_2(t) \tag{4.10a}$$

$$\dot{\hat{\underline{x}}}_2(t) = F_{22}(t)\ \hat{\underline{x}}_2(t) + K_2(t)(\underline{z}(t) - H_2(t)\ \hat{\underline{x}}_2(t)) \ . \tag{4.10b}$$

Find the parameters of the feedback matrices $K_1(t)$, $K_2(t)$ and $K_3(t)$ such that the cost functional

$$J = E\{[\underline{x}_1^T(t_f)\quad \tilde{\underline{x}}_2^T(t_f)]\ \Pi(t_f) \begin{bmatrix} \underline{x}_1(t_f) \\ \tilde{\underline{x}}_2(t_f) \end{bmatrix}\} \tag{4.11}$$

with the disturbance compensation error $\tilde{\underline{x}}_2(t) = \underline{x}_2(t) - \hat{\underline{x}}_2(t)$ and the terminal-time $t_o \leq t \leq t_f$ is minimized.

The choice of the controller structure (eq. 4.10) can be motivated by the following considerations:

1. The simplest structure of the controller would be the static output feedback $K_1(t)\ \underline{z}(t)$. By the assumption of an invertible control input matrix $D_1(t)$ this type of control is sufficient for stabilizing the physical system.

2. The disturbance vector $\underline{w}^*(t)$ (eq. 2.5b) contains information by the time-correlation of $\underline{x}_2(t)$. It seems to be useful to construct an estimator for $\underline{x}_2(t)$. The estimate $\hat{\underline{x}}_2(t)$ serves to com-

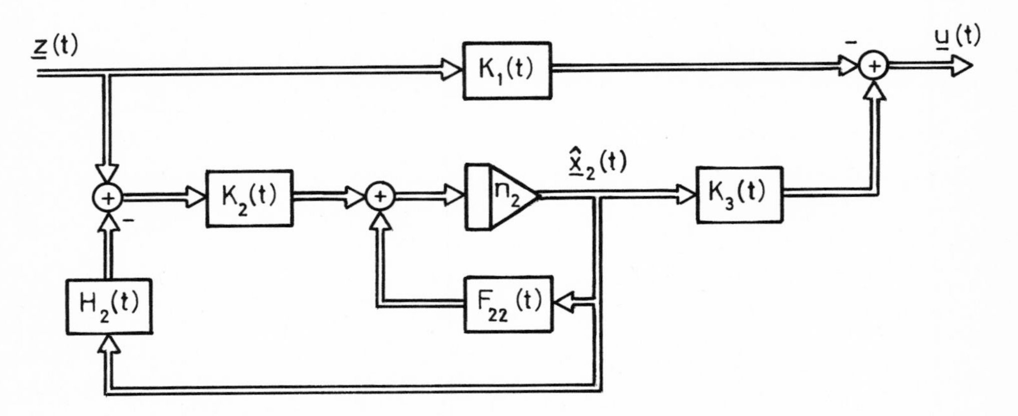

Fig. 4.7 Structure of the Output Feedback Controller

pensate the correlated noise $\underline{x}_2(t)$ in the mean and provides for a more accurate behaviour of the closed-loop system.

The solution of this parameter optimization problem is given by [18]

$$K_1(t) = D_1^{-1}\, K_1^*(t) \tag{4.12a}$$

$$K_2(t) = K_2^*(t)$$

$$K_3(t) = -\, D_1^{-1}(t)\, F_{12}(t) + K_1(t)\, H_2(t)$$

with

$$K^*(t) = \begin{pmatrix} K_1^*(t) \\ K_2^*(t) \end{pmatrix} = P^*(t)\, H^T\, R^{-1}(t) \tag{4.12b}$$

where the error covariance matrix $P^*(t)$ is solution of the n-th order, $n = n_1 + n_2$, matrix Riccati differential equation

$$\begin{aligned} \dot{P}^*(t) &= F(t)\, P^*(t) + P^*(t)\, F^T(t) \\ &\quad - P^*(t)\, H^T(t)\, R^{-1}(t)\, H(t)\, P^*(t) \\ &\quad + G(t)\, Q(t)\, G^T(t) \end{aligned} \tag{4.12c}$$

$$P^*(t_o) = P_o \quad .$$

By the kind of solution it is to be expected that there exists an equivalence between linear optimal filtering and linear output vector feedback. This is shown and discussed in detail in [18]. As a consequence properties and results of optimal filtering (chapter 3) can directly be applied to the proposed controller design procedure. Some main results are repeated:

1. The stability behaviour is governed by (chapter 3, T1) or in the constant gain design case by (chapter 3, T2).
2. The controller provides for optimality uniformly in t, i.e. it is independent of the terminal time t_f and the weighting matrix $\Pi(t_f)$.
3. The accuracy behaviour of the closed-loop system can be judged off-line by inspection of the diagonal elements of the error covariance matrix $P^*(t)$.

By the equivalence it is also possible to use the already known numerical results from the above discussed filters. The optimal output vector feedback controller involves a dynamical compensator of order n_2 -- the order of the disturbance model (eq. 4.6a) -- and time-varying controller gain matrices. The resulting accuracy of the closed-loop system -- equal to that of the Kalman-Bucy filter -- is to be seen from (fig. 4.4). According to the design considerations of the suboptimal filter algorithm the design of a suboptimal controller can be performed yielding a static output vector feedback control

$$\underline{u}(t) = K\,\underline{z}(t) \quad . \tag{4.13}$$

The feedback matrix K is equal to the filter gain matrix in (eq. 4.8). The accuracy results of the suboptimally controlled system are to be seen from (fig. 4.6). The considerations about loss of accuracy and simplicity of realization are the same as in the filter case.

From these results it is to be seen that the proposed controller design procedure for systems of the assumed structure leads to simple controllers and most accurate closed-loop system behaviour. All results and experiences of linear optimal filtering can be applied directly. Moreover the procedure is independent of the weighting matrix of the cost functional such that the optimal solution is maintained within one step and a satisfying optimal solution needs not to be determined by trial and error.

5. SUMMARY

The -- at least approximate -- mathematical description of dynamical systems by means of linear vector differential equations is discussed with special regard to random system and measurement noise modelling. The derivation of the Kalman-Bucy filter using a parameter optimization approach and the properties of the linear optimal filter solution, especially the stability conditions, are reviewed. The practical problem of controlling a Doppler-aided inertial navigation system in an as accurate and as simple as possible way is investigated. To provide for a lower bound of accuracy of state vector feedback based controllers the linear optimal filter and a suboptimal estimation algorithm, derived in a general and typical way regarding system properties and realization requirements, are examined. An optimal output vector feedback controller is proposed which enables the design of low-order random disturbance compensators with most accurate closed-loop system behaviour. By the equivalence of linear optimal filtering and the proposed optimal output vector feedback control all properties and results of estimation can be applied directly to the control problem.

LITERATURE

[1] COURANT, R., HILBERT, D. "Methods of Mathematical Physics", Vol. I, II. Interscience Publishers, New York, 1962.

[2] KWAKERNAAK, H. SIVAN, R. "Linear Optimal Control Systems". Wiley-Interscience, New York, 1972.

[3] ANDERSON, B.D.O., MOORE, J.B. "Linear Optimal Control". Prentice-Hall, Englewood Cliffs, 1971.

[4] JAZWINSKI, A.H. "Stochastic Processes and Filtering Theory". Academic Press, New York, 1970.

[5] PAPOULIS, A. "Probability, Random Variables, and Stochastic Processes". McGraw Hill, New York, 1965.

[6] GELB, A., e.a. "Applied Optimal Estimation". The M.I.T. Press, Cambridge, 1974.

[7] -- "Theory and Applications of Kalman Filtering". AGARDograph 139, 1970.

[8] ATHANS, M. "The Matrix Minimum Principle". J. on Information and Control, vol. 11, p. 592-606, 1968.

[9] HOFMANN, W., KORTÜM, W. "Reduced-Order State Estimators with Application in Rotational Dynamics". "Gyrodynamics", Springer-Verlag, Berlin, 1974.

[10] BROCKETT, R.W. "Finite Dimensional Linear Systems". John Wiley & Sons, New York, 1970.

[11] -- "A Short Course on Kalman Filtering and Application". The Analytic Science Corporation, Reading Mass.

[12] KUCERA, V. "A Contribution to Matrix Quadratic Equations". IEEE Trans. on Automatic Control, vol. AC-17 (June 1972), pp. 344-347.

[13] VAUGHAN, D.R. "A Negative Exponential Solution for the Matrix Riccati Equation". IEEE Trans. Automatic Control, vol. AC-14, (Feb. 1969), pp. 72-75.

[14] MÜLLER, P.C. "Special Problems of Gyrodynamics". International Centre of Mechanical Sciences, Courses and Lectures No. 63, Springer-Verlag, Udine, 1970.

[15] HOFMANN, W. "Die Anwendung der Kalman-Bucy-Filtertheorie beim Entwurf von Rückkopplungsnetzwerken für Dopplergestützte Trägheitsnavigationssysteme im Flug". DFVLR-IB Nr. 552-75/18.

[16] BRYSON, A.E., HO, Y. "Applied Optimal Control". Ginn and Company, London, 1969.

[17] HOFMANN, W. "Application of Inners to the Design of Optimal State Estimators Under Stability Constraints". Proceedings of the Conference on Information Sciences and Systems, Baltimore, 1976.

[18] HOFMANN, W. "Optimal Stochastic Disturbance Compensation by Output Vector Feedback". to appear as DLR-Forschungsbericht.

INDEX